Lecture Notes in Mathematics

Editors:
J.-M. Morel, Cachan
F. Takens, Groningen
B. Teissier, Paris

Martin C. Olsson

Compactifying Moduli
Spaces for Abelian Varieties

 Springer

Martin C. Olsson
Department of Mathematics
University of California
Berkeley, CA 94720-3840
USA
molsson@math.berkeley.edu

ISBN 978-3-540-70518-5   ISBN 978-3-540-70519-2 (eBook)
DOI: 10.1007/978-3-540-70519-2

Lecture Notes in Mathematics ISSN print edition: 0075-8434
                    ISSN electronic edition: 1617-9692

Library of Congress Control Number: 2008931163

Mathematics Subject Classification (2000): 14K10, 14K25

Cover design: SPi Publishing Services

Printed on acid-free paper

9 8 7 6 5 4 3 2 1

springer.com

Contents

Summary VII

0   Introduction 1

1   A Brief Primer on Algebraic Stacks 7
    1.1 S-groupoids 7
    1.2   Stacks 17
    1.3   Comparison of Topologies 25
    1.4   Coarse Moduli Spaces 26
    1.5   Rigidification of Stacks 26

2   Preliminaries 31
    2.1   Abelian Schemes and Torsors 31
    2.2   Biextensions 34
    2.3   Logarithmic Geometry 43
    2.4   Summary of Alexeev's Results 51

3   Moduli of Broken Toric Varieties 57
    3.1   The Basic Construction 57
    3.2   Automorphisms of the Standard Family over a Field 67
    3.3   Deformation Theory 70
    3.4   Algebraization 75
    3.5   Approximation 76
    3.6   Automorphisms over a General Base 78
    3.7   The Stack $\mathscr{K}_Q$ 80

4   Moduli of Principally Polarized Abelian Varieties 85
    4.1   The Standard Construction 85
    4.2   Automorphisms over a Field 93
    4.3   Deformation Theory 98
    4.4   Isomorphisms over Artinian Local Rings 110

4.5   Versal Families 113
4.6   Definition of the Moduli Problem 121
4.7   The Valuative Criterion for Properness 121
4.8   Algebraization 125
4.9   Completion of Proof of 4.6.2 130

5   **Moduli of Abelian Varieties with Higher Degree
    Polarizations** 135
5.1   Rethinking $\mathscr{A}_{g,d}$ 135
5.2   The Standard Construction 138
5.3   Another Interpretation of $\widetilde{\mathscr{P}} \to \mathscr{P}$ 142
5.4   The Theta Group 144
5.5   Deformation Theory 158
5.6   Isomorphisms without Log Structures 160
5.7   Algebraization of Formal Log Structures 163
5.8   Description of the Group H_S^{gp} 166
5.9   Specialization 178
5.10 Isomorphisms in $\overline{\mathscr{T}}_{g,d}$ 201
5.11 Rigidification 202
5.12 Example: Higher Dimensional Tate Curve 207
5.13 The Case $g = 1$ 225

6   **Level Structure** 231
6.1   First Approach Using Kummer étale Topology 231
6.2   Second Approach using the Theta Group 237
6.3   Resolving Singularities of Theta Functions 241

References 273

Index of Terminology 277

Index of Notation 279

Summary

The problem of compactifying the moduli space $\mathscr{A}_g$ of principally polarized abelian varieties has a long and rich history. The majority of recent work has focused on the toroidal compactifications constructed over $\mathbb{C}$ by Mumford and his coworkers, and over $\mathbb{Z}$ by Chai and Faltings. The main drawback of these compactifications is that they are not canonical and do not represent any reasonable moduli problem on the category of schemes. The starting point for this work is the realization of Alexeev and Nakamura that there is a canonical compactification of the moduli space of principally polarized abelian varieties. Indeed Alexeev describes a moduli problem representable by a proper algebraic stack over $\mathbb{Z}$ which contains $\mathscr{A}_g$ as a dense open subset of one of its irreducible components.

In this text we explain how, using logarithmic structures in the sense of Fontaine, Illusie, and Kato, one can define a moduli problem "carving out" the main component in Alexeev's space. We also explain how to generalize the theory to higher degree polarizations and discuss various applications to moduli spaces for abelian varieties with level structure.

If d and g are positive integers we construct a proper algebraic stack with finite diagonal $\overline{\mathscr{A}}_{g,d}$ over $\mathbb{Z}$ containing the moduli stack $\mathscr{A}_{g,d}$ of abelian varieties with a polarization of degree d as a dense open substack. The main features of the stack $\overline{\mathscr{A}}_{g,d}$ are that (i) over $\mathbb{Z}[1/d]$ it is log smooth (i.e. has toroidal singularities), and (ii) there is a canonical extension of the kernel of the universal polarization over $\mathscr{A}_{g,d}$ to $\overline{\mathscr{A}}_{g,d}$. The stack $\overline{\mathscr{A}}_{g,d}$ is obtained by a certain "rigidification" procedure from a solution to a moduli problem. In the case $d = 1$ the stack $\overline{\mathscr{A}}_{g,1}$ is equal to the normalization of the main component in Alexeev's compactification. In the higher degree case, our study should be viewed as a higher dimensional version of the theory of generalized elliptic curves introduced by Deligne and Rapoport.

0

Introduction

In attempting to study any moduli space M, one of the basic first steps is to find a good compactification $M \subset \overline{M}$. Preferably the compactification $\overline{M}$ should have reasonable geometric properties (i.e. smooth with $\overline{M} - M$ a divisor with normal crossings), and the space $\overline{M}$ should also have a reasonable moduli interpretation with boundary points corresponding to degenerate objects.

Probably the most basic example of this situation is the moduli space $\mathcal{M}_{1,1}$ classifying elliptic curves and variant spaces classifying elliptic curves with level structure. For $\mathcal{M}_{1,1}$ the compactification $\overline{\mathcal{M}}_{1,1}$ is the stack which to any scheme T associates the groupoid of pairs $(f : E \to T, e)$, where $f : E \to T$ is a proper flat morphism and $e : T \to E^{\mathrm{sm}}$ is a section into the smooth locus of f such that for every geometric point $\bar{t} \to T$ the fiber $E_{\bar{t}}$ is either a genus 1 smooth curve or a rational nodal curve.

In order to generalize this compactification $\mathcal{M}_{1,1} \subset \overline{\mathcal{M}}_{1,1}$ to moduli spaces Y_Γ classifying elliptic curves with Γ–level structure for some arithmetic subgroup $\Gamma \subset SL_2(\mathbb{Z})$, Deligne and Rapoport introduced the notion of "generalized elliptic curves" in [15]. The main difficulty is that for a scheme T, an integer $N \geq 1$, and an object $(f : E \to T, e) \in \overline{\mathcal{M}}_{1,1}(T)$ there is no good notion of the N–torsion subgroup of E. More precisely, if there exists a dense open subset $U \subset T$ such that the restriction $f_U : E_U \to U$ is smooth, then the finite flat U–group scheme $E_U[N]$ does not extend to a finite flat group scheme over T. Deligne and Rapoport solve this problem by introducing "N-gons" which enable them to define a reasonable notion of N–torsion group for degenerate objects.

The moduli space $\mathcal{M}_{1,1}$ has two natural generalizations. First one can let the genus and number of marked points vary which leads to the moduli spaces $\mathcal{M}_{g,n}$ of genus g curves with n-marked points. These spaces of course have modular compactifications $\mathcal{M}_{g,n} \subset \overline{\mathcal{M}}_{g,n}$ defined by Deligne, Mumford and Knudsen. The second generalization of $\mathcal{M}_{1,1}$ is moduli spaces for higher dimensional (polarized) abelian varieties. Constructing compactifications of moduli spaces for polarized abelian varieties has historically been a much more difficult problem.

M.C. Olsson, *Compactifying Moduli Spaces for Abelian Varieties*. Lecture Notes in Mathematics 1958.
© Springer-Verlag Berlin Heidelberg 2008

Let $\mathscr{A}_g$ denote the moduli space of principally polarized abelian varieties of dimension g for some integer $g \geq 1$. The first compactification of $\mathscr{A}_g$ over $\mathbb{C}$ is the so-called *Satake* or *minimal* compactification $\mathscr{A}_g \subset \mathscr{A}_g^*$ constructed by Satake in [47]. The space $\mathscr{A}_g^*$ is normal but in general singular at the boundary. The basic question following the construction of the Satake compactification is then how to resolve the singularities of $\mathscr{A}_g^*$ and to generalize the theory to one over $\mathbb{Z}$.

Over the complex numbers such resolutions of $\mathscr{A}_g^*$ were constructed by Ash, Mumford, Rapoport, and Tai in [9] where they constructed the so-called toroidal compactifications of $\mathscr{A}_g$. These compactifications are smooth with boundary a divisor with normal crossings. Unfortunately, these compactifications are not canonical and there is no simple modular interpretation (though recently Kajiwara, Kato, and Nakayama [22] have given a modular interpretation of these compactifications using their theory of *log abelian varieties*). Later Chai and Faltings [13] extended the toroidal compactifications to $\mathbb{Z}$.

The problem remained however to define a compactification of $\mathscr{A}_g$ with a simple modular interpretation, and to generalize the theory of Deligne–Rapoport to also give modular compactifications of moduli spaces for abelian varieties with level structure and higher degree polarizations. This is the purpose of this text.

The starting point for our work is the paper [3] in which Alexeev studied moduli of varieties with action of semi–abelian schemes (Alexeev's work in turn built on the work of several people including Namikawa [39] and work with Nakamura [4]). He constructed compact moduli spaces for two basic moduli problems, one of which leads to a functorial compactification of the moduli space of principally polarized abelian varieties $\mathscr{A}_g$. One feature of his approach, is that the resulting moduli spaces have many irreducible components with one "distinguished" component containing $\mathscr{A}_g$. One of the main ideas in this text is that using logarithmic geometry in the sense of Fontaine and Illusie ([24]) one can give a relatively simple functorial description of the normalizations of the main components. In fact this idea can also be applied to give a modular interpretation of Alexeev's moduli spaces of "broken toric varieties".

In the principally polarized case our work yields an Artin stack $\mathscr{K}_g$ with the following properties:

(i) The diagonal of $\mathscr{K}_g$ is finite and $\mathscr{K}_g$ is proper of $\mathrm{Spec}(\mathbb{Z})$.

(ii) There is a natural open immersion $\mathscr{A}_g \hookrightarrow \mathscr{K}_g$ identifying $\mathscr{A}_g$ with a dense open substack of $\mathscr{K}_g$.

(iii) There is a good "analytic theory" at the boundary of $\mathscr{A}_g$ in $\mathscr{K}_g$ generalizing the theory of the Tate curve for elliptic curves.

(iv) The stack $\mathscr{K}_g$ has only toroidal singularities (in fact the complement $\mathscr{K}_g \backslash \mathscr{A}_g$ defines a fine saturated log structure $M_{\mathscr{K}_g}$ on $\mathscr{K}_g$ such that the log stack $(\mathscr{K}_g, M_{\mathscr{K}_g})$ is log smooth over $\mathrm{Spec}(\mathbb{Z})$ with the trivial log structure).

In order to study moduli of abelian varieties with higher degree polarizations and level structure, we need a different point of view on how to classify abelian schemes with polarization. Let g and d be positive integers and let $\mathscr{A}_{g,d}$ denote the moduli stack classifying pairs (A, λ) where A is an abelian scheme and $\lambda : A \to A^t$ is a polarization of degree d (by convention this means that the kernel of λ is a finite flat group scheme of rank d^2). The stack $\mathscr{A}_{g,d}$ can be viewed as follows. Let $\mathscr{T}_{g,d}$ denote the stack over $\mathbb{Z}$ associating to any scheme S the groupoid of triples (A, P, L), where A is an abelian scheme over S of relative dimension g, P is a A-torsor, and L is an ample line bundle on P such that the map

$$\lambda_L : A \to \underline{\mathrm{Pic}}^0(P), \quad a \mapsto [t_a^* L \otimes L^{-1}] \qquad (0.0.0.1)$$

has kernel a finite flat group scheme of rank d^2, where $t_a : P \to P$ denotes the action on P of a (scheme-valued) point $a \in A$. We will show that $\mathscr{T}_{g,d}$ is in fact an algebraic stack over $\mathbb{Z}$.

For such a triple (A, P, L) over a scheme S, let $\mathscr{G}_{(A,P,L)}$ denote the group of automorphisms of the triple (A, P, L) which are the identity on A. That is, $\mathscr{G}_{(A,P,L)}$ is the group scheme classifying pairs (β, ι), where $\beta : P \to P$ is an automorphism commuting with the A–action and $\iota : \beta^* L \to L$ is an isomorphism of line bundles on P. We call this group $\mathscr{G}_{(A,P,L)}$ the *theta group* of (A, P, L). There is a natural inclusion $\mathbb{G}_m \hookrightarrow \mathscr{G}_{(A,P,L)}$ sending $u \in \mathbb{G}_m$ to the element with $\beta = \mathrm{id}$ and $\iota = u$. This inclusion identifies $\mathbb{G}_m$ with a central subgroup of $\mathscr{G}_{(A,P,L)}$ and we write $H_{(A,P,L)}$ for the quotient. If the torsor P is trivial, then the group $\mathscr{G}_{(A,P,L)}$ is the theta group in the sense of Mumford [36, part I, §1]. In particular, by descent theory the group scheme $H_{(A,P,L)}$ is a finite flat group scheme of rank d^2 over S.

As explained in 2.1.5 there is a canonical isomorphism $A^t \simeq \underline{\mathrm{Pic}}^0(P)$. The map 0.0.0.1 therefore induces a polarization of degree d on A. This defines a map

$$\pi : \mathscr{T}_{g,d} \to \mathscr{A}_{g,d}, \quad (A, P, L) \mapsto (A, \lambda_L).$$

For any object $(A, P, L) \in \mathscr{T}_{g,d}(S)$ (for some scheme S), the kernel of the morphism of group schemes

$$\underline{\mathrm{Aut}}_{\mathscr{T}_{g,d}}(A, P, L) \to \underline{\mathrm{Aut}}_{\mathscr{A}_{g,d}}(\pi(A, P, L))$$

is precisely the group scheme $\mathscr{G}_{(A,P,L)}$. This implies that one can obtain $\mathscr{A}_{g,d}$ by a purely stack-theoretic construction called *rigidification* which "kills off" the extra automorphisms provided by $\mathscr{G}_{(A,P,L)}$ (the reader not famililar with the notion of rigidification may wish to consult the examples 1.5.7 and 1.5.8). Thus in many ways the stack $\mathscr{T}_{g,d}$ is a more basic object than $\mathscr{A}_{g,d}$.

With this in mind, our approach to compactifying $\mathscr{A}_{g,d}$ is to first construct an open immersion $\mathscr{T}_{g,d} \hookrightarrow \overline{\mathscr{T}}_{g,d}$ and an extension of the theta group over $\mathscr{T}_{g,d}$ to an extension of a finite flat group scheme of rank d^2 by $\mathbb{G}_m$ over the stack $\overline{\mathscr{T}}_{g,d}$. The stack $\overline{\mathscr{T}}_{g,d}$ should be viewed as a compactification of $\mathscr{T}_{g,d}$, though of course it is not compact (not even separated) since the diagonal is not proper. We can then apply the rigidification construction to $\overline{\mathcal{T}}_{g,d}$ with respect to the extension of the theta group to get a compactification $\mathscr{A}_{g,d} \hookrightarrow \overline{\mathscr{A}}_{g,d}$. The stack $\overline{\mathscr{A}}_{g,d}$ is proper over $\mathbb{Z}$ with finite diagonal, $\mathscr{A}_{g,d} \hookrightarrow \overline{\mathscr{A}}_{g,d}$ is a dense open immersion, and over $\mathbb{Z}[1/d]$ the stack $\overline{\mathscr{A}}_{g,d}$ is log smooth (so it has toroidal singularities). Moreover, over $\overline{\mathscr{A}}_{g,d}$ there is a tautological finite flat group scheme $\mathscr{H} \to \overline{\mathscr{A}}_{g,d}$ whose restriction to $\mathscr{A}_{g,d}$ is the kernel of the universal polarization of degree d^2.

Remark 0.0.1. The stack $\overline{\mathscr{A}}_{g,1}$ is canonically isomorphic to $\mathscr{K}_g$. However, because the moduli interpretation of these two stacks are different (the definition of $\mathscr{K}_g$ avoids the rigidification procedure by a trick that only works in the principally polarized case) we make the notational distinction.

The text is organized as follows. Because of the many technical details involved with our construction in full generality, we have chosen to present the theory by studying in turn three moduli problems of increasing technical difficulty (broken toric varieties, principally polarized abelian varieties, abelian varieties with higher degree polarizations).

We also make heavy use of stack-theoretic techniques. For this reason we have included in chapter 1 a brief primer on algebraic stacks. The reader familiar with stacks can skip this chapter and refer back as needed.

In chapter 2 we summarize the necessary background material for the rest of the text. We state our conventions about semi-abelian schemes, review the necessary theory of biextensions, and summarize the material we need from logarithmic geometry [24] and for the convenience of the reader we recall some of Alexeev's results from [3].

Chapter 3 is devoted to moduli of "broken toric varieties". This chapter is independent of the other chapters. We have included it here because it illustrates many of the key ideas used for the moduli of abelian varieties without many of the technical details. The moduli problems considered in this chapter have also been studied extensively in other contexts (see for example [26] and [30]), so this chapter may be of independent interest.

In chapter 4 we turn to the problem of compactifying the moduli space of principally polarized abelian varieties. Though we subsequently will also study higher degree polarizations and level structure, we first consider the principally polarized case which does not require the more intricate theory of the theta group and is more closely related to Alexeev's work.

In chapter 5 we then turn to the full theory. The main ingredient needed to generalize the principally polarized case is to study in detail degenerations of the theta group.

Finally in chapter 6 we explain how to construct compact moduli spaces for abelian varieties with level structure. We present two approaches. One using the theory of logarithmic étale cohomology, and the second using the theory of the theta group developed in 5. The second approach has the advantage that it lends itself to a study of reductions of moduli spaces at primes dividing the level. We intend to discuss this in future writings. We also discuss in detail how to construct modular compactifications of the moduli spaces for abelian varieties with "theta level structure" defined by Mumford in [36].

0.0.2 (Acknowledgements). The author is grateful to V. Alexeev and S. Keel for several helpful conversations. The author also would like to thank the Institute for Advanced Study where part of this work was done for its excellent working conditions, and the American Institute of Mathematics which hosted an excellent workshop which helped initiate this project. Finally the author thanks the referees for helpful suggestions. The author was partially supported by an NSF post–doctoral research fellowship, NSF grant DMS-0714086, and an Alfred P. Sloan research fellowship.

1

A Brief Primer on Algebraic Stacks

We recall in this chapter a few basic definitions and results (without proofs) about algebraic stacks. For a more complete treatment (and in particular for proofs of many of the results stated here) see the book of Laumon and Moret-Bailly [31] and Vistoli's paper [49].

Throughout this chapter we work over a fixed a base scheme S, and consider the étale topology on the category (Sch/S) of schemes over S, unless otherwise mentioned.

1.1 S-groupoids

1.1.1. If F is a category and $p : F \to$ (Sch/S) is a functor, then for any S-scheme X we define $F(X)$ to be the category whose objects are objects $x \in F$ such that $p(x) = X$ and whose morphisms are morphisms $\rho : x \to x'$ in F such that the induced map

$$X = p(x) \xrightarrow{p(\rho)} p(x') = X$$

is equal to id_X.

If $f : X \to Y$ is a morphism of S-schemes and $y \in F(Y)$, then *a pullback of y along f* is an arrow $h : x \to y$ in F with $p(h) = f$ such that the following hold: given an object $z \in F$ with $p(z) = Z$ for some S-scheme Z, a morphism $g : z \to y$ in F, and a factorization

$$Z \xrightarrow{u} X \xrightarrow{f} Y$$

of $p(g)$, there exists a unique morphism $v : z \to x$ such that $p(v) = u$ and $h \circ v = g$. In a diagram:

M.C. Olsson, *Compactifying Moduli Spaces for Abelian Varieties*. Lecture Notes in Mathematics 1958.
© Springer-Verlag Berlin Heidelberg 2008

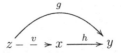

$$Z \xrightarrow{u} X \xrightarrow{f} Y.$$

The arrow $h : x \to y$ is unique up to unique isomorphism. Abusively we often refer to x as the *pullback of y* and usually denote it by f^*y.

Definition 1.1.2. A *fibered category over S* is a category F with a functor $p : F \to (\mathrm{Sch}/S)$ such that for every $y \in F$ with image $Y \in (\mathrm{Sch}/S)$ and morphism $f : X \to Y$ there exists a pullback of y along f.

Definition 1.1.3. A *groupoid over S* (or *S-groupoid*) is a functor $p : F \to (\mathrm{Sch}/S)$ such that the following hold:

(i) For every $y \in F$ with $p(y) = Y$ and morphism of S-schemes $f : X \to Y$, there exists a pullback $h : x \to y$ of y along f (so F is a fibered category over S).

(ii) For every $X \in (\mathrm{Sch}/S)$ the category $F(X)$ is a groupoid (which by definition means that all morphisms in $F(X)$ are isomorphisms).

Remark 1.1.4. Conditions (i) and (ii) imply that every arrow in F is a pullback arrow.

Example 1.1.5. Let G/S be a flat group scheme, and define BG to be the following category:

Objects. Pairs (T, P), where T is an S-scheme and P is a G-torsor over T.

Morphisms. A morphisms $(T', P') \to (T, P)$ in BG is a commutative (in fact necessarily cartesian) diagram

$$\begin{array}{ccc} P' & \xrightarrow{a} & P \\ \downarrow & & \downarrow \\ T' & \longrightarrow & T, \end{array}$$

where the morphism a is G-equivariant. There is a functor

$$F : BG \to (\mathrm{Sch}/S), \quad (T, P) \mapsto T,$$

which makes BG an S-groupoid.

Example 1.1.6. The preceding example can be generalized as follows. Let G/S be as in 1.1.5, and let X/S be a scheme with G-action. Define $[X/G]$ to be the following category.

Objects. Triples (T, P, ρ), where T is an S-scheme, P is a G-torsor over T, and $\rho : P \to X$ is a G-equivariant map.

Morphisms. A morphism $(T', P', \rho') \to (T, P, \rho)$ is a commutative diagram

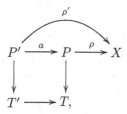

where a is G-equivariant. Then $[X/G]$ is an S-groupoid with projection

$$F : [X/G] \to (\mathrm{Sch}/S), \quad (T, P, \rho) \mapsto T.$$

Note that $BG = [S/G]$, where G acts trivially on S.

Example 1.1.7. Fix an integer $g \geq 1$, and let $\mathscr{A}_g$ denote the following category.

Objects. Triples (T, A, λ), where T is an S-scheme, A/T is an abelian scheme of relative dimension g, and $\lambda : A \to A^t$ is a principal polarization (where A^t denotes the dual abelian scheme of A).

Morphisms. A morphism $(T', A', \lambda') \to (T, A, \lambda)$ is a cartesian diagram

$$
\begin{array}{ccc}
A' & \xrightarrow{\;a\;} & A \\
\downarrow & & \downarrow \\
T' & \longrightarrow & T,
\end{array}
$$

where a is a homomorphism of abelian schemes such that the induced diagram

$$
\begin{array}{ccc}
A' & \xrightarrow{\;a\;} & T' \times_T A \\
{\scriptstyle \lambda'}\downarrow & & \downarrow{\scriptstyle \lambda} \\
A'^t & \xleftarrow{\;a^*\;} & T' \times_T A^t
\end{array}
$$

commutes. Then $\mathscr{A}_g$ is an S-groupoid with projection

$$F : \mathscr{A}_g \to (\mathrm{Sch}/S), \quad (T, A, \lambda) \mapsto T.$$

1.1.8. If $p : F \to (\mathrm{Sch}/S)$ and $q : H \to (\mathrm{Sch}/S)$ are groupoids over S, then a morphism $F \to H$ of S-groupoids is a functor $\rho : F \to H$ such that $p = q \circ \rho$ ('on-the-nose' equality of functors). If $\rho, \rho' : F \to H$ are two morphisms of S-groupoids, then a morphism $\gamma : \rho \to \rho'$ is a natural transformation of functors such that the induced automorphism

$$p = q \circ \rho \xrightarrow{\;\gamma\;} q \circ \rho' = p$$

is the identity. Note that γ is necessarily an isomorphism. We denote by

$$\mathrm{HOM}(F, H)$$

the groupoid of morphisms of S-groupoids $F \to H$.

Example 1.1.9. Let $F : (\mathrm{Sch}/S)^{\mathrm{op}} \to \mathrm{Set}$ be a functor. Then F defines a fibered category $p_F : F^* \to (\mathrm{Sch}/S)$ as follows:

   Objects. Pairs (T, f), where $T \in (\mathrm{Sch}/S)$ and $f \in F(T)$.
   Morphisms. A morphism $(T', f') \to (T, f)$ is a morphism $h : T' \to T$ such that the two elements $f', h^*(f) \in F(T')$ are equal.
   The functor p_F sends (T, f) to T. A natural transformation of functors $\rho : F \to G$ induces a morphism of S-groupoids $F^* \to G^*$ and the induced map

$$\mathrm{Hom}(F, G) \to \mathrm{HOM}(F^*, G^*)$$

is an equivalence of categories [49, 3.26], where the set $\mathrm{Hom}(F, G)$ is viewed as a category with only identity morphisms.

   In what follows, we will usually not distinguish between the functor F and the corresponding S-groupoid F^* and simply write F for both. In particular, if $X \in (\mathrm{Sch}/S)$ then the functor of points h_X of X can be viewed as an S-groupoid, which we usually denote (abusively) again by X.

Remark 1.1.10. There is a version of Yoneda's lemma for S-groupoids. Let X be an S-scheme and let $p : F \to (\mathrm{Sch}/S)$ be an S-groupoid. For any morphism $\rho : h_X \to F$ of S-groupoids, we get by looking at the fiber over X a functor

$$\rho_X : h_X(X) \to F(X).$$

Evaluating this map on $\mathrm{id}_X \in h_X(X)$ we get an object $\rho_X(\mathrm{id}_X) \in F(X)$. The Yoneda lemma in this context [49, 3.6.2] says that the induced functor

$$\mathrm{HOM}(h_X, F) \to F(X)$$

is an equivalence of categories. We will therefore often view objects of $F(X)$ as morphisms $h_X \to F$ and vice versa.

1.1.11. It is often convenient to think about S-groupoids F together with a choice of a pullback arrow $h : f^*x \to x$ for every morphism $f : Y \to X$ in (Sch/S) and object $x \in F(X)$. The choice of such a collection K of pullback arrows is called a *cleavage* of F [49, 3.9]. It follows from the axiom of choice that a cleavage always exists.

   If K is a cleavage of F we obtain a functor

$$f^* : F(X) \to F(Y), \quad x \mapsto f^*x,$$

for every morphism $f : Y \to X$ in (Sch/S). An arrow $\alpha : x' \to x$ in $F(X)$ is sent to the unique dotted arrow (which exists since $f^*x \to x$ is a pullback) filling in the following diagram in F

$$\begin{array}{ccc} f^*x' & \longrightarrow & x' \\ \downarrow & & \downarrow{\scriptstyle\alpha} \\ f^*x & \longrightarrow & x. \end{array}$$

It is therefore tempting to think of an S-groupoid as a functor from $(\mathrm{Sch}/S)^{\mathrm{op}}$ to groupoids sending X to $F(X)$. Here the fact that the collection of groupoids forms a 2-category, and not a category, becomes an issue.

The main point is that in general one cannot choose the cleavage K in such a way that for a composite

$$Z \xrightarrow{\ g\ } Y \xrightarrow{\ f\ } X$$

the two functors

$$(fg)^*, g^* \circ f^* : F(X) \to F(Z)$$

are equal (they are, however, canonically isomorphic by the universal property of pullback). A simple counterexample is given in [49, 3.14]. This leads to the following definition.

Definition 1.1.12 ([49, 3.10]). A *pseudo-functor* $\mathcal{F}$ on (Sch/S) consists of the following data:

(i) For $X \in (\mathrm{Sch}/S)$ a groupoid $\mathcal{F}(X)$.
(ii) For every S-morphism $f : Y \to X$ a functor

$$f^* : \mathcal{F}(X) \to \mathcal{F}(Y).$$

(iii) For every $X \in (\mathrm{Sch}/S)$ an isomorphism $\epsilon_X : \mathrm{id}_X^* \simeq \mathrm{id}_{\mathcal{F}(X)}$.
(iv) For every composite

$$Z \xrightarrow{\ g\ } Y \xrightarrow{\ f\ } X$$

an isomorphism

$$\alpha_{g,f} : g^* f^* \simeq (fg)^* : \mathcal{F}(X) \to \mathcal{F}(Z).$$

This data is in addition required to satisfy the following conditions.

(a) For any S-morphism $f : Y \to X$ and $x \in \mathcal{F}(X)$ the diagrams

$$\begin{array}{ccc} \mathrm{id}_Y^* f^*(x) & & \\ {\scriptstyle\alpha_{\mathrm{id}_Y,f}}\downarrow & \searrow{\scriptstyle\epsilon_Y} & \\ (f \circ \mathrm{id}_Y)^*(x) & = \!\!= & f^*x \end{array}$$

and

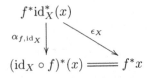

commute.

(b) For a composite

$$W \xrightarrow{h} Z \xrightarrow{g} Y \xrightarrow{f} Z$$

and $x \in \mathcal{F}(X)$ the diagram

$$
\begin{array}{ccc}
h^*g^*f^*(x) & \xrightarrow{\alpha_{h,g}} & (gh)^*f^*(x) \\
{\scriptstyle\alpha_{g,f}}\downarrow & & \downarrow{\scriptstyle\alpha_{gh,f}} \\
h^*(fg)^*(x) & \xrightarrow{\alpha_{h,fg}} & (fgh)^*(x)
\end{array}
$$

commutes.

Remark 1.1.13. In [49, 3.10] Vistoli defines a notion of pseudo-functor with (Sch/S) replaced by an arbitrary category and not necessarily taking values in groupoids. However, the above definition suffices for our purposes.

1.1.14. An S-groupoid F with a cleavage defines a pseudo-functor $\mathcal{F}$ as in 1.1.11.

Conversely a pseudo-functor $\mathcal{F}$ defines an S-groupoid F as follows. The objects of F are pairs (X, x), where $X \in (Sch/S)$ and $x \in \mathcal{F}(X)$. A morphism $(Y, y) \to (X, x)$ is a pair (f, f^b), where $f : Y \to X$ is a morphism in (Sch/S) and $f^b : y \to f^*x$ is a morphism in $\mathcal{F}(Y)$. The composition of two arrows

$$(g, g^b) : (Z, z) \to (Y, y), \quad (f, f^b) : (Y, y) \to (X, x)$$

is defined to be gf together with the composite morphisms

$$(fg)^*x \xrightarrow{\alpha_{g,f}^{-1}} g^*f^*x \xrightarrow{f^b} g^*y \xrightarrow{g^b} z.$$

As explained in [49, 3.1.3] this defines a category which is an S-groupoid with projection to (Sch/S) given by $(X, x) \mapsto X$. Note also that F comes equipped with a cleavage: if $f : Y \to X$ is a morphism in (Sch/S) and $x \in F(X) = \mathcal{F}(X)$ the we get a pullback morphism

$$(Y, f^*x) \to (X, x).$$

In this way one sees as in [49, 3.1.3] that the theory of fibered categories with cleavages is equivalent to the theory of pseudo-functors.

Example 1.1.15. Choose (using again the axiom of choice) for every morphism $f : X \to Y$ a pullback functor

$$(\text{Sch}/Y) \to (\text{Sch}/X), \quad (Z \to Y) \mapsto (Z \times_Y X \xrightarrow{\text{pr}} X).$$

Such a choice defines a cleavage of $\mathscr{A}_g$. Namely, if X is an S-scheme and (A, λ) is a principally polarized abelian scheme of relative dimension g over X, then for any morphism $f : Y \to X$ we define the pullback $f^*(A, \lambda)$ to be the fiber product $A \times_X Y$ with the principal polarization

$$f^*\lambda := \lambda \times \text{id} : A \times_X Y \to A \times_X Y.$$

The commutative diagram

$$
\begin{array}{ccc}
A \times_X Y & \longrightarrow & A \\
\downarrow & & \downarrow \\
Y & \xrightarrow{\ f\ } & X
\end{array}
$$

then defines a morphism $(A \times_X Y, f^*\lambda) \to (A, \lambda)$ in $\mathscr{A}_g$ over f. In this way we obtain a functor

$$f^* : \mathscr{A}_g(X) \to \mathscr{A}_g(Y)$$

together with isomorphisms ϵ_X and $\alpha_{f,g}$ defining a pseudo-functor.

Following customary abuse of terminology, we will usually define an S-groupoid by specifying the corresponding pseudo-functor. For example, $\mathscr{A}_g$ will be described as the fibered category which to any S-scheme X associates the groupoid of principally polarized abelian schemes over X.

1.1.16. In general, if $\mathscr{G}$ is a groupoid then $\mathscr{G}$ is characterized by the following data.

(i) The set of objects G_0.

(ii) The set of morphisms G_1.

(iii) The maps $s, t : G_1 \to G_0$ sending a morphism ρ to its source and target respectively, the map $e : G_0 \to G_1$ sending an object $g \in G_0$ to id $: g \to g$, and the map $\iota : G_1 \to G_1$ sending a morphism to its inverse.

(iv) The map

$$m : G_1 \times_{s,G_0,t} G_1 \to G_1 \quad (\rho, \eta) \mapsto \rho \circ \eta$$

defining the composition of arrows.

The morphisms s, t, m, ι, and e satisfy the following relations:

(v) $s \circ e = t \circ e = \text{id}$, $s \circ \iota = t$, $t \circ \iota = s$, $s \circ m = s \circ \text{pr}_2$, $t \circ m = t \circ \text{pr}_1$.

(vi) The diagram

$$
\begin{array}{ccc}
G_1 \times_{s,G_0,t} G_1 \times_{s,G_0,t} G_1 & \xrightarrow{\ m \times \text{id}\ } & G_1 \times_{s,G_0,t} G_1 \\
\downarrow{\scriptstyle \text{id} \times m} & & \downarrow{\scriptstyle m} \\
G_1 \times_{s,G_0,t} G_1 & \xrightarrow{\ \ \ m\ \ \ } & G_1
\end{array}
\qquad (1.1.16.1)
$$

commutes.

(vii) The diagrams

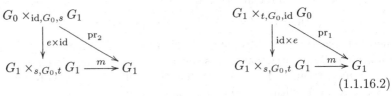

$$(1.1.16.2)$$

commute.

(viii) The squares

are cartesian.

Remark 1.1.17. The map ι is determined by m and e and condition (viii). Indeed if P denotes the fiber product of the diagram

$$
\begin{array}{ccc}
 & & G_1 \times_{s,G_0,t} G_1 \\
 & & \downarrow m \\
G_0 & \xrightarrow{\ e\ } & G_1
\end{array}
$$

then the composite

$$P \longrightarrow G_1 \times_{s,G_0,t} G_1 \xrightarrow{\ \mathrm{pr}_2\ } G_1$$

is an isomorphism and identifies P with G_1. The composite

$$G_1 \xrightarrow{\ \sim\ } P \longrightarrow G_1 \times_{s,G_0,t} G_1 \xrightarrow{\ \mathrm{pr}_1\ } G_1$$

is then the map ι. Thus the groupoid $\mathscr{G}$ is specified by the data (G_0, G_1, s, t, m).

1.1.18. Let $\mathcal{C}$ be any category with fiber products, and suppose given objects $G_0, G_1 \in \mathcal{C}$ together with morphisms in $\mathcal{C}$

$$s, t : G_1 \to G_0, \quad m : G_1 \times_{s,G_0,t} G_1 \to G_1, \quad e : G_0 \to G_1, \quad \iota : G_1 \to G_1.$$

We call $(G_0, G_1, s, t, m, e, \iota)$ a *groupoid in $\mathcal{C}$* if the axioms (now in $\mathcal{C}$) (v)-(viii) hold. As noted in 1.1.17 such a groupoid is determined by the data (G_0, G_1, s, t, m, e).

1.1.19. Let (X_0, X_1, s, t, m, e) be a groupoid in (Sch/S). We then obtain an S-groupoid

$$p : \{X_0/X_1\} \to (\mathrm{Sch}/S)$$

defined as follows.

Objects. The objects of $\{X_0/X_1\}$ are pairs (T, t), where $T \in (\mathrm{Sch}/S)$ and $t : T \to X_0$ is a morphism of S-schemes.

Morphisms. A morphism $(T', t') \to (T, t)$ is a pair $(h, \tilde{h})$, where $h : T' \to T$ is a morphism of S-schemes and $\tilde{h} : T' \to X_1$ is a morphism such that $s \circ \tilde{h} = t'$ and $t \circ \tilde{h} = t \circ h$.

The functor p sends (T, t) to T.

Note that if T is a scheme, then the groupoid $\{X_0/X_1\}(T)$ has objects $X_0(T)$ and morphisms the set $X_1(T)$ and the composition law is given by the morphism m.

Remark 1.1.20. One can generalize the definition of $\{X_0/X_1\}$ by taking X_0 and X_1 in 1.1.19 to be just functors $(\mathrm{Sch}/S)^{\mathrm{op}} \to \mathrm{Set}$ and s, t, e, and m to be morphisms of functors. Again one obtains an S-groupoid $\{X_0/X_1\}$.

1.1.21. Let F be an S-groupoid, and fix a cleavage K for F. For a morphism $f : X \to Y$ in (Sch/S) and object $y \in F(Y)$ the pullback arrow given by K will be denoted by

$$f^* y \to y.$$

Recall that for composable morphisms

$$Z \xrightarrow{\ g\ } X \xrightarrow{\ f\ } Y$$

there is a unique isomorphism $g^* f^* y \simeq (fg)^* y$ in $F(Z)$ over y.

Now let $t_1, t_2 \in F(T)$ be objects over some scheme T. Define

$$\underline{\mathrm{Isom}}(t_1, t_2) : (\mathrm{Sch}/T)^{\mathrm{op}} \to \mathrm{Set}$$

to be the functor sending $f : T' \to T$ to the set of isomorphisms $f^* t_1 \to f^* t_2$ in $F(T')$. If $g : T'' \to T'$ is a morphism and $\iota : f^* t_1 \to f^* t_2$ is an isomorphism in $F(T')$ then $g^*(\iota) \in \underline{\mathrm{Isom}}(t_1, t_2)(T'')$ is defined to be the isomorphism

$$(fg)^* t_1 \xrightarrow{\ \mathrm{can}\ } g^* f^* t_1 \xrightarrow{\ \iota\ } g^* f^* t_2 \xrightarrow{\ \mathrm{can}\ } (fg)^* t_2.$$

Remark 1.1.22. The definition of the sheaf $\underline{\mathrm{Isom}}(t_1, t_2)$ depends on the choice of a cleavage. However, a straightforward verification shows that two different cleavages define canonically isomorphic sheaves. In what follows we will therefore write $\underline{\mathrm{Isom}}(t_1, t_2)$ without reference to a particular choice of cleavage.

1.1.23. Given a diagram of S-groupoids

$$
\begin{array}{ccc}
 & & F \\
 & & \downarrow a \\
G & \xrightarrow{\;b\;} & H
\end{array}
$$

one can form the fiber product $G \times_H F$ as follows. The objects of $G \times_H F$ are triples (g, f, ι), where $g \in G$ and $f \in F$ are objects such that $p_G(g) = p_F(f)$ (where $p_G : G \to (\mathrm{Sch}/S)$ and $p_F : F \to (\mathrm{Sch}/S)$ are the projections) and $\iota : b(g) \to a(f)$ is an isomorphism in H such that the map

$$p_G(g) = p_H(b(g)) \xrightarrow{\iota} p_H(a(f)) = p_F(f) = p_G(g)$$

is the identity map. A morphism $(g', f', \iota') \to (g, f, \iota)$ is a pair (σ, γ), where $\sigma : g' \to g$ is a morphism in G, $\gamma : f' \to f$ is a morphism in F, the two projections $p_G(\sigma)$ and $p_F(\gamma)$ are equal, and such that the diagram

$$
\begin{array}{ccc}
b(g') & \xrightarrow{\;\sigma\;} & b(g) \\
\downarrow{\iota'} & & \downarrow{\iota} \\
a(f') & \xrightarrow{\;\gamma\;} & a(f)
\end{array}
$$

commutes. There is a natural diagram

$$
\begin{array}{ccc}
G \times_H F & \xrightarrow{\;\mathrm{pr}_2\;} & F \\
\downarrow{\mathrm{pr}_1} & & \downarrow a \\
G & \xrightarrow{\;b\;} & H,
\end{array}
\qquad (1.1.23.1)
$$

where pr_1 (resp. pr_2) is the morphism sending (g, f, ι) to g (resp. f). In addition, there is an isomorphism of functors $\sigma : b \circ \mathrm{pr}_1 \to a \circ \mathrm{pr}_2$ defined by

$$b \circ \mathrm{pr}_1(g, f, \iota) = b(g) \xrightarrow{\;\iota\;} a(f) = a \circ \mathrm{pr}_2(g, f, \iota).$$

The square 1.1.23.1 is universal in the sense that for any S-groupoid Z the natural functor

$$\mathrm{HOM}(Z, G \times_H F) \to \mathrm{HOM}(Z, G) \times_{\mathrm{HOM}(Z,H)} \mathrm{HOM}(Z, F)$$

induced by the diagram 1.1.23.1 and σ is an equivalence.

Example 1.1.24. If in 1.1.23 we take H to be the trivial S-groupoid defined by h_S, then we get the product $G \times F$.

1.1.25. For an S-groupoid F, the diagonal map

$$\Delta : F \to F \times F$$

is the functor sending $f \in F$ to (f, f). If $T \in (\mathrm{Sch}/S)$ and $t_1, t_2 \in F(T)$ are two objects, then the fiber product of the diagram

$$
\begin{array}{c}
h_T \\
\downarrow {\scriptstyle t_1 \times t_2} \\
F \xrightarrow{\ \Delta\ } F \times F
\end{array}
$$

is canonically isomorphic to the S-groupoid associated to the functor $\underline{\mathrm{Isom}}(t_1, t_2)$.

Example 1.1.26. Let $\{X_0/X_1\}$ be the S-groupoid associated to morphisms of schemes

$$
s, t : X_1 \to X_0, e : X_0 \to X_1, m : X_1 \times_{s, X_0, t} X_1 \to X_1
$$

as in 1.1.19. If T is a scheme, and $t_1, t_2 \in X_0(T)$ are elements defining objects of $\{X_0/X_1\}(T)$ then the functor

$$
\underline{\mathrm{Isom}}(t_1, t_2) : (\mathrm{Sch}/T)^{\mathrm{op}} \to \mathrm{Set}
$$

is represented by the fiber product of the diagram

$$
\begin{array}{c}
X_1 \\
\downarrow {\scriptstyle s \times t} \\
T \xrightarrow{\ t_1 \times t_2\ } X_0 \times_S X_0.
\end{array}
$$

1.2 Stacks

Definition 1.2.1. An S-groupoid F is called a *prestack* if for every S-scheme T and objects $t_1, t_2 \in F(T)$ the functor $\underline{\mathrm{Isom}}(t_1, t_2)$ is a sheaf on (Sch/T) with respect to the étale topology.

Example 1.2.2. With notation as in 1.1.26, the S-groupoid $\{X_0/X_1\}$ is a prestack, as representable functors are sheaves with respect to the étale topology [49, 2.55].

1.2.3. Let F be an S-groupoid with a cleavage K, and consider a scheme T together with an étale cover $\{f_i : T_i \to T\}$. Define a category $F(\{T_i \to T\})$ as follows.

Objects. Collections $(\{t_i, \sigma_{ij}\})$, where $t_i \in F(T_i)$ and $\sigma_{ij} : \mathrm{pr}_1^* t_i \to \mathrm{pr}_2^* t_j$ is an isomorphism in $F(T_i \times_T T_j)$ such that for every triple i, j, k the diagram

$$
\begin{array}{ccc}
\mathrm{pr}_{12}^* \mathrm{pr}_1^* t_i \xrightarrow{\mathrm{pr}_{12}^* \sigma_{ij}} \mathrm{pr}_{12}^* \mathrm{pr}_2^* t_j \simeq \mathrm{pr}_{23}^* \mathrm{pr}_1^* t_j \xrightarrow{\sigma_{jk}} \mathrm{pr}_{23}^* \mathrm{pr}_2^* t_k \\
\downarrow {\scriptstyle \mathrm{can}} \qquad\qquad\qquad\qquad\qquad\qquad\qquad\qquad \downarrow {\scriptstyle \mathrm{can}} \\
\mathrm{pr}_{13}^* \mathrm{pr}_1^* t_i \xrightarrow{\qquad\qquad \mathrm{pr}_{13}^* \sigma_{ik} \qquad\qquad} \mathrm{pr}_{13}^* \mathrm{pr}_2^* t_k
\end{array}
$$

commutes, where $\mathrm{pr}_{\alpha\beta}$ denotes the projection to the (α, β)-factors of $T_i \times_T T_j \times_T T_k$.

Morphisms. A morphism $(\{t_i, \sigma_{ij}\}) \to (\{t'_i, \sigma'_{ij}\})$ is a collection of morphisms $\gamma_i : t_i \to t'_i$ in $F(T_i)$ such that the diagrams

$$
\begin{array}{ccc}
\mathrm{pr}_1^* t_i & \xrightarrow{\gamma_i} & \mathrm{pr}_1^* t'_i \\
\downarrow{\scriptstyle\sigma_{ij}} & & \downarrow{\scriptstyle\sigma'_{ij}} \\
\mathrm{pr}_2^* t_j & \xrightarrow{\gamma_j} & \mathrm{pr}_1^* t_j
\end{array}
$$

in $F(T_i \times_T T_j)$ commute. Note that there is a natural functor

$$F(T) \to F(\{f_i : T_i \to T\})$$

sending an object $t \in F(T)$ to the $\{f_i^* t_i\}$ with the canonical isomorphisms

$$\mathrm{pr}_1^* f_i^* t_i \xrightarrow{\mathrm{can}} q^* t \xrightarrow{\mathrm{can}} \mathrm{pr}_2^* f_j^* t_j,$$

where $q : T_i \times_T T_j \to T$ is the structure morphism.

Definition 1.2.4. A prestack F is a *stack* if for every étale covering $\{f_i : T_i \to T\}$ of a scheme T the functor

$$F(T) \to F(\{T_i \to T\})$$

is an equivalence of categories.

Remark 1.2.5. As explained in [49, 4.1.2] two different cleavages on F define canonically equivalent categories $F(\{T_i \to T\})$. From this one sees that the condition that a prestack F is a stack is independent of the choice of a cleavage.

Example 1.2.6. The S-groupoid $\mathscr{A}_g$ is a stack. This does not follow immediately from descent theory as we only consider polarizations $\lambda : A \to A^t$ and not a specific line bundle representing the polarization. Nonetheless one can proceed as follows.

Let T be a scheme and let $(A, \lambda) \in \mathscr{A}_g(T)$ be a principally polarized abelian scheme over T. We claim that there is a canonical relatively very ample line bundle M_λ on A whose associated map $A \to A^t$ is 4λ together with a trivialization $e^* M_\lambda \simeq \mathscr{O}_T$. Once this is shown the fact that $\mathscr{A}_g$ is a stack follows from descent theory.

To construct the line bundle M_λ, first recall that a line bundle L on A is called *symmetric* (see also 6.3.2) if there exists an isomorphism $\iota^* L \to L$, where $\iota : A \to A$ is the map sending a to $-a$. Étale locally on T, there exists a symmetric line bundle L on A representing λ (see for example 6.3.18). The line bundle L is unique up to tensoring L with a degree 0 line bundle N which is also symmetric. Now if N is a degree 0 symmetric line bundle on A we have

$\iota^* N \simeq N^{-1}$ so the condition that N is symmetric is equivalent to $N^2 \simeq \mathscr{O}_A$. It follows that the line bundle $M_\lambda := L^{\otimes 4}$ is independent of the choice of L up to isomorphism. If we further fix an isomorphism $\chi : e^* M_\lambda \to \mathscr{O}_T$ then it follows that for any morphism $g : T' \to T$, object $(A', \lambda') \in \mathscr{A}_g(T')$, and isomorphism $\tilde{g} : g^*(A, \lambda) \to (A', \lambda')$ there is a unique isomorphism of line bundles $\tilde{g}^* M_\lambda \to M_{\lambda'}$ on A' compatible with the isomorphisms

$$\tilde{g}^* \chi : e'^* \tilde{g}^* M_\lambda \to \mathscr{O}_{T'}, \quad \chi' : e'^* M_{\lambda'} \to \mathscr{O}_{T'}.$$

Finally the fact that M_λ is relatively very ample follows from [35, Theorem on p. 163].

1.2.7. If F is an S-groupoid, one can form its associated stack F^a, which is a stack together with a morphism $F \to F^a$ such that for any other stack G the natural functor

$$\mathrm{HOM}(F^a, G) \to \mathrm{HOM}(F, G)$$

is an equivalence of categories [31, 3.2]. As one would expect, the construction of F^a is rather formal. However, it should be noted that the construction shows that if F is a prestack then the functor $F \to F^a$ is fully faithful.

In particular, for the S-groupoid $\{X_0/X_1\}$ defined in 1.1.19 there is an associated stack which we will denote by $[X_0/X_1]$.

Example 1.2.8. If F is a functor $(\mathrm{Sch}/S)^{\mathrm{op}} \to$ Set with associated S-groupoid F^*, then we have $(F^*)^a = (F^a)^*$, where F^a denotes the sheaf associated to F. In particular, F^* is a stack if and only if F is a sheaf.

Example 1.2.9. If

$$s, t : X_1 \to X_0, \quad e : X_0 \to X_1, \quad m : X_1 \times_{s, X_0, t} X_1 \to X_1$$

is a collection of morphisms of schemes as in 1.1.19 such that $s \times t : X_1 \to X_0 \times_S X_0$ is a monomorphism, then for every S-scheme T the subset $X_1(T) \subset X_0(T) \times X_0(T)$ is an equivalence relation on $X_0(T)$, and $\{X_0/X_1\}$ is the S-groupoid corresponding to the presheaf

$$T \mapsto X_0(T)/X_1(T).$$

The stack $[X_0/X_1]$ is then the stack associated to the sheaf associated to this presheaf.

Definition 1.2.10. An *algebraic space* is a sheaf

$$X : (\mathrm{Sch}/S)^{\mathrm{op}} \to \mathrm{Set}$$

whose associated S-groupoid is isomorphic to $[X_0/X_1]$ for some collection of data as in 1.1.19 with the morphisms

$$s, t : X_1 \to X_0$$

étale, and

$$s \times t : X_1 \to X_0 \times_S X_0$$

a quasi-compact monomorphism.

Remark 1.2.11. The above definition of algebraic space coincides with the one in [29]. One can relax the condition on the diagonal, but to develop the theory in this greater generality requires some foundational work and is not needed in this monograph.

1.2.12. A morphism of stacks $f : \mathscr{X} \to \mathscr{Y}$ is *schematic* if for every scheme Y and morphism $y : Y \to \mathscr{Y}$ (equivalently object $y \in \mathscr{Y}(Y)$) the fiber product

$$\mathscr{X} \times_{\mathscr{Y}} Y$$

is representable by a scheme.

Let P is a property of morphisms of schemes which is stable under base change (i.e. if $g : Z \to W$ is a morphism of schemes having property P then for every morphism of schemes $W' \to W$ the morphism $Z \times_W W' \to W'$ also has property P). For example, P could be the property of being a closed immersion, immersion, étale, smooth etc. We say that a schematic morphism of stacks $f : \mathscr{X} \to \mathscr{Y}$ has *property P* if for every scheme Y and morphism $y : Y \to \mathscr{Y}$ the morphism of schemes

$$\mathscr{X} \times_{\mathscr{Y}} Y \to Y$$

has property P.

If $\mathscr{X}$ is a stack, then the condition that for every scheme T and objects $t_1, t_2 \in \mathscr{X}(T)$ the functor $\underline{\mathrm{Isom}}(t_1, t_2)$ is representable by a scheme is equivalent to the statement that the diagonal

$$\Delta : \mathscr{X} \to \mathscr{X} \times \mathscr{X}$$

is schematic by 1.1.25. This also implies that for any diagram with T and T' schemes

$$
\begin{array}{c}
T \\
\downarrow a \\
T' \xrightarrow{\ b\ } \mathscr{X}
\end{array}
$$

the fiber product is representable by a scheme, as this fiber product is isomorphic to the fiber product of the diagram

$$
\begin{array}{c}
T \times T' \\
\downarrow a \times b \\
\mathscr{X} \xrightarrow{\ \Delta\ } \mathscr{X} \times \mathscr{X}.
\end{array}
$$

Therefore we can talk about a morphism $T \to \mathscr{X}$ being étale, flat etc., when T is a scheme and the diagonal of $\mathscr{X}$ is schematic (after we introduce algebraic spaces this condition on the diagonal will be relaxed).

Proposition 1.2.13. *A functor*

$$X : (\mathrm{Sch}/S)^{op} \to \mathrm{Set}$$

is an algebraic space if and only if the following hold:

(i) X *is a sheaf with respect to the étale topology.*
(ii) *The diagonal* $\Delta : X \to X \times X$ *is schematic, and quasi-compact.*
(iii) *There exists a scheme* U *and an étale surjective morphism* $U \to X$.

Proof. See [29, II.1.7]. $\square$

1.2.14. Most of the usual theory of schemes as developed in EGA [17], can be extended to algebraic spaces.

First of all, if X is an algebraic space then since the diagonal $\Delta : X \to X \times X$ is schematic, we can talk about the diagonal being separated or proper. We say that X is *separated* if the diagonal of X is proper (this is equivalent to saying that the diagonal is a closed immersion).

As we now explain, we can also talk about a morphism of algebraic spaces being smooth, étale, locally of finite type etc.

Let $f : X \to Y$ be a morphism of algebraic spaces. Let $q : U \to Y$ be an étale surjection with U a scheme, and let X_U denote the fiber product $X \times_Y U$. Then X_U is again an algebraic space, and therefore we can also choose an étale surjection $V \to X_U$ with V a scheme. We then have a commutative diagram

$$
\begin{array}{ccccc}
V & \xrightarrow{\;r\;} & X_U & \xrightarrow{\;\rho\;} & U \\
& \searrow{\scriptstyle p} & \downarrow & & \downarrow{\scriptstyle q} \\
& & X & \xrightarrow{\;f\;} & Y,
\end{array}
\qquad (1.2.14.1)
$$

where the square is cartesian and r and q (and hence also p) are étale surjections.

Let P be a property of morphisms of schemes. We say that P is *local in the étale topology on source and target* if for any morphism of schemes $f : X \to Y$ and diagram 1.2.14.1 the morphism f has property P if and only if the morphism $\rho r : V \to U$ has property P.

Definition 1.2.15. Let P be a property of morphisms of schemes which is local in the étale topology on source and target. We say that a morphism of algebraic spaces $f : X \to Y$ *has property* P if for every diagram 1.2.14.1 the morphism of schemes ρr has property P.

For example, we can talk about a morphism of algebraic spaces being smooth, étale, flat etc.

Remark 1.2.16. A straightforward verification shows that if $f : X \to Y$ is a morphism of algebraic spaces and if for some diagram 1.2.14.1 the morphism ρr has property P, then f has property P.

Definition 1.2.17. A morphism of stacks $f : \mathscr{X} \to \mathscr{Y}$ is *representable* if for every scheme Y and morphism $y : Y \to \mathscr{Y}$ the fiber product $\mathscr{X} \times_{\mathscr{Y}} Y$ is an algebraic space.

Remark 1.2.18. By descent theory, if $f : \mathscr{X} \to \mathscr{Y}$ is a representable morphism of stacks then for every algebraic space Y and morphism $y : Y \to \mathscr{Y}$ the fiber product $\mathscr{X} \times_{\mathscr{Y}} Y$ is an algebraic space.

1.2.19. Let P be a property of morphisms of algebraic spaces. We say that P is *stable under base change and local in the étale topology on target* if the following hold:

(i) If $f : X \to Y$ has property P then for any morphism $Y' \to Y$ the base change $X \times_Y Y' \to Y'$ has property P.
(ii) If $f : X \to Y$ is a morphism and $\{Y_i \to Y\}$ is an étale covering of Y, then f has property P if and only if for every i the base change $X \times_Y Y_i \to Y_i$ has property P.

For example, P could be the property of being étale, surjective, flat etc. (see [31, 3.10] for a more exhaustive list).

Definition 1.2.20. Let $f : \mathscr{X} \to \mathscr{Y}$ be a representable morphism of stacks, and let P be a property of morphisms of algebraic spaces which is stable under base change and local in the étale topology on target. We say that f *has property P* if for every morphism of schemes $Y \to \mathscr{Y}$ the morphism

$$\mathscr{X} \times_{\mathscr{Y}} Y \to Y$$

has property P.

1.2.21. There are a few properties of morphisms that are not étale local on source and target but still can be extended to algebraic spaces. Most notably, we say that an algebraic space X is *quasi-compact* if there exists an étale surjection $U \to X$ with U a quasi-compact scheme. A morphism $f : X \to Y$ of algebraic spaces is called *quasi-compact* if for every quasi-compact scheme U and morphism $U \to Y$ the fiber product $X \times_Y U$ is a quasi-compact algebraic space.

Definition 1.2.22. A stack $\mathscr{X}$ is *algebraic* if the following conditions hold:

(i) The diagonal $\Delta : \mathscr{X} \to \mathscr{X} \times \mathscr{X}$ is representable and quasi-compact.
(ii) There exists a smooth surjection $X \to \mathscr{X}$ with X a scheme.

An algebraic stack $\mathscr{X}$ is *Deligne-Mumford* if the diagonal Δ is in addition separated and if there exists an étale surjection $X \to \mathscr{X}$ with X a scheme.

Remark 1.2.23. We call a smooth surjection $X \to \mathscr{X}$ as in (ii) a *presentation*.

Remark 1.2.24. By [29, II.6.16] any algebraic space quasi-finite and separated over a scheme is again a scheme. It follows from this that if $\mathscr{X}$ is a Deligne-Mumford stack over S then the diagonal $\Delta : \mathscr{X} \to \mathscr{X} \times_S \mathscr{X}$ is schematic (or more generally if $\mathscr{X}$ has quasi-finite and separated diagonal).

1.2.25. Let $\mathscr{X}/S$ be an algebraic stack, and let $X_0 \to \mathscr{X}$ be a presentation. Set

$$X_1 := X_0 \times_{\mathscr{X}} X_0,$$

and let $s, t : X_1 \to X_0$ be the two projections. Also let $e : X_0 \to X_1$ be the diagonal and define

$$m : X_1 \times_{t, X_0, s} X_1 \to X_1$$

to be the map obtained as the composite

$$X_1 \times_{t, X_0, s} X_1 \xrightarrow{\simeq} X_0 \times_{\mathscr{X}} \times X_0 \times_{\mathscr{X}} X_0 \xrightarrow{\mathrm{pr}_{13}} X_0 \times_{\mathscr{X}} X_0 = X_1,$$

where the first map is the canonical isomorphism and the second map is the projection onto the first and third factors. Then one verifies that the axioms in 1.1.16 (v)–(viii) hold, and moreover there is a natural isomorphism

$$[X_0/X_1] \simeq \mathscr{X}.$$

1.2.26. Let $f : \mathscr{X} \to \mathscr{Y}$ be a morphism of algebraic S-stacks, and let $Q : Y \to \mathscr{Y}$ be a presentation. Then $\mathscr{X}_Y := \mathscr{X} \times_{\mathscr{Y}} Y$ is again an algebraic S-stack, and the projection $\mathscr{X}_Y \to \mathscr{X}$ is smooth and surjective. Choose a presentation $R : X \to \mathscr{X}_Y$. Then the composite $X \to \mathscr{X}_Y \to \mathscr{X}$ is also a presentation, and we have a commutative diagram

$$
\begin{array}{ccccc}
X & \xrightarrow{\;R\;} & \mathscr{X}_Y & \xrightarrow{\;\rho\;} & Y \\
 & \searrow & \downarrow & & \downarrow Q \\
 & & \mathscr{X} & \xrightarrow{\;f\;} & \mathscr{Y},
\end{array}
\qquad (1.2.26.1)
$$

where the square is cartesian.

1.2.27. Let P be a property of morphisms of schemes. We say that P is *local in the smooth topology on source and target* if for every commutative diagram of schemes

$$
\begin{array}{ccccc}
X'' & \xrightarrow{\;p'\;} & X' & \longrightarrow & X \\
 & f'' \searrow & \downarrow & & \downarrow f \\
 & & Y' & \xrightarrow{\;q\;} & Y,
\end{array}
$$

where the square is cartesian and the morphisms q and p' are smooth and surjective, the morphism f has property P if and only if the morphism f'' has property P. For example, P could be the property of being surjective, locally of finite type, flat, or smooth.

By the discussion in 1.2.14, we can then also talk about a morphism of algebraic spaces having property P.

If P is a property of morphisms of schemes which is local in the smooth topology on source and target, then we can extend the notion to morphisms of algebraic stacks as follows. Let $f : \mathscr{X} \to \mathscr{Y}$ be a morphism of algebraic stacks. We say that f has property P if for every diagram 1.2.26.1 the morphism of algebraic spaces $\rho R : X \to Y$ has property P. As in 1.2.16, this equivalent to verifying that for a single diagram 1.2.26.1 the morphism ρR has property P.

1.2.28. As in the case of algebraic spaces 1.2.21, we can also talk about a morphism of algebraic S-stacks $f : \mathscr{X} \to \mathscr{Y}$ being quasi-compact. First of all, an algebraic S-stack $\mathscr{X}$ is called *quasi-compact* if there exists a presentation $P : X \to \mathscr{X}$ with X a quasi-compact scheme. A morphism $f : \mathscr{X} \to \mathscr{Y}$ is called *quasi-compact* if for every quasi-compact scheme Y and morphism $Y \to \mathscr{Y}$ the fiber product $\mathscr{X} \times_{\mathscr{Y}} Y$ is a quasi-compact algebraic stack.

1.2.29. A morphism $f : \mathscr{X} \to \mathscr{Y}$ of algebraic stacks is called a *closed immersion* (resp. *open immersion, immersion*) if it is schematic and if for every morphisms $Y \to \mathscr{Y}$ with Y a scheme the map of schemes

$$\mathscr{X} \times_{\mathscr{Y}} Y \to Y$$

is a closed immersion (resp. open immersion, immersion).

This enables us to associate to an algebraic stack $\mathscr{X}$ a topological space $|\mathscr{X}|$ as follows. First of all, define an S-field to be a field K together with a morphism $\mathrm{Spec}(K) \to S$. Morphisms of S-fields are defined to be S-morphisms. For an S-field K, let $|\mathscr{X}(K)|$ denote the set of isomorphism classes in the groupoid $\mathscr{X}(K)$. We define $|\mathscr{X}|$ to be the quotient of the set

$$\coprod_{S\text{-fields } K} |\mathscr{X}(K)|$$

by the equivalence relation obtained by declaring $x_K \in |\mathscr{X}(K)|$ and $x_{K'} \in |\mathscr{X}(K')|$ (for two S-fields K and K') to be equivalent if there exists a diagram of S-fields

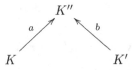

such that the images of x_K and $x_{K'}$ in $|\mathscr{X}(K'')|$ are equal.

Note that if $f : \mathscr{X} \to \mathscr{Y}$ is a morphism of algebraic stacks, then there is a natural induced morphism $|f| : |\mathscr{X}| \to |\mathscr{Y}|$. If f is an immersion, then this map is an injection.

In particular, if $\mathscr{X}$ is an algebraic S-stack then for any open substack $\mathscr{U} \subset \mathscr{X}$ we obtain a subset $|\mathscr{U}| \subset |\mathscr{X}|$. We define a topology on $|\mathscr{X}|$ by declaring the subsets $|\mathscr{U}| \subset |\mathscr{X}|$ to be open. One verifies immediately that if $f : \mathscr{X} \to \mathscr{Y}$ is a morphism of algebraic stacks then the induced map $|f| : |\mathscr{X}| \to |\mathscr{Y}|$ is continuous.

Remark 1.2.30. We leave it to the reader to show that in the case when $\mathscr{X}$ is a scheme the space $|\mathscr{X}|$ coincides with the underlying topological space of $\mathscr{X}$ in the usual sense.

Definition 1.2.31. A morphism of algebraic S-stacks $f : \mathscr{X} \to \mathscr{Y}$ is *closed* if the map $|f| : |\mathscr{X}| \to |\mathscr{Y}|$ is a closed map of topological spaces. The morphism f is called *universally closed* if for every morphism of algebraic stacks $\mathscr{Y}' \to \mathscr{Y}$ the base change $\mathscr{X} \times_{\mathscr{Y}} \mathscr{Y}' \to \mathscr{Y}'$ is closed.

1.2.32. Once we have a notion of a universally closed map we can talk about proper morphisms of stacks as follows.

First of all, if $f : X \to Y$ is a morphism of algebraic spaces then the diagonal $X \to X \times_Y X$ is schematic so it makes sense to say that it is proper. We say that f is *separated* if this is the case. A morphism $f : X \to Y$ of algebraic spaces is called *proper* if it is locally of finite type, quasi-compact, separated, and universally closed. One checks that the property of a morphism of algebraic spaces being proper stable under base change and local on the target in the sense of 1.2.19. Therefore is also makes sense to say that a representable morphisms of stacks is proper.

A morphism of algebraic S-stacks $f : \mathscr{X} \to \mathscr{Y}$ is called *separated* if the (representable) diagonal

$$\mathscr{X} \to \mathscr{X} \times_{\mathscr{Y}} \mathscr{X}$$

is proper. The morphism $f : \mathscr{X} \to \mathscr{Y}$ is called *proper* if it is separated, locally of finite type and quasi-compact, and universally closed.

1.3 Comparison of Topologies

1.3.1. In the above, we have used throughout the étale topology on (Sch/S). It is often convenient to also consider the fppf topology on (Sch/S). The following theorem of Artin ensures that one gets an equivalent notion of algebraic stack.

Theorem 1.3.2 ([7, 6.1]) *Let (X_0, X_1, s, t, m) be a groupoid in (Sch/S) and assume the following hold:*

(i) X_0 and X_1 are locally of finite type over S.
(ii) The morphisms $s, t : X_1 \to X_0$ are flat and locally of finite presentation.
(iii) The map $s \times t : X_1 \to X_0 \times X_0$ is of finite type.

Let $\mathscr{X}$ be the fibered category obtained by taking the stack associated to $\{X_0/X_1\}$ with respect to the fppf topology. Then $\mathscr{X}$ is an algebraic stack in the sense of 1.2.22.

1.4 Coarse Moduli Spaces

1.4.1. Let S be a scheme and $\mathscr{X}/S$ an Artin stack locally of finite presentation with finite diagonal. A *coarse moduli space* for $\mathscr{X}$ is a morphism $\pi : \mathscr{X} \to X$ to an algebraic space such that:

(i) π is initial for maps to algebraic spaces.
(ii) For every algebraically closed field k the map $|\mathscr{X}(k)| \to X(k)$ is bijective, where $|\mathscr{X}(k)|$ denotes the set of isomorphism classes in $\mathscr{X}(k)$.

Theorem 1.4.2 ([27, 1.3], [14, 1.1]) *There exists a coarse moduli space $\pi :$ $\mathscr{X} \to X$. In addition:*

(i) X/S is separated and locally of finite type if S is noetherian.
(ii) π is proper.
(iii) If $X' \to X$ is a flat morphism of algebraic spaces, then

$$\pi' : \mathscr{X}' := \mathscr{X} \times_X X' \to X'$$

is a coarse moduli space for $\mathscr{X}'$.

1.4.3. The key case to keep in mind is when S is affine and $\mathscr{X}$ admits a finite flat surjection $U \to \mathscr{X}$ with U an affine scheme. In this case since the diagonal of $\mathscr{X}$ is finite the algebraic space

$$V := U \times_{\mathscr{X}} U$$

is in fact an affine scheme finite over U. Then the proof of 1.4.2 shows that the coarse space X is equal to the spectrum of the equalizer of the two rings homomorphisms

$$\mathrm{pr}_1^*, \mathrm{pr}_2^* : \Gamma(U, \mathscr{O}_U) \to \Gamma(V, \mathscr{O}_V).$$

In particular, if G is a finite group acting on an affine scheme U and $\mathscr{X}$ denotes the stack quotient $[U/G]$ then the coarse space is the spectrum of the ring of invariants $\Gamma(U, \mathscr{O}_U)^G$.

In fact, one can show that étale locally on the coarse space any Deligne-Mumford stack is of the form $[U/G]$ [2, 2.2.3].

Remark 1.4.4. Assume S is locally noetherian and that $\mathscr{X}/S$ is an algebraic stack of finite type with finite diagonal. Then using the valuative criterion and the fact that $\pi : \mathscr{X} \to X$ is proper and surjective one sees that $\mathscr{X}$ is proper over S if and only if X is proper over S.

1.5 Rigidification of Stacks

The results of this section are only used in chapters 5 and 6.

1.5.1. Let $\mathscr{X}$ be an Artin stack and let $\mathscr{I} \to \mathscr{X}$ denote the inertia stack. By definition $\mathscr{I}$ is the fiber product of the diagram

$$\mathcal{X}$$
$$\downarrow \scriptstyle{\Delta}$$
$$\mathcal{X} \xrightarrow{\;\Delta\;} \mathcal{X} \times \mathcal{X}.$$
(1.5.1.1)

More concretely, for any scheme S the groupoid $\mathscr{I}(S)$ is the groupoid of pairs (x, α), where $x \in \mathcal{X}(S)$ and α is an automorphism of x in $\mathcal{X}(S)$. In particular $\mathscr{I}$ is a group algebraic space over $\mathcal{X}$.

Assume given a group scheme $\mathcal{G}$ over $\mathcal{X}$ and a closed immersion $\mathcal{G} \subset \mathscr{I}$ of group spaces over $\mathcal{X}$. Assume that for any scheme S and $x \in \mathcal{X}(S)$ the subgroup scheme $x^*\mathcal{G} \subset \underline{\mathrm{Aut}}(x)$ is a normal subgroup scheme flat over S, where $x^*\mathcal{G}$ denotes $\mathcal{G} \times_{\mathcal{X}, x} S$. This implies that for any second object $x' \in \mathcal{X}(S)$ and isomorphism $\iota : x \to x'$ the induced isomorphism

$$\underline{\mathrm{Aut}}(x) \to \underline{\mathrm{Aut}}(x'), \quad g \mapsto \iota \circ g \circ \iota^{-1} \qquad (1.5.1.2)$$

sends $x^*\mathcal{G}$ to $x'^*\mathcal{G}$.

A scheme valued section $\alpha \in x^*\mathcal{G}$ acts on the space $\underline{\mathrm{Hom}}(x, x')$ by

$$\iota \mapsto \iota \circ \alpha. \qquad (1.5.1.3)$$

This defines an action of $x^*\mathcal{G}$, and we can form the quotient

$$x^*\mathcal{G} \backslash \underline{\mathrm{Hom}}(x, x'). \qquad (1.5.1.4)$$

This quotient is an algebraic space since $x^*\mathcal{G}$ is flat over S. Observe also that since 1.5.1.2 sends $x^*\mathcal{G}$ to $x'^*\mathcal{G}$ the quotient 1.5.1.4 is also isomorphic to

$$\underline{\mathrm{Hom}}(x, x')/x'^*\mathcal{G}, \qquad (1.5.1.5)$$

where $\alpha \in x'^*\mathcal{G}$ acts by

$$\iota \mapsto \alpha \circ \iota. \qquad (1.5.1.6)$$

Let $\underline{\mathrm{Hom}}_{\overline{\mathcal{X}}}(x, x')$ denote the space 1.5.1.4.

Lemma 1.5.2 *Let* $x, x', x'' \in \mathcal{X}(S)$ *be three object. Then there exists a unique morphism*

$$\underline{\mathrm{Hom}}_{\overline{\mathcal{X}}}(x, x') \times_S \underline{\mathrm{Hom}}_{\overline{\mathcal{X}}}(x', x'') \to \underline{\mathrm{Hom}}_{\overline{\mathcal{X}}}(x, x'') \qquad (1.5.2.1)$$

such that the diagram

$$\begin{array}{ccc}
\underline{\mathrm{Hom}}_{\mathcal{X}}(x, x') \times_S \underline{\mathrm{Hom}}_{\mathcal{X}}(x', x'') & \xrightarrow{\;\text{composition}\;} & \underline{\mathrm{Hom}}_{\mathcal{X}}(x, x'') \\
\downarrow & & \downarrow \\
\underline{\mathrm{Hom}}_{\overline{\mathcal{X}}}(x, x') \times_S \underline{\mathrm{Hom}}_{\overline{\mathcal{X}}}(x', x'') & \xrightarrow{\hspace{2cm}} & \underline{\mathrm{Hom}}_{\overline{\mathcal{X}}}(x, x'')
\end{array} \qquad (1.5.2.2)$$

commutes.

Proof. We leave this to the reader. □

1.5.3. Let $\overline{\mathscr{X}}^{\mathrm{ps}}$ denote the prestack which to any scheme S associates the category whose objects are the objects of $\mathscr{X}(S)$ and whose morphisms $x \to x'$ are the elements of

$$\underline{\mathrm{Hom}}_{\overline{\mathscr{X}}}(x, x'). \tag{1.5.3.1}$$

Using 1.5.2 we obtain a category $\overline{\mathscr{X}}^{\mathrm{ps}}(S)$. Let $\overline{\mathscr{X}}$ denote the associated stack.

Proposition 1.5.4. *The stack $\overline{\mathscr{X}}$ is an algebraic stack.*

Proof. Let $x : U \to \mathscr{X}$ be a smooth covering by a scheme, and let $R = \underline{\mathrm{Hom}}_{\mathscr{X}}(x, x)$. Also let $\overline{R}$ denote $\underline{\mathrm{Hom}}_{\overline{\mathscr{X}}}(x, x)$. Then $\overline{\mathscr{X}}$ is the stack associated to a groupoid $\overline{R} \rightrightarrows U$. Thus it suffices to show that the projections $\overline{R} \to U$ are flat (here we also use 1.3.2). By construction the natural map $R \to \overline{R}$ is faithfully flat, and hence the projections $\overline{R} \to U$ are also flat since $R \to U$ is smooth. □

Definition 1.5.5. The stack $\overline{\mathscr{X}}$ in 1.5.4 is called the *rigidification* of $\mathscr{X}$ with respect to $\mathscr{G}$.

Remark 1.5.6. Note that the projection $\mathscr{X} \to \overline{\mathscr{X}}$ is faithfully flat. Moreover, if $U \to \overline{\mathscr{X}}$ is a morphism and $s : U \to \mathscr{X}$ is a section, then $U \times_{\overline{\mathscr{X}}} \mathscr{X}$ is isomorphic to the classifying stack $Bs^*\mathscr{G}$ over U of the group scheme $s^*\mathscr{G}$. In particular, if $\mathscr{G}$ is smooth over $\mathscr{X}$ then the morphism $\mathscr{X} \to \overline{\mathscr{X}}$ is also smooth.

Example 1.5.7. Let $f : X \to S$ be a proper morphism of algebraic spaces such that the natural map $\mathscr{O}_S \to f_*\mathscr{O}_X$ is an isomorphism universally (i.e. the same holds after base change $S' \to S$). Let $\mathscr{P}ic_{X/S}$ denote the fibered category over (Sch/S) whose fiber over a scheme $S' \to S$ is the groupoid of line bundles on $X_{S'} := X \times_S S'$. By [7, Appendix 2] the fibered category $\mathscr{P}ic_{X/S}$ is an algebraic stack. The assumption on f implies that for any S-scheme $S' \to S$ and line bundle L on $X_{S'}$ the automorphism group of L is canonically isomorphic to $\Gamma(S', \mathscr{O}_{S'}^*)$. This implies that the inertia stack of $\mathscr{P}ic_{X/S}$ is isomorphic to $\mathbb{G}_m \times \mathscr{P}ic_{X/S}$.

We can therefore apply the rigidification construction with $\mathscr{G}$ equal to the inertia stack to get a morphism

$$\mathscr{P}ic_{X/S} \to \underline{\mathrm{Pic}}_{X/S}$$

of algebraic stacks. Now by construction, the objects of $\underline{\mathrm{Pic}}_{X/S}$ have no non-trivial automorphisms, so $\underline{\mathrm{Pic}}_{X/S}$ is in fact an algebraic space [31, 8.1.1]. The algebraic space $\underline{\mathrm{Pic}}_{X/S}$ is called the *relative Picard space* of X/S. It follows from the construction of the rigidification that $\underline{\mathrm{Pic}}_{X/S}$ is the sheaf on the category of S-schemes with the fppf topology associated to the presheaf

$$(S' \to S) \mapsto \{\text{isom. classes of line bundles on } X \times_S S'\}. \tag{1.5.7.1}$$

Example 1.5.8. Let $\mathscr{X}$ be the stack over the category of schemes whose fiber over any scheme S is the groupoid of pairs (E, L), where E is an ellliptic curve over S and L is a line bundle on E which has degree 0 on all geometric fibers. For any such object (E, L) there is an inclusion

$$\mathbb{G}_m \hookrightarrow \underline{\mathrm{Aut}}_{\mathscr{X}}(E, L) \tag{1.5.8.1}$$

sending a unit $u \in \mathscr{O}_S^*$ to the automorphism of (E, L) which is the identity on E and multiplication by u on L. We can therefore form the rigidification $\mathscr{X} \to \overline{\mathscr{X}}$ with respect to $\mathbb{G}_m$. By the preceding example and using the canonical identification $E \simeq \underline{\mathrm{Pic}}^0_{E/S}$, it follows that $\overline{\mathscr{X}}$ can be identified with the stack which to any scheme S associates the groupoid of pairs (E, p), where E/S is an elliptic curve and $p : S \to E$ is a section. Equivalently the rigidification $\overline{\mathscr{X}}$ is isomorphic to the universal elliptic curve $\mathscr{E}$ over $\mathscr{M}_{1,1}$ (the moduli stack of elliptic curves).

2

Preliminaries

In this chapter we review some background material used in the main part of the text. The experienced reader may wish to just skim this chapter for our notational conventions and then proceed to chapter 3.

2.1 Abelian Schemes and Torsors

2.1.1. If S is an algebraic space, an *abelian algebraic space* over S is a proper smooth group algebraic space A/S with geometrically connected fibers. An important fact due to Raynaud [13, 1.9] is that when S is a scheme any abelian algebraic space over S is in fact a scheme.

A *semi-abelian scheme* over a scheme S is a smooth commutative group scheme G/S such that every geometric fiber of G is an extension of an abelian scheme by a torus.

2.1.2. If $f : P \to S$ is a proper morphism of algebraic spaces with geometrically connected and reduced fibers, then the map $\mathscr{O}_S \to f_* \mathscr{O}_P$ is an isomorphism. From a general theorem of Artin [5, 7.3] it follows that the Picard functor $\underline{\mathrm{Pic}}(P)$ defined to be the sheaf with respect to the fppf topology of the presheaf

$$T/S \mapsto \{\text{isomorphism classes of invertible sheaves on } P_T\}$$

is an algebraic space locally of finite presentation over S (see also 1.5.7)

The case we will be interested in is when P is a torsor under an abelian algebraic space A/S. In this case define a subfunctor $\underline{\mathrm{Pic}}^0(P) \subset \underline{\mathrm{Pic}}(P)$ as follows. The points of $\underline{\mathrm{Pic}}^0(P)$ over a scheme-valued point $\bar{t} : \mathrm{Spec}(\Omega) \to S$ with Ω an algebraically closed field is the subgroup of $\underline{\mathrm{Pic}}(P)(\Omega)$ consisting of isomorphism classes of line bundles $\mathscr{L}$ such that for every $a \in A_{\bar{t}}(\Omega)$ the line bundles $t_a^* \mathscr{L}$ and $\mathscr{L}$ are isomorphic (note that this depends only on the isomorphism class of $\mathscr{L}$). The subfunctor $\underline{\mathrm{Pic}}^0(P) \subset \underline{\mathrm{Pic}}(P)$ is defined by

M.C. Olsson, *Compactifying Moduli Spaces for Abelian Varieties*. Lecture Notes in Mathematics 1958.
© Springer-Verlag Berlin Heidelberg 2008

associating to any scheme T/S the subset of $\underline{\mathrm{Pic}}(P)(T)$ of isomorphism classes of line bundles $\mathscr{L}$ such that for every algebraically closed field Ω and point $\bar{t} : \mathrm{Spec}(\Omega) \to S$ the image of $[\mathscr{L}]$ in $\underline{\mathrm{Pic}}(P)(\Omega)$ is in $\underline{\mathrm{Pic}}^0(P)(\Omega)$.

In the case when $P = A$ is the trivial torsor, the subfunctor $\underline{\mathrm{Pic}}^0(P) \subset \underline{\mathrm{Pic}}(P)$ is the dual abelian scheme, denoted A^t.

Proposition 2.1.3. *The subfunctor $\underline{\mathrm{Pic}}^0(P) \subset \underline{\mathrm{Pic}}(P)$ is a smooth proper algebraic space over S. Tensor product defines the structure of an abelian algebraic space on $\underline{\mathrm{Pic}}^0(P)$.*

Proof. Since $\underline{\mathrm{Pic}}^0(P) \subset \underline{\mathrm{Pic}}(P)$ is in any case a subsheaf (with respect to the étale topology) it suffices to prove the proposition after replacing S by an étale cover. We may therefore assume that P is a trivial torsor and that S is a scheme. In this case $\underline{\mathrm{Pic}}^0(P)$ is the dual abelian scheme of A as mentioned above. $\square$

2.1.4. For any S–scheme T, the set of isomorphisms of A_T–torsors $\iota : A_T \to P_T$ is canonically in bijection with the set $P(T)$. Any such isomorphism ι defines an isomorphism $\iota_* : A^t \to \underline{\mathrm{Pic}}^0(P)$. This defines a morphism of schemes

$$A^t \times P \to \underline{\mathrm{Pic}}^0(P), \quad ([\mathscr{L}], \iota) \mapsto [\iota_* \mathscr{L}]. \tag{2.1.4.1}$$

Proposition 2.1.5. *The morphism 2.1.4.1 factors uniquely as*

$$A^t \times P \xrightarrow{\mathrm{pr}_1} A^t \xrightarrow{\sigma} \underline{\mathrm{Pic}}^0(P), \tag{2.1.5.1}$$

where σ is an isomorphism.

Proof. The uniqueness of the factorization is clear. To prove the existence we can by descent theory work étale locally on S and may therefore assume that S is a scheme and that $P \simeq A$ is the trivial torsor. In this case the map 2.1.4.1 is identified with the translation action

$$A^t \times A \to A^t, \quad ([\mathscr{L}], a) \mapsto [t_a^* \mathscr{L}].$$

It is well known that this action is trivial (see for example [3, 4.1.12]). $\square$

2.1.6. Note in particular that an invertible sheaf $\mathscr{L}$ on P defines a homomorphism $\lambda_{\mathscr{L}} : A \to A^t$ by

$$A \to \underline{\mathrm{Pic}}^0(P) \simeq A^t, \quad a \mapsto [t_a^* \mathscr{L} \otimes \mathscr{L}^{-1}].$$

2.1.7. The dual abelian scheme A^t/S has a very useful description which does not require the sheafification involved in the definition of $\underline{\mathrm{Pic}}$ in general. Namely define a *rigidified line bundle* on A to be a pair $(\mathscr{L}, \iota)$, where $\mathscr{L}$ is a line bundle on A and $\iota : \mathscr{O}_S \to e^* \mathscr{L}$ is an isomorphism, where $e : S \to A$ is the identity section. Any isomorphism $(\mathscr{L}, \iota) \to (\mathscr{L}', \iota')$ between two such pairs is unique if it exists. Using this one shows that $\underline{\mathrm{Pic}}(A)$ can also be viewed as representing the functor

$T/S \mapsto \{$isomorphism classes of rigidified line bundles on $A_T\}$.

In particular, over $A \times_S A^t$ there is a tautological line bundle $\mathscr{B}$, called the *Poincare bundle*, with a trivialization of the restriction of $\mathscr{B}$ to $\{e\} \times A^t$.

If $\mathscr{M}$ is a line bundle on A, then for any scheme-valued point $a \in A$ the line bundle

$$t_a^* \mathscr{M} \otimes \mathscr{M}^{-1} \otimes_{\mathscr{O}_S} \mathscr{M}^{-1}(a) \otimes_{\mathscr{O}_S} \mathscr{M}(e) \qquad (2.1.7.1)$$

has a canonical rigidification. We obtain a map $\lambda_{\mathscr{M}} : A \to A^t$ by sending $a \in A$ to 2.1.7.1.

2.1.8. Let B be a scheme and A/B an abelian scheme over B. Fix a finitely generated free abelian group X with associated torus T. A semi-abelian scheme G sitting in an exact sequence

$$0 \to T \to G \to A \to 0 \qquad (2.1.8.1)$$

defines a homomorphism $c : X \to A^t$ as follows. The group X is the character group of T, so any $x \in X$ defines an extension

$$0 \to \mathbb{G}_m \to E_x \to A \to 0 \qquad (2.1.8.2)$$

by pushing 2.1.8.1 out along the homomorphism $x : T \to \mathbb{G}_m$. Let $\mathscr{L}_x$ denote the corresponding line bundle on A. The identity element of G induces a trivialization of $\mathscr{L}_x(0)$ and hence $\mathscr{L}_x$ is a rigidified line bundle. Moreover it follows from the construction that there is a canonical isomorphism of rigidied line bundles

$$\mathscr{L}_x \otimes \mathscr{L}_{x'} \simeq \mathscr{L}_{x+x'} \qquad (2.1.8.3)$$

for $x, x' \in X$. In particular, the $\mathscr{O}_A$–module

$$\oplus_{x \in X} \mathscr{L}_x \qquad (2.1.8.4)$$

has a natural algebra structure and there is a canonical isomorphism over A

$$G \to \underline{\mathrm{Spec}}_A (\oplus_{x \in X} \mathscr{L}_x). \qquad (2.1.8.5)$$

For any $a \in A(B)$ there exists étale locally on B an isomorphism $t_a^* \mathscr{L}_x \to \mathscr{L}_x$. Indeed étale locally on B there exists a lifting $\tilde{a} \in G(B)$ of a. Then the map

$$t_{\tilde{a}}^* : \mathscr{O}_G \to \mathscr{O}_G$$

induces an isomorphism $t_{\tilde{a}}^* \mathscr{L}_x \to \mathscr{L}_x$ since G is commutative.

In terms of the isomorphism 2.1.8.5, the group structure on G can be described as follows. A (scheme-valued) point $g \in G(S)$ is given by a point $a \in A(S)$ together with trivializations $\iota_x : a^* \mathscr{L}_x \to \mathscr{O}_S$ which are compatible in the sense that for any two $x, x' \in X$ the diagram

$$a^* \mathscr{L}_x \otimes a^* \mathscr{L}_{x'} \xrightarrow{\iota_x \otimes \iota_{x'}} \mathscr{L}_x \otimes \mathscr{L}_{x'}$$

$$2.1.8.3 \downarrow \qquad\qquad \downarrow 2.1.8.3 \qquad\qquad (2.1.8.6)$$

$$a^* \mathscr{L}_{x+x'} \xrightarrow{\iota_{x+x'}} \mathscr{L}_{x+x'}$$

commutes. Such trivializations ι_x are equivalent to the structure of rigidified line bundles on the $t_a^* \mathscr{L}_x$. The translation action of a point $g = (a, \{\iota_x\})$ on G can be described as follows.

Namely, since the translation action of A on A^t is trivial, there exists a unique isomorphism of rigidified line bundles $t_a^* \mathscr{L}_x \to \mathscr{L}_x$ (where $t_a^* \mathscr{L}_x$ is rigidified using ι_x). This defines a morphism of algebras

$$t_a^*(\oplus_x \mathscr{L}_x) \to \oplus_{x \in X} \mathscr{L}_x \qquad\qquad (2.1.8.7)$$

which gives the translation action of g on G.

Conversely any homomorphism $c : X \to A^t$ defines a semi-abelian scheme G sitting in an extension 2.1.8.1. Indeed for every $x \in X$ let $\mathscr{L}_x$ be the rigidified line bundle corresponding to $c(x)$, and set

$$\mathscr{O}_G := \oplus_{x \in X} \mathscr{L}_x.$$

Then $\mathscr{O}_G$ is given an algebra structure using the unique isomorphism of rigidified line bundles

$$\mathscr{L}_x \otimes \mathscr{L}_{x'} \to \mathscr{L}_{x+x'}$$

corresponding to the fact that c is a homomorphism. The group structure on G is defined as in the preceding paragraph.

2.2 Biextensions

We only review the aspects of the theory we need in what follows. For a complete account see [19, VII] and [11].

2.2.1. Let S be a scheme and F and G abelian sheaves in the topos S_{Et} of all S–schemes with the big étale topology. In the applications the sheaves F and G will be either abelian schemes over S or constant sheaves associated to some abelian group.

A $\mathbb{G}_m$–*biextension of* $F \times G$ is a sheaf of sets E with a map $\pi : E \to F \times G$ and the following additional structure:

(i) A faithful action of $\mathbb{G}_m$ on E over $F \times G$ such that the quotient sheaf $[E/\mathbb{G}_m]$ is isomorphic to $F \times G$ via the map induced by π.

(ii) For a local section $q \in G$, let $E_{-,q} \subset E$ denote $\pi^{-1}(F \times q)$. Then we require for any local section $q \in G$ a structure on $E_{-,q}$ of an extension of abelian groups

$$0 \to \mathbb{G}_m \to E_{-,q} \to F \to 0 \qquad\qquad (2.2.1.1)$$

compatible with the $\mathbb{G}_m$–action from (i). This structure is determined by isomorphisms of $\mathbb{G}_m$–torsors over F

$$\varphi_{p,p';q} : E_{p,q} \wedge E_{p',q} \to E_{p+p',q} \qquad (2.2.1.2)$$

satisfying certain compatibilities (see [19, VII.2.1]). Here p and p' are local sections of F and $E_{p,q} \subset E$ denotes the subsheaf of elements mapping to $(p,q) \in F \times G$.

(iii) Similarly, for a local section $p \in F$, let $E_{p,-} \subset E$ denote $\pi^{-1}(p \times G)$. Then we require for any local section $p \in F$ a structure on $E_{p,-}$ of an extension of abelian groups

$$0 \to \mathbb{G}_m \to E_{p,-} \to G \to 0 \qquad (2.2.1.3)$$

compatible with the $\mathbb{G}_m$–action from (i). This structure is determined by isomorphisms of $\mathbb{G}_m$–torsors

$$\psi_{p;q,q'} : E_{p,q} \wedge E_{p,q'} \to E_{p,q+q'} \qquad (2.2.1.4)$$

satisfying certain compatibilities [19, VII.2.1], where q and q' are local sections of G.

(iv) For any local sections $p, p' \in F$ and $q, q' \in G$ the diagram

$$
\begin{array}{ccc}
E_{p,q}\wedge E_{p,q'}\wedge E_{p',q}\wedge E_{p',q'} & \xrightarrow{\text{flip}} & E_{p,q}\wedge E_{p',q}\wedge E_{p,q'}\wedge E_{p',q'} \\
\downarrow{\scriptstyle \psi_{p;q,q'}\wedge\psi_{p';q,q'}} & & \downarrow{\scriptstyle \varphi_{p,p';q}\wedge\varphi_{p,p';q'}} \\
E_{p,q+q'} \wedge E_{p',q+q'} & & E_{p+p',q} \wedge E_{p+p';q'} \\
& \searrow{\scriptstyle \varphi_{p,p';q+q'}} \quad \swarrow{\scriptstyle \psi_{p+p';q,q'}} & \\
& E_{p+p',q+q'} &
\end{array}
$$

$$(2.2.1.5)$$

commutes.

If $\pi : E \to F \times G$ and $\pi' : E' \to F \times G$ are two $\mathbb{G}_m$–biextensions of $F \times G$, then a morphism of $\mathbb{G}_m$–biextensions $f : E \to E'$ is a map of sheaves over $F \times G$ compatible with the $\mathbb{G}_m$–actions and the maps $\varphi_{p,p';q}$ and $\psi_{p;q,q'}$. Note that (i) implies that any morphism of $\mathbb{G}_m$–biextensions of $F \times G$ is automatically an isomorphism. The collection of $\mathbb{G}_m$–biextensions of $F \times G$ therefore form a groupoid denoted $\underline{\text{Biext}}(F, G; \mathbb{G}_m)$.

The *trivial* $\mathbb{G}_m$–biextension of $F \times G$ is defined to be the product $\mathbb{G}_m \times F \times G$ with $\mathbb{G}_m$–action on the first factor, and the map to $F \times G$ given by projection to the last two factors. The maps $\varphi_{p,p';q}$ and $\psi_{p;q,q'}$ are induced by the group law on $\mathbb{G}_m \times F \times G$. If E is a $\mathbb{G}_m$–biextension of $F \times G$, then a *trivialization* of E is an isomorphism from the trivial $\mathbb{G}_m$–biextension of $F \times G$ to E.

If $f : F' \to F$ and $g : G' \to G$ are morphisms of abelian sheaves and $\pi : E \to F \times G$ is a $\mathbb{G}_m$–biextension of $F \times G$, then the pullback $\pi' : E \times_{F \times G} (F' \times G') \to F' \times G'$ has a natural structure of a $\mathbb{G}_m$-biextension of $F' \times G'$ induced by the structure on E. We therefore have a pullback functor

$$(f \times g)^* : \underline{\text{Biext}}(F, G; \mathbb{G}_m) \to \underline{\text{Biext}}(F', G'; \mathbb{G}_m). \qquad (2.2.1.6)$$

For later use, let us discuss some examples of pullbacks.

2.2.2. Let $E \in \underline{\text{Biext}}(F, G; \mathbb{G}_m)$ be a biextension, and let $f : F \to F$ be the zero map and $g : G \to G$ the identity map. Denote by E' the pullback

$$E' := (f \times g)^* E. \qquad (2.2.2.1)$$

For any sections $(p, q) \in F \times G$ we have $E'_{p,q} = E_{0,q}$. Let $e_q \in E_{0,q}$ denote the section corresponding to the identity element in the group $E_{-,q}$.

Lemma 2.2.3 *The sections e_q define a trivialization of E'.*

Proof. Let $s_{p,q} \in E'_{p,q}$ denote the section $e_q \in E'_{p,q} = E_{0,q}$.

Let $\psi'_{p;q,q'}$ and $\varphi'_{p,p';q}$ be the maps giving E' the biextension structure. Then we need to show the following:

(i) For every $p, p' \in F$ and $q \in G$ the map

$$\varphi'_{p,p';q} : E'_{p,q} \wedge E'_{p',q} \to E'_{p+p',q} \qquad (2.2.3.1)$$

sends $s_{p,q} \wedge s_{p',q}$ to $s_{p+p',q}$;
(ii) For every $p \in F$ and $q, q' \in G$ the map

$$\psi'_{p;q,q'} : E'_{p,q} \wedge E'_{p,q'} \to E_{p,q+q'} \qquad (2.2.3.2)$$

sends $s_{p,q} \wedge s_{p,q'}$ to $s_{p,q+q'}$.

Statement (i) is immediate as $\varphi'_{p,p';q}$ is simply the map

$$E_{0,q} \wedge E_{0,q} \to E_{0,q} \qquad (2.2.3.3)$$

induced by the group law on $E_{-,q}$.

For statement (ii), note that the map $\psi'_{p;q,q'}$ is given by the map

$$\psi_{0;q,q'} : E_{0,q} \wedge E_{0,q'} \to E_{0,q+q'}. \qquad (2.2.3.4)$$

Now the identity element $e_{q+q'} \in E_{0,q+q'}$ is characterized by the condition that

$$\varphi_{0,0;q+q'}(e_{q+q'} \wedge e_{q+q'}) = e_{q+q'}. \qquad (2.2.3.5)$$

We show that $\psi_{0;q,q'}(e_q \wedge e_{q'})$ also has this property.

By the commutativity of 2.2.1.5, the diagram

$$
\begin{array}{ccc}
E_{0,q} \wedge E_{0,q'} \wedge E_{0,q} \wedge E_{0,q'} & \xrightarrow{\;\text{flip}\;} & E_{0,q} \wedge E_{0,q} \wedge E_{0,q'} \wedge E_{0,q'} \\
\Big\downarrow{\scriptstyle \psi_{0;q,q'} \wedge \psi_{0;q,q'}} & & \Big\downarrow{\scriptstyle \varphi_{0,0;q} \wedge \varphi_{0,0;q'}} \\
E_{0,q+q'} \wedge E_{0,q+q'} & & E_{0,q} \wedge E_{0;q'} \\
\end{array}
$$

with $\varphi_{0,0;q+q'}$ and $\psi_{0;q,q'}$ converging to $E_{0,q+q'}$

$$(2.2.3.6)$$

commutes. Chasing the section

$$
e_q \wedge e_{q'} \wedge e_q \wedge e_{q'} \in E_{0,q} \wedge E_{0,q'} \wedge E_{0,q} \wedge E_{0,q'} \tag{2.2.3.7}
$$

along the two paths to $E_{0,q+q'}$ we obtain

$$
\varphi_{0,0;q+q'}((\psi_{0;q,q'}(e_q \wedge e_{q'})) \wedge (\psi_{0;q,q'}(e_q \wedge e_{q'}))) = \psi_{0;q,q'}(e_q \wedge e_{q'}) \tag{2.2.3.8}
$$

as desired. □

2.2.4. Similarly, if we take f to be the identity map and g to be the zero morphism, then the sections $f_p \in E_{p,0}$ corresponding to the identity elements of the groups $E_{p,-}$ define a trivialization of the biextension

$$
E'' := (\mathrm{id} \times 0)^* E. \tag{2.2.4.1}
$$

2.2.5. We can also pullback a biextension E of $F \times G$ by $\mathbb{G}_m$ along the map

$$
(-1) : F \times G \to F \times G, \quad (a,b) \mapsto (-a,-b). \tag{2.2.5.1}
$$

Concretely for local sections $(p,q) \in F \times G$ we have

$$
((-1)^* E)_{p,q} = E_{-p,-q}. \tag{2.2.5.2}
$$

Let

$$
\sigma_{p,q} : E_{p,q} \to ((-1)^* E)_{p,q} = E_{-p,-q} \tag{2.2.5.3}
$$

by the isomorphism of $\mathbb{G}_m$-torsors characterized by the condition that the diagram

$$
\begin{array}{ccc}
E_{p,q} \wedge E_{p,-q} & \xrightarrow{\;\sigma_{p,q} \times \mathrm{id}\;} & E_{-p,-q} \wedge E_{p,-q} \\
\Big\downarrow{\scriptstyle \psi_{p;q,-q}} & & \Big\downarrow{\scriptstyle \varphi_{-p,p;-q}} \\
E_{p,0} & \xrightarrow{\;\text{can}\;} & E_{0,-q} \\
\end{array}
\tag{2.2.5.4}
$$

commutes, where can denotes the unique isomorphism of $\mathbb{G}_m$-torsors sending the identity element of $E_{p,-}$ (an element of $E_{p,0}$) to the identity element of $E_{-,q}$ (an element of $E_{0,-q}$).

Lemma 2.2.6 *The maps* $\sigma_{p,q}$ *define an isomorphism of* $\mathbb{G}_m$*-biextensions of* $F \times G$

$$\sigma : E \to (-1)^* E. \tag{2.2.6.1}$$

Proof. We can rewrite the diagram 2.2.5.4 as the diagram

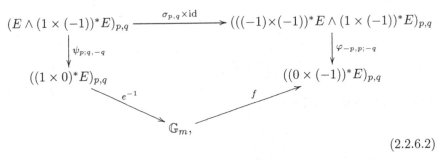

$$\tag{2.2.6.2}$$

where e and f are the maps of biextensions defined in 2.2.2 and 2.2.4. Since the maps $\psi_{p;q,-q}$ and $\varphi_{-p,p;-q}$ induce morphisms of biextensions

$$E \wedge (1 \times (-1))^* E \to (1 \times 0)^* E \tag{2.2.6.3}$$

and

$$((-1) \times (-1))^* E \wedge (1 \times (-1))^* E \to (0 \times -1)^* E \tag{2.2.6.4}$$

by [11, 1.2] this proves that the maps $\sigma_{p,q}$ are obtained from a composition of morphisms of biextensions (which is also a morphism of biextensions). $\square$

2.2.7. If $F = G$, then the "flip" map $\iota : F \times F \to F \times F$ sending (a, b) to (b, a) induces a functor

$$\iota^* : \underline{\mathrm{Biext}}(F, F; \mathbb{G}_m) \to \underline{\mathrm{Biext}}(F, F; \mathbb{G}_m). \tag{2.2.7.1}$$

A *symmetric* $\mathbb{G}_m$*-biextension of* F is defined to be a $\mathbb{G}_m$–biextension E of $F \times F$ together with an isomorphism $\lambda : \iota^* E \to E$ such that $\lambda \circ i^*(\lambda) = \mathrm{id}$ (note that $(\iota^*)^2 = \mathrm{id}$). If (E', λ') is a second symmetric $\mathbb{G}_m$–biextension of F, then a morphism $(E', \lambda') \to (E, \lambda)$ is a morphism $h : E' \to E$ of biextensions such that the diagram

$$\tag{2.2.7.2}$$

$$\begin{array}{ccc} \iota^* E' & \xrightarrow{\lambda'} & E' \\ \iota^*(h) \downarrow & & \downarrow h \\ \iota^* E & \xrightarrow{\lambda} & E \end{array}$$

commutes. We denote by $\underline{\mathrm{Biext}}^{\mathrm{sym}}(F, \mathbb{G}_m)$ the category of symmetric $\mathbb{G}_m$–biextensions of F. The "flip" map $\mathbb{G}_m \times F \times F \to \mathbb{G}_m \times F \times F$ sending (u, a, b) to (u, b, a) induces the structure of a symmetric $\mathbb{G}_m$–biextension on the trivial $\mathbb{G}_m$–biextension of $F \times F$. As above, we therefore have a notion of a trivialization of a symmetric $\mathbb{G}_m$–biextension of F.

2.2.8. Let X be a free abelian group of finite rank, and view X as a constant sheaf. In this case the group of automorphisms of a symmetric $\mathbb{G}_m$–biextension $\pi : E \to X \times X$ of X is canonically isomorphic to $\mathrm{Hom}(S^2X, \mathbb{G}_m)$, where S^2X denotes the second symmetric power of X. To see this let $h : E \to E$ be such an automorphism. Since h is a morphism over $X \times X$ and E is a $\mathbb{G}_m$–torsor over $X \times X$, for any local section $e \in E$ there exists a unique element $u \in \mathbb{G}_m$ such that $h(e) = u(e)$. Furthermore, since h is compatible with the $\mathbb{G}_m$–action the element u depends only on $\pi(e)$. We therefore obtain a set map $b : X \times X \to \mathbb{G}_m$ by associating to any pair (x, y) the section of $\mathbb{G}_m$ obtained by locally choosing a lifting $e \in E$ of (x, y) and sending (x, y) to the corresponding unit $u \in \mathbb{G}_m$. Compatibility with (2.2.1 (ii) and (iii)) implies that this map b in fact is bilinear. Furthermore, the commutativity of 2.2.7.2 amounts to the condition $b(x, y) = b(y, x)$. Thus any automorphism h of E is determined by a map $b : S^2X \to \mathbb{G}_m$. Conversely any such map b induces an automorphism by the formula

$$e \mapsto b(\pi(e)) \cdot e. \tag{2.2.8.1}$$

2.2.9. Let A/S be an abelian scheme, A^t/S the dual abelian scheme, and $\lambda : A \to A^t$ a principal polarization defined by an invertible sheaf $\mathscr{M}$ on A. Via the isomorphism λ, the Poincare bundle $\mathscr{B} \to A \times A^t$ defines a $\mathbb{G}_m$–torsor (denoted by the same letter) $\pi : \mathscr{B} \to A \times A$. This torsor can be described as follows. For an integer n and subset $I \subset \{1, \ldots, n\}$, let $m_I : A^{\times n} \to A$ be the map

$$(a_1, \ldots, a_n) \mapsto \sum_{i \in I} a_i, \tag{2.2.9.1}$$

where $A^{\times n}$ denotes the n–fold fiber product over S of A with itself. If I is the empty set then m_I sends all points of A^n to the identity element of A. Then $\mathscr{B} \to A \times A$ is canonically isomorphic to

$$\Lambda(\mathscr{M}) := \bigotimes_{I \subset \{1,2\}} m_I^* \mathscr{M}^{(-1)^{\mathrm{card}(I)}}. \tag{2.2.9.2}$$

In other words, for any two scheme–valued points $a, b \in A$, the fiber of $\mathscr{B}$ over $(a, b) \in A \times A$ is equal to

$$\mathscr{M}(a + b) \otimes \mathscr{M}(a)^{-1} \otimes \mathscr{M}(b)^{-1} \otimes \mathscr{M}(0). \tag{2.2.9.3}$$

Note also that the definition of $\Lambda(\mathscr{M})$ is symmetric in the two factors of $A \times A$ so there is a canonical isomorphism $\iota : \mathscr{B} \to \mathscr{B}$ over the flip map $A \times A \to A \times A$.

The theorem of the cube [11, 2.4] provides a canonical isomorphism

$$\rho : \mathscr{O}_{A^3} \to \Theta(\mathscr{M}) := \bigotimes_{I \subset \{1,2,3\}} m_I^* \mathscr{M}^{(-1)^{\mathrm{card}(I)}}. \tag{2.2.9.4}$$

For any three scheme–valued points $a, b, c \in A$ this gives a canonical isomorphism

$$\mathcal{M}(a+b) \otimes \mathcal{M}^{-1}(a) \otimes \mathcal{M}^{-1}(b) \otimes \mathcal{M}(0) \otimes \mathcal{M}(a+c) \otimes \mathcal{M}^{-1}(a) \otimes \mathcal{M}^{-1}(c)$$

$$\downarrow$$

$$\mathcal{M}(a+b+c) \otimes \mathcal{M}^{-1}(b+c) \otimes \mathcal{M}^{-1}(a).$$

$$(2.2.9.5)$$

For points $p, p', q \in A$ this induces an isomorphism

$$\psi_{p;q,q'} : \mathcal{B}_{p,q} \otimes \mathcal{B}_{p,q'} \to \mathcal{B}_{p,q+q'}, \qquad (2.2.9.6)$$

and also by symmetry for $p, q, q' \in A$ an isomorphism

$$\varphi_{p,p';q} : \mathcal{B}_{p,q} \otimes \mathcal{B}_{p',q} \to \mathcal{B}_{p+p',q}. \qquad (2.2.9.7)$$

It is shown in [11, 2.4] that these maps together with the above defined map ι give $\mathcal{B}$ the structure of a symmetric $\mathbb{G}_m$–biextension of A.

2.2.10. Let X be a free abelian group of finite rank, let $(A, \mathcal{M})$ be an abelian scheme with an invertible sheaf defining a principal polarization over some base S with $\mathrm{Pic}(S) = 0$, and let $c : X \to A(S)$ be a homomorphism. We fix a rigidification of $\mathcal{M}$. Pulling back $\mathcal{M}$ along c we obtain a $\mathbb{G}_m$–torsor W over X (viewed as a constant sheaf on the category of S–schemes). Let

$$\lambda_{\mathcal{M}} : A \to A^t, \quad a \in A \mapsto t_a^* \mathcal{M} \otimes \mathcal{M}^{-1} \otimes \mathcal{M}^{-1}(a) \otimes \mathcal{M}(0) \qquad (2.2.10.1)$$

be the isomorphism defined by $\mathcal{M}$ (where we view A^t as classifying invertible sheaves rigidified along 0), and let $\mathcal{B} \to A \times A$ denote the symmetric $\mathbb{G}_m$–biextension defined by the Poincare bundle. Pulling back along $c \times c : X \times X \to A \times A$ we also obtain a symmetric $\mathbb{G}_m$–biextension E of X. From above we know that for $(x, y) \in X \times X$ the fiber $E_{x,y}$ is equal to

$$\mathcal{M}(c(x) + c(y)) \otimes \mathcal{M}(c(x))^{-1} \otimes \mathcal{M}(c(y))^{-1} \otimes \mathcal{M}(0). \qquad (2.2.10.2)$$

Now let $\psi : X \to c^* \mathcal{M}^{-1}$ be a trivialization. This trivialization ψ defines a trivialization τ of the $\mathbb{G}_m$–torsor over $X \times X$ underlying E by sending $(x, y) \in X \times X$ to

$$\psi(x+y) \otimes \psi(x)^{-1} \otimes \psi(y)^{-1} \otimes \psi(0).$$

In what follows it will be important to make explicit the additional conditions on ψ needed for τ to be compatible with the symmetric biextension structure.

For a point $a \in A$, let L_a denote the rigidified invertible sheaf corresponding to $\lambda_{\mathcal{M}}(a)$. The sheaf L_a is equal to the restriction of the Poincare bundle to $A \simeq A \times \{a\} \subset A \times A$. In particular, for a point $b \in A$ we have $L_a(b) \simeq \mathcal{B}_{(b,a)}$. For $x \in X$, we also sometimes write L_x for $L_{c(x)}$ if no confusion seems likely to arise.

Lemma 2.2.11 *For any integer $d \geq 0$ and $x, y \in X$ the sheaves $t_{c(y)}^*(\mathscr{M}^d \otimes L_x)$ and $\mathscr{M}^d \otimes L_{x+dy}$ on A are non–canonically isomorphic, where $t_{c(y)} : A \to A$ denotes translation by $c(y)$.*

Proof. View A^t as the connected component of the space classifying isomorphism classes of line bundles on A. Then the isomorphism $\lambda : A \to A^t$ sends a scheme–valued point $a \in A$ to the isomorphism class of $L_a := t_a^*\mathscr{M} \otimes \mathscr{M}^{-1}$. Since λ is a homomorphism, there exists an isomorphism

$$(t_x^*\mathscr{M}) \otimes (t_y^*\mathscr{M}) \simeq \mathscr{M} \otimes t_{x+y}^*\mathscr{M}. \tag{2.2.11.1}$$

It follows that

$$t_y^*(\mathscr{M}^d \otimes L_x) \simeq t_y^*(\mathscr{M}^{d-1} \otimes t_x^*\mathscr{M}) \simeq t_y^*\mathscr{M}^{d-1} \otimes t_{x+y}^*\mathscr{M} \simeq \mathscr{M}^{d-1} \otimes t_{x+dy}^*\mathscr{M}, \tag{2.2.11.2}$$

and also

$$\mathscr{M}^d \otimes L_{x+dy} \simeq \mathscr{M}^{d-1} \otimes t_{x+dy}^*\mathscr{M}. \tag{2.2.11.3}$$

It follows that étale locally on S the two line bundles in the lemma are isomorphic. It follows that the functor on S–schemes

$$T/S \mapsto \{\text{isomorphisms } t_{c(y)}^*(\mathscr{M}^d \otimes L_x) \to \mathscr{M}^d \otimes L_{x+dy} \text{ over } A_T\} \tag{2.2.11.4}$$

is a $\mathbb{G}_m$–torsor. Since $\mathrm{Pic}(S) = 0$ this torsor is trivial so there exists an isomorphism over S. □

2.2.12. It follows that to give an isomorphism

$$t_{c(y)}^*(\mathscr{M}^d \otimes L_x) \to \mathscr{M}^d \otimes L_{x+dy} \tag{2.2.12.1}$$

is equivalent to giving an isomorphism of $\mathscr{O}_S$–modules

$$(\mathscr{M}^d \otimes L_x)(c(y)) \to \mathscr{M}^d(0) \otimes L_{x+dy}(0). \tag{2.2.12.2}$$

Since $\mathscr{M}^d \otimes L_{x+dy}$ is rigidified at 0 this in turn is equivalent to a trivialization of $(\mathscr{M}^d \otimes L_x)(c(y))$. On the other hand, there is a canonical isomorphism

$$(\mathscr{M}^d \otimes L_x)(c(y)) \simeq \mathscr{M}(c(y))^d \otimes \mathscr{B}_{(c(y), c(x))}. \tag{2.2.12.3}$$

It follows that ψ and τ define an isomorphism

$$\psi(y)^d \tau(y, x) : t_{c(y)}^*(\mathscr{M}^d \otimes L_x) \to \mathscr{M}^d \otimes L_{x+dy}. \tag{2.2.12.4}$$

Proposition 2.2.13. *The trivialization τ is compatible with the symmetric biextension structure on $(c \times c)^*\mathscr{B}$ if and only if the following two conditions hold (see the proof for explanation of the numbering):*
(iii)' *For any $x, x', y \in X$ and $d \geq 0$ the diagram*

$$t^*_{c(y)}(\mathcal{M}^d \otimes L_x) \otimes t^*_{c(y)}(\mathcal{M}^{d'} \otimes L_{x'}) \xrightarrow{\quad can \quad} t^*_{c(y)}(\mathcal{M}^{d+d'} \otimes L_{x+x'})$$

$$\psi(y)^d \tau(y,x) \otimes \psi(y)^{d'} \tau(y,x') \downarrow \qquad\qquad\qquad \downarrow \psi(y)^{d+d'} \tau(y,x+x')$$

$$(\mathcal{M}^d \otimes L_{x+dy}) \otimes (\mathcal{M}^{d'} \otimes L_{x'+d'y}) \xrightarrow{\quad can \quad} \mathcal{M}^{d+d'} \otimes L_{x+x'+(d+d')y}$$

$$(2.2.13.1)$$

commutes, where "can" denotes the canonical isomorphisms described in
4.1.10.2.

(ii)' *For any* $x, y, y' \in X$, *the diagram*

$$t^*_{c(y+y')}(\mathcal{M}^d \otimes L_x) \xrightarrow{\quad \psi(y)^d \tau(y,x) \quad} t^*_{c(y')}(\mathcal{M}^d \otimes L_{x+dy})$$

$$\psi(y+y')^d \tau(y+y',x) \searrow \qquad \swarrow \psi(y')^d \tau(y',x+dy)$$

$$\mathcal{M}^d \otimes L_{x+d(y+y')}$$

$$(2.2.13.2)$$

commutes.

Proof. First we claim that (iii)' is equivalent to compatibility with the structure in (2.2.1 (iii)). To see this note first that 2.2.13.1 clearly commutes when $x = x' = 0$. From this it follows that it suffices to consider the case when $d = 0$. In this case the commutativity of 2.2.13.1 amounts to the statement that the image of $\tau(y,x) \otimes \tau(y,x')$ under the canonical map

$$\mathcal{B}_{(y,x)} \otimes \mathcal{B}_{(y,x')} = L_x(y) \otimes L_{x'}(y) \to L_{x+x'}(y) = \mathcal{B}_{(y,x+x')} \qquad (2.2.13.3)$$

is equal to $\tau(y, x + x')$. This is precisely compatibility with the structure in (2.2.1 (iii)).

Next we claim that condition (ii)' in the case when $d = 0$ is equivalent to compatibility of τ with (2.2.1 (ii)). Indeed in this case the composite map

$$(\psi(y')^d \tau(y', x + dy)) \circ (\psi(y)^d \tau(y, x)) \qquad (2.2.13.4)$$

is equal to the map induced by the image of $\tau(y,x) \otimes \tau(y',x)$ under the map

$$L_x(y) \otimes L_x(y') = \mathcal{B}_{(y,x)} \otimes \mathcal{B}_{(y',x)} \to \mathcal{B}_{(y+y',x)} \simeq L_x(y+y'). \qquad (2.2.13.5)$$

Note also that (ii)' holds in the case when $x = 0$ by the definition of $\tau(y,y')$.

Condition (ii)' in general follows from these two special cases and (iii)'. To see this note that there is a commutative diagram

$$t^*_{c(y+y')}(\mathcal{M}^d) \otimes t^*_{c(y+y')} L_x \xrightarrow{\quad can \quad} t^*_{c(y+y')}(\mathcal{M}^d \otimes L_x) \qquad , \qquad (2.2.13.6)$$

$$\downarrow \qquad\qquad\qquad\qquad \downarrow \psi(y)^d \tau(y,x)$$

$$t^*_{c(y')}(\mathcal{M}^d \otimes L_{dy}) \otimes t^*_{c(y')}(L_x) \xrightarrow{\quad can \quad} t^*_{c(y')}(\mathcal{M}^d \otimes L_{x+dy})$$

$$\downarrow \qquad\qquad\qquad\qquad \downarrow \psi(y')^d \tau(y',x+dy)$$

$$\mathcal{M}^d \otimes L_{d(y+y')} \otimes L_x \xrightarrow{\quad can \quad} \mathcal{M}^d \otimes L_{x+d(y+y')}$$

where the left column is obtained by taking the tensor product of the maps 2.2.13.4 in the cases $x = 0$ and $d = 0$. Then (iii)' implies that it suffices to show that the composite of the left column is equal to the tensor product of $\psi(y + y')^d$ and $\tau(y + y', x)$ which follows from the above special cases.

Finally compatibility with (2.2.1 (iv)) is automatic by the definition of τ, as is the compatibility with the isomorphism ι giving the symmetric structure.
$\square$

2.3 Logarithmic Geometry

In this section we review the necessary parts of the theory of logarithmic geometry developed by Fontaine, Illusie, and Kato. For complete treatments of the theory we recommend [24] and [41].

2.3.1. Let us start by reviewing some terminology from the theory of monoids.

We only consider commutative monoids with unit, and morphisms of monoids are required to preserve the unit element. We usually write the monoid law additively (the main exception being a ring R viewed as a monoid under multiplication).

The inclusion functor

$$(\text{abelian groups}) \subset (\text{monoids})$$

has a left adjoint sending a monoid M to the group M^{gp} which is the quotient of the set of pairs $\{(a, b) \in M \times M\}$ by the equivalence relation

$(a, b) \sim (c, d)$ if there exists $s \in M$ such that $s + a + d = s + b + c$.

The group structure on M^{gp} is induced by the addition

$$(a, b) + (c, d) = (a + c, b + d).$$

The adjunction map $\pi : M \to M^{gp}$ sends $a \in M$ to $(a, 0)$.

A monoid is called *integral* if for every $a \in M$ the translation map

$$M \to M, \quad b \mapsto a + b$$

is injective. The monoid M is called *coherent* if M is a finitely generated monoid, and *fine* if M is coherent and integral.

Let X be a scheme.

Definition 2.3.2. (i) A *pre-log structure* on X is a pair (M, α), where M is a sheaf of monoids on the étale site of X and $\alpha : M \to \mathscr{O}_X$ is a morphism of sheaves of monoids (where $\mathscr{O}_X$ is viewed as a monoid under multiplication).

(ii) A pre-log structure (M, α) is called a *log structure* if the map α induces a bijection $\alpha^{-1}(\mathscr{O}_X^*) \to \mathscr{O}_X^*$.

(iii) A *log scheme* is a pair (X, M_X) consisting of a scheme X and a log structure M_X on X.

Remark 2.3.3. As in (iii) above, when dealing with (pre-)log structures we usually omit the map α from the notation and write simply M for the pair (M, α).

Remark 2.3.4. If (X, M) is a log scheme then the units $M^* \subset M$ are by the definition of log structure identified with $\mathscr{O}_X^*$ via the map $\alpha : M \to \mathscr{O}_X$. We let $\lambda : \mathscr{O}_X^* \hookrightarrow M$ be the resulting inclusion. The monoid law on M defines an action of $\mathscr{O}_X^*$ on M by translation. The quotient $\overline{M} := M/\mathscr{O}_X^*$ has a natural monoid structure induced by the monoid structure on M.

Remark 2.3.5. The notion of log structure makes sense in any ringed topos. Using the étale topology on algebraic spaces and Deligne-Mumford stacks and the lisse-étale topology on Artin stacks (see for example [31, §12]) we can therefore also talk about log algebraic spaces and log algebraic stacks.

2.3.6. The natural inclusion functor

$$(\text{log structures on } X) \hookrightarrow (\text{pre-log structures on } X)$$

has a left adjoint $M \mapsto M^a$. The log structure M^a is obtained from M by setting M^a equal to the pushout $M \oplus_{\alpha^{-1}\mathscr{O}_X^*} \mathscr{O}_X^*$ in the category of sheaves of monoids of the diagram

$$\begin{array}{ccc} \alpha^{-1}\mathscr{O}_X^* & \xrightarrow{\lambda} & M \\ \downarrow & & \\ \mathscr{O}_X^*, & & \end{array} \qquad (2.3.6.1)$$

with the map to $\mathscr{O}_X$ induced by the map $M \to \mathscr{O}_X$ and the inclusion $\mathscr{O}_X^* \hookrightarrow \mathscr{O}_X$. We refer to M^a as the *log structure associated to* M.

The basic example of this construction is the following. If P is a finitely generated integral monoid and $\beta : P \to \Gamma(X, \mathscr{O}_X)$ is a morphism of monoids, we obtain a pre-log structure by viewing P as a constant sheaf with the map to $\mathscr{O}_X$ defined by β. By passing to the associated log structure we obtain a log structure on X.

Definition 2.3.7. A log structure M on X is *fine* if there exists an étale cover $\{U_i \to X\}_{i \in I}$ and finitely generated integral monoids $\{P_i\}_{i \in I}$ with maps $\beta_i : P_i \to \Gamma(U_i, \mathscr{O}_{U_i})$ such that the restriction $M|_{U_i}$ is isomorphic to the log structure defined by the pair (P_i, β_i).

Definition 2.3.8. A *chart* for a fine log structure M on a scheme X is a map $\beta : P \to \Gamma(X, M)$ from a fine monoid P such that the induced map

$$P^a \to M$$

is an isomorphism, where P^a is the log structure associated to the prelog structure defined by the composite

$$P \xrightarrow{\beta} \Gamma(X, M) \xrightarrow{\alpha} \Gamma(X, \mathscr{O}_X).$$

Example 2.3.9. Let R be a ring. If P is a finitely generated integral monoid, we write $\text{Spec}(P \to R[P])$ for the log scheme whose underlying scheme is $\text{Spec}(R[P])$ (where $R[P]$ is the monoid algebra on P) and whose log structure is associated to the prelog structure given by the natural map of monoids $P \to R[P]$.

2.3.10. If $f : X \to Y$ is a morphism of schemes and M_Y is a log structure on Y, then the pullback f^*M_Y of M_Y to X is defined to be the log structure associated to the prelog structure

$$f^{-1}M_Y \to f^{-1}\mathscr{O}_Y \to \mathscr{O}_X. \tag{2.3.10.1}$$

One checks immediately that if M_Y is fine then f^*M_Y is also fine.

This construction enables one to define a category of log schemes: A morphism $(X, M_X) \to (Y, M_Y)$ is a pair (f, f^b), where $f : X \to Y$ is a morphism of schemes and $f^b : f^*M_Y \to M_X$ is a morphism of log structures on X.

Example 2.3.11. Let $\theta : Q \to P$ be a morphism of fine monoids, and let R be a ring. Then θ induces a natural morphism of fine log schemes

$$\text{Spec}(P \to R[P]) \to \text{Spec}(Q \to R[Q]).$$

Many of the classical notions (e.g. smooth, flat, local complete intersection...) have logarithmic analogues. The key notions we need in this paper are the following:

Definition 2.3.12. (i) A morphism $(f, f^b) : (X, M_X) \to (Y, M_Y)$ of log schemes is *strict* if the map $f^b : f^*M_Y \to M_X$ is an isomorphism.

(ii) A morphism $(f, f^b) : (X, M_X) \to (Y, M_Y)$ is a *closed immersion* (resp. *strict closed immersion*) if $f : X \to Y$ is a closed immersion and $f^b : f^*M_X \to M_Y$ is surjective (resp. an isomorphism).

(iii) A morphism $(f, f^b) : (X, M_X) \to (Y, M_Y)$ is *log smooth* (resp. *log étale*) if $f : X \to Y$ is locally of finite presentation and for every commutative diagram

$$
\begin{array}{ccc}
(T_0, M_{T_0}) & \xrightarrow{\ a\ } & (X, M_X) \\
{\scriptstyle j}\downarrow & & \downarrow \\
(T, M_T) & \longrightarrow & (Y, M_Y)
\end{array}
\tag{2.3.12.1}
$$

with j a strict closed immersion defined by a nilpotent ideal, there exists (resp. there exists a unique) morphism $(T, M_T) \to (X, M_X)$ filling in the diagram.

Remark 2.3.13. If (X, M) is a fine log scheme (i.e. a scheme with fine log structure) then giving a chart $P \to \Gamma(X, M)$ is equivalent to giving a strict morphism of log schemes

$$(X, M) \to \text{Spec}(P \to \mathbb{Z}[P]). \tag{2.3.13.1}$$

Lemma 2.3.14 *Let* $(f, f^b) : (X, M_X) \to (Y, M_Y)$ *be a morphism of fine log schemes, and let* $h : U \to X$ *be a smooth surjection. Denote by* M_U *the pullback of* M_X *to* U *so we have a commutative diagram of log schemes*

$$(U, M_U) \xrightarrow{(h, h^b)} (X, M_X) \xrightarrow{(f, f^b)} (Y, M_Y). \tag{2.3.14.1}$$
$$\underbrace{\phantom{(U, M_U) \xrightarrow{(h, h^b)} (X, M_X) \xrightarrow{}}}_{(g, g^b)}$$

Then (f, f^b) *is log smooth if and only if* (g, g^b) *is log smooth.*

Proof. This follows immediately from the definition of a log smooth morphism. □

2.3.15. One of the most remarkable aspects of the logarithmic theory is that the notion of log smoothness behaves so much like the usual notion of smoothness for schemes (a stack–theoretic "explanation" for this phenomenon is given in [42]). In particular, as we now explain the étale local structure of log smooth morphisms is very simple and there is a good deformation theory of log smooth morphisms.

Theorem 2.3.16 ([24, 3.5]) *Let* $f : (X, M_X) \to (Y, M_Y)$ *be a log smooth morphism of fine log schemes, let* $\bar{x} \to X$ *be a geometric point and let* $\bar{y} \to Y$ *be the composite* $\bar{x} \to X \to Y$. *Then after replacing* X *and* Y *by étale neighborhoods of* $\bar{x}$ *and* $\bar{y}$ *respectively, there exists charts* $\beta_X : P \to M_X$, $\beta_Y : Q \to M_Y$, *and a morphism* $\theta : Q \to P$ *such that the following hold:*
(i) *The diagram of fine log schemes*

$$
\begin{array}{ccc}
(X, M_X) & \xrightarrow{\beta_X} & \mathrm{Spec}(P \to \mathbb{Z}[P]) \\
f \downarrow & & \downarrow \theta \\
(Y, M_Y) & \xrightarrow{\beta_Y} & \mathrm{Spec}(Q \to \mathbb{Z}[Q])
\end{array}
\tag{2.3.16.1}
$$

commutes.
(ii) *The induced map*

$$X \to Y \times_{\mathrm{Spec}(\mathbb{Z}[Q])} \mathrm{Spec}(\mathbb{Z}[P]) \tag{2.3.16.2}$$

is étale.
(iii) *The kernel of* $\theta^{gp} : Q^{gp} \to P^{gp}$ *is a finite group and the orders of* $\mathrm{Ker}(\theta^{gp})$ *and the torsion part of* $\mathrm{Coker}(\theta^{gp})$ *are invertible in* $k(x)$.

Conversely if étale locally there exists charts satisfying the above conditions then the morphism $(X, M_X) \to (Y, M_Y)$ *is log smooth.*

Example 2.3.17. If S is a scheme which we view as a log scheme with the trivial log structure $\mathcal{O}_S^* \hookrightarrow \mathcal{O}_S$, then a morphism of fine log schemes

$(X, M_X) \to (S, \mathscr{O}_S^*)$ is log smooth if and only if étale locally on S and X the log scheme (X, M_X) is isomorphic to $\mathrm{Spec}(P \to \mathscr{O}_S[P])$ for some finitely generated integral monoid P. Thus in the case of trivial log structure on the base log smoothness essentially amounts to "toric singularities".

Example 2.3.18. Probably the most important example from the point of view of degenerations is the following. Let k be a field and M_k the log structure on k associated to the map $\mathbb{N} \to k$ sending all nonzero elements to 0. Let $\Delta : \mathbb{N} \to \mathbb{N}^2$ be the diagonal map and set $X = \mathrm{Spec}(k \otimes_{k[\mathbb{N}]} k[\mathbb{N}^2]) = \mathrm{Spec}(k[x, y]/(xy))$ with log structure M_X induced by the natural map $\mathbb{N}^2 \to k \otimes_{k[\mathbb{N}]} k[\mathbb{N}^2]$. Then the morphism

$$(X, M_X) \to (\mathrm{Spec}(k), M_k) \tag{2.3.18.1}$$

is log smooth.

2.3.19. One technical difficulty that arises when dealing with log smoothness is that in general the underlying morphism of schemes of a log smooth morphism need not be flat. All the examples considered in this text will satisfy an additional property that ensures that the underlying morphism of schemes is flat. A morphism of integral monoids $\theta : P \to Q$ is called *integral* if the map of algebras $\mathbb{Z}[P] \to \mathbb{Z}[Q]$ induced by θ is flat (see [24, 4.1] for several other characterizations of this property). A morphism of log schemes $f : (X, M_X) \to (Y, M_Y)$ is called *integral* if for every geometric point $\bar{x} \to X$ the map $f^{-1}\overline{M}_{Y,f(\bar{x})} \to \overline{M}_{X,\bar{x}}$ is an integral morphism of monoids. By [24, 4.5], if $f : (X, M_X) \to (Y, M_Y)$ is a log smooth and integral morphism of fine log schemes then the underlying morphism of schemes $X \to Y$ is flat.

2.3.20. As in the case of schemes without log structures, the notion of log smoothness is intimately tied to differentials.

Let $f : (X, M_X) \to (Y, M_Y)$ be a morphism of fine log schemes locally of finite presentation. For a scheme T and a quasi–coherent sheaf I on T, let $T[I]$ denote the scheme with same underlying topological space as that of T, but with structure sheaf the $\mathscr{O}_T$–algebra $\mathscr{O}_T \oplus I$ with algebra structure given by $(a + i)(c + j) = ac + (aj + ci)$. The ideal I defines a closed immersion $j : T \hookrightarrow T[I]$ for which the natural map $T[I] \to T$ induced by $\mathscr{O}_T \to \mathscr{O}_T[I]$ sending a to a is a retraction. If M_T is a fine log structure on T, let $M_{T[I]}$ denote the log structure on $T[I]$ obtained by pullback along $\mathscr{O}_T \to \mathscr{O}_T[I]$ so that we have a diagram of fine log schemes

$$(T, M_T) \xrightarrow{\ j\ } (T[I], M_{T[I]}) \xrightarrow{\ \pi\ } (T, M_T). \tag{2.3.20.1}$$

The above construction is functorial in the pair (T, I).

Consider now the functor F on the category of quasi–coherent sheaves on X associating to any quasi-coherent sheaf I the set of morphisms of fine log schemes $(X[I], M_{X[I]}) \to (X, M_X)$ filling in the commutative diagram

$$(X, M_X) \xrightarrow{\ \text{id}\ } (X, M_X)$$

$$j \downarrow \qquad\qquad\qquad \downarrow f \qquad\qquad (2.3.20.2)$$

$$(X[I], M_{X[I]}) \xrightarrow{\ f \circ \pi\ } (Y, M_Y).$$

Theorem 2.3.21 ([24, 3.9]) *There exists a (necessarily unique) quasi–coherent sheaf $\Omega^1_{(X,M_X)/(Y,M_Y)}$ on X and an isomorphism of functors*

$$F \simeq \mathrm{Hom}(\Omega^1_{(X,M_X)/(Y,M_Y)}, -). \qquad (2.3.21.1)$$

2.3.22. The sheaf $\Omega^1_{(X,M_X)/(Y,M_Y)}$ is called the sheaf of *logarithmic differentials* of (X, M_X) over (Y, M_Y). Note that the identity map

$$\Omega^1_{(X,M_X)/(Y,M_Y)} \to \Omega^1_{(X,M_X)/(Y,M_Y)} \qquad (2.3.22.1)$$

defines a morphism

$$\rho : (X[\Omega^1_{(X,M_X)/(Y,M_Y)}], M_{X[\Omega^1_{(X,M_X)/(Y,M_Y)}]}) \to (X, M_X). \qquad (2.3.22.2)$$

This defines in particular a morphism $\rho^* : \mathscr{O}_X \to \mathscr{O}_X[\Omega^1_{(X,M_X)/(Y,M_Y)}]$. Taking the difference of this morphism and the morphism $\pi^* : \mathscr{O}_X \to \mathscr{O}_X[\Omega^1_{(X,M_X)/(Y,M_Y)}]$ we obtain a derivation $d : \mathscr{O}_X \to \Omega^1_{(X,M_X)/(Y,M_Y)}$. This defines in particular a morphism of quasi–coherent sheaves

$$\Omega^1_{X/Y} \to \Omega^1_{(X,M_X)/(Y,M_Y)}. \qquad (2.3.22.3)$$

Remark 2.3.23. Note that in the case when the log structures M_X and M_Y are trivial, we recover the usual Kähler differentials.

Remark 2.3.24. The sheaf of logarithmic differentials $\Omega^1_{(X,M_X)/(Y,M_Y)}$ has the following more concrete description (see [24, 1.7]). It is the quotient of the $\mathscr{O}_X$-module

$$\Omega^1_{X/Y} \oplus (\mathscr{O}_X \otimes_{\mathbb{Z}} M_X^{\mathrm{gp}})$$

by the $\mathscr{O}_X$-submodule generated locally by sections of the following form:

(i) $(d\alpha(a), 0) - (0, \alpha(a) \otimes a)$, where $a \in M_X$.
(ii) $(0, 1 \otimes a)$ for $a \in M_X$ in the image of $f^{-1}M_Y$.

The following summarizes the basic properties of logarithmic differentials:

Theorem 2.3.25 ([41, 3.2.1 and 3.2.3]) (i) *If $f : (X, M_X) \to (Y, M_Y)$ is a log smooth morphism of fine log schemes, then $\Omega^1_{(X,M_X)/(Y,M_Y)}$ is a locally free $\mathscr{O}_X$–module of finite type.*
(ii) *For any composite*

$$(X, M_X) \xrightarrow{\ f\ } (Y, M_Y) \xrightarrow{\ g\ } (S, M_S) \qquad (2.3.25.1)$$

there is an associated exact sequence

$$f^*\Omega^1_{(Y,M_Y)/(S,M_S)} \xrightarrow{\ s\ } \Omega^1_{(X,M_X)/(S,M_S)} \longrightarrow \Omega^1_{(X,M_X)/(Y,M_Y)} \longrightarrow 0.$$
$$(2.3.25.2)$$

If gf is log smooth, then f is log smooth if and only if s is injective and the image is locally a direct summand.

Example 2.3.26. Let P be a finitely generated integral monoid, and let (X, M_X) denote the log scheme $\mathrm{Spec}(P \to \mathbb{Z}[P])$. Then one can show that $\Omega^1_{(X,M_X)/\mathbb{Z}} \simeq \mathcal{O}_X \otimes_{\mathbb{Z}} P^{gp}$ [24, 1.8]. The differential $d : \mathcal{O}_X \to \Omega^1_{(X,M_X)/\mathbb{Z}}$ sends a section $e_p \in \mathbb{Z}[P]$ which is the image of an element $p \in P$ to $e_p \otimes p$. In this case when $P = \mathbb{N}^r$ with standard generators e_i ($1 \leq i \leq r$) the module $\Omega^1_{(X,M_X)/\mathbb{Z}}$ is isomorphic to the classically defined module of logarithmic differentials on $\mathbb{A}^r \simeq \mathrm{Spec}(\mathbb{Z}[X_1, \ldots, X_r])$ with the section $1 \otimes e_i \in \mathcal{O}_X \otimes_{\mathbb{Z}} \mathbb{Z}^r$ playing the role of $d\log(X_i)$.

2.3.27. If $f : (X, M_X) \to (Y, M_Y)$ is log smooth the dual of the vector bundle $\Omega^1_{(X,M_X)/(Y,M_Y)}$, denoted $T_{(X,M_X)/(Y,M_Y)}$, is called the *log tangent bundle*.

As in the classical case of schemes, the cohomology of the log tangent bundle controls the deformation theory of log smooth morphisms. Let $i : (Y_0, M_{Y_0}) \hookrightarrow (Y, M_Y)$ be a strict closed immersion defined by a square–zero quasi–coherent ideal $I \subset \mathcal{O}_Y$, and let $f_0 : (X_0, M_{X_0}) \to (Y_0, M_{Y_0})$ be a log smooth and integral morphism. A *log smooth deformation of f_0* is a commutative diagram of log schemes

$$
\begin{array}{ccc}
(X_0, M_{X_0}) & \xrightarrow{\ j\ } & (X, M_X) \\
{\scriptstyle f_0}\downarrow & & \downarrow{\scriptstyle f} \\
(Y_0, M_{Y_0}) & \xrightarrow{\ i\ } & (Y, M_Y),
\end{array}
\qquad (2.3.27.1)
$$

where j is a strict closed immersion, and the underlying diagram of schemes is cartesian. Note that since f_0 is assumed integral any deformation f is also integral from which it follows that $X \to Y$ is a flat deformation of $X_0 \to Y_0$.

Theorem 2.3.28 ([24, 3.14]) (i) *There is a canonical obstruction*

$$o \in H^2(X_0, I \otimes_{\mathcal{O}_{Y_0}} T_{(X_0,M_{X_0})/(Y_0,M_{Y_0})}) \qquad (2.3.28.1)$$

whose vanishing is necessary and sufficient for there to exists a log smooth deformation of f_0.
(ii) *If $o = 0$ then the set of isomorphism classes of log smooth deformations of f_0 is canonically a torsor under $H^1(X_0, I \otimes_{\mathcal{O}_{Y_0}} T_{(X_0,M_{X_0})/(Y_0,M_{Y_0})})$.*
(iii) *For any log smooth deformation f of f_0, the group of automorphisms of f is canonically isomorphic to $H^0(X_0, I \otimes_{\mathcal{O}_{Y_0}} T_{(X_0,M_{X_0})/(Y_0,M_{Y_0})})$.*

2.3.29. For technical reasons we will also need the notion of saturated log scheme.

A fine monoid P is called *saturated* if for any $p \in P^{\mathrm{gp}}$ for which there exists $n \geq 1$ such that $np \in P$ we have $p \in P$.

A fine log structure M on a scheme X is called *saturated* if for every geometric point $\bar{x} \to X$ the monoid $\overline{M}_{\bar{x}}$ is saturated. We also say that the log scheme (X, M) is *saturated* if M is a saturated log structure.

Proposition 2.3.30 ([41, Chapter II, 2.4.5]). *The inclusion functor*

$$\text{(saturated fine log schemes)} \hookrightarrow \text{(fine log schemes)} \tag{2.3.30.1}$$

has a right adjoint $(X, M) \mapsto (X^{\mathrm{sat}}, M^{\mathrm{sat}})$.

Remark 2.3.31. The log scheme $(X^{\mathrm{sat}}, M^{\mathrm{sat}})$ is called the *saturation* of (X, M).

Remark 2.3.32. Locally the saturation of a fine log scheme (X, M) can be described as follows. Let P be a finitely generated integral monoid, and assume given a strict morphism

$$(X, M) \to \mathrm{Spec}(P \to \mathbb{Z}[P]). \tag{2.3.32.1}$$

Let $P' \subset P^{gp}$ be the submonoid of elements $p \in P^{gp}$ for which there exists an integer $n > 0$ such that $np \in P$. Then P' is a saturated monoid. The saturation $(X^{\mathrm{sat}}, M^{\mathrm{sat}})$ is then equal to the scheme

$$X^{\mathrm{sat}} := X \times_{\mathrm{Spec}(\mathbb{Z}[P])} \mathrm{Spec}(\mathbb{Z}[P']) \tag{2.3.32.2}$$

with the log structure M^{sat} equal to the pullback of the log structure on $\mathrm{Spec}(P' \to \mathbb{Z}[P'])$.

Remark 2.3.33. Following standard practice, we usually write 'fs' for 'fine and saturated'.

2.3.34. The category of log schemes (resp. fine log schemes, fs log schemes) has fiber products. Given a diagram of log schemes

$$(X, M_X) \tag{2.3.34.1}$$

$$\downarrow a$$

$$(Y, M_Y) \xrightarrow{\ b\ } (Z, M_Z)$$

the fiber product in the category of log schemes is equal to the scheme $X \times_Z Y$ with log structure the pushout

$$\mathrm{pr}_1^* M_X \oplus_{p^* M_Z} \mathrm{pr}_2^* M_Y \tag{2.3.34.2}$$

with the natural map to $\mathcal{O}_{X \times_Z Y}$. Here we write $p : X \times_Z Y \to Z$ for the projection.

For fiber products in the categories of fine and fs log schemes more care has to be taken, as the inclusion functors

$$(\text{fs log schemes}) \hookrightarrow (\text{fine log schemes}) \hookrightarrow (\text{log schemes}) \qquad (2.3.34.3)$$

do not preserve fiber products. To understand this, the key case to consider is the following. Let

$$
\begin{array}{c}
P \\
\uparrow \\
Q \longleftarrow R
\end{array}
\qquad (2.3.34.4)
$$

by a diagram of fine (resp. fs) monoids, and consider the resulting diagram of log schemes

$$
\begin{array}{c}
\mathrm{Spec}(P \to \mathbb{Z}[P]) \ . \qquad (2.3.34.5) \\
\downarrow \\
\mathrm{Spec}(Q \to \mathbb{Z}[Q]) \longrightarrow \mathrm{Spec}(R \to \mathbb{Z}[R])
\end{array}
$$

Let $P \oplus_R Q$ be the pushout in the category of monoids. Then this pushout need not be fine (resp. fs). Let $(P \oplus_R Q)^{\mathrm{int}}$ be the image of $P \oplus_R Q$ in $(P \oplus_R Q)^{\mathrm{gp}}$. Then $(P \oplus_R Q)^{\mathrm{int}}$ is the pushout of the diagram 2.3.34.4 in the category of integral monoids, and in the fs case the pushout of 2.3.34.4 in the category of fs monoids is given by the saturation $(P \oplus_R Q)^{\mathrm{sat}}$ of $(P \oplus_R Q)^{\mathrm{int}}$. The fiber product of 2.3.34.5 in the category of fine (resp. fs) log schemes is then equal to

$$\mathrm{Spec}((P \oplus_R Q)^{\mathrm{int}} \to \mathrm{Spec}(\mathbb{Z}[(P \oplus_R Q)^{\mathrm{int}}]) \qquad (2.3.34.6)$$

$$(\text{resp. } \mathrm{Spec}((P \oplus_R Q)^{\mathrm{sat}} \to \mathrm{Spec}(\mathbb{Z}[(P \oplus_R Q)^{\mathrm{sat}}])). \qquad (2.3.34.7)$$

2.4 Summary of Alexeev's Results

For the convenience of the reader we summarize in this section the main results of Alexeev [3]. At various points in the work that follows we have found it convenient to reduce certain proofs to earlier results of Alexeev instead of proving everything "from scratch".

2.4.1. Recall [3, 1.1.6] that a reduced scheme P is called *seminormal* if for every reduced scheme P' and proper bijective morphism $f : P' \to P$ such that for every $p' \in P'$ mapping to $p \in P$ the map $k(p) \to k(p')$ is an isomorphism, the morphism f is an isomorphism.

In his paper, Alexeev considers two kinds of moduli problems. The first concerns polarized toric varieties and the second abelian varities. We summarize here the main results from both problems. The reader only interested in abelian varieties can skip to 2.4.7.

Broken Toric Varieties.

2.4.2. Let X be a finitely generated free abelian group, and let $T = \mathrm{Spec}(\mathbb{Z}[X])$ be the corresponding torus. If S is a scheme, we write T_S for the base change of T to S.

2.4.3. Let $B = \mathrm{Spec}(k)$ be the spectrum of an algebraically closed field k, and let P/B be an affine integral scheme with action of the torus T_B such that the action has only finitely many orbits. Write $P = \mathrm{Spec}(R)$, and assume that for every $p \in P(k)$ the stabilizer group scheme $H_p \subset T_B$ of p is connected and reduced.

The T_B–action on P defines (and is defined by) an X–grading $R = \oplus_{\chi \in X} R_\chi$. Since R is an integral domain the set

$$S := \{\chi \in X | R_\chi \neq 0\} \subset X \tag{2.4.3.1}$$

is a submonoid of X. Since the T_B–action has only finitely many orbits, there exists an orbit which is dense in P. Pick a point $p \in P(k)$ in this orbit so that the map $T_B \to P$ sending a scheme-valued point $g \in T_B$ to $g(p) \in P$ is dominant. This map defines a T_B–invariant inclusion $R \hookrightarrow k[X]$. This shows in particular that each R_χ is a k–vector space of dimension 1 or 0. Furthermore, if $X' \subset X$ denotes the subgroup generated by S and $T \to T'$ the associated quotient torus of T, then the map $T_B \to P$ defines by $p \in P(k)$ factors through a dense open immersion $T'_B \hookrightarrow P$. Furthermore we obtain a morphism $\mathrm{Spec}(k[S]) \to P$ which by [3, 2.3.13] is an isomorphism if P is seminormal.

More generally, if P is reduced, affine, but not necessarily irreducible, we obtain a collection of cones $S_i \subset X$ as follows. Let $\{P_i \subset P\}$ be the closures of the orbits of the T–action. Since each P_i is an irreducible scheme we get a collection of cones $S_i \subset X$. If $P_i \subset P_j$ then one sees from the construction that S_i is a face of S_j.

2.4.4. We can also use this to study the projective situation. Let $(P, \mathscr{L})$ be a projective scheme with T–action over k with a linearized ample line bundle $\mathscr{L}$. Assume that P is seminormal and connected, the T–action has only finitely many orbits, and that for every point $p \in P$ the stabilizer group scheme is reduced.

Set $R = \oplus_{i \geq 0} H^0(P, \mathscr{L}^{\otimes i})$. Let $\mathbb{T}$ denote the torus $T \times \mathbb{G}_m$. The $\mathbb{Z}$–grading on R defines an action of $\mathbb{G}_m$ on R which one can show commutes with the T–action thereby defining an action of $\mathbb{T}$ on R. Applying the above discussion to R with this $\mathbb{T}$–action we obtain a collection of cones $S_i \subset \mathbb{Z} \oplus X$. Let $\Delta \subset X_{\mathbb{R}}$ be the union of the intersections of the cones $S_{i,\mathbb{R}}$ with the set

$(1, X_{\mathbb{R}}) \subset \mathbb{Z} \oplus X_{\mathbb{R}}$. Then one can show that Δ is a polytope in $X_{\mathbb{R}}$ and that the $S_{i,\mathbb{R}} \cap \Delta$ define a paving of $X_{\mathbb{R}}$ in the sense of 3.1.1 below.

2.4.5. The association of the polytope Δ to $(P, \mathscr{L})$ in the preceding paragraph behaves well in families. If S is a scheme and $f : P \to S$ is a projective flat morphism with $T = \mathrm{Spec}(\mathbb{Z}[X])$ a torus acting on P over S, and $\mathscr{L}$ is a T–linearized ample invertible sheaf on P then for any geometric point $\bar{s} \to S$ the fiber $P_{\bar{s}}$ defines a polytope $\Delta_{\bar{s}} \subset X_{\mathbb{R}}$. As explained in [3, 2.10.1], if S is connected then the polytopes $\Delta_{\bar{s}}$ are all equal.

For a fixed polytope $Q \subset X_{\mathbb{R}}$, this enables one to define a stack $\mathcal{TP}^{\mathrm{fr}}[Q]$ which to any scheme S associates the groupoid of triples $(f : P \to S, \mathscr{L}, \theta \in f_*\mathscr{L})$ as follows:

(i) $f : P \to S$ is a proper flat morphism of schemes, and T acts on P over S.
(ii) For every geometric point $\bar{s} \to S$ the fiber $P_{\bar{s}}$ is seminormal and connected with associated polytope $Q \subset X_{\mathbb{R}}$.
(iii) $\mathscr{L}$ is a relatively ample invertible sheaf on P.
(iv) $\theta \in f_*\mathscr{L}$ is a section such that for every geometric point $\bar{s} \to S$ the zero locus of the section $\theta_{\bar{s}} \in H^0(P_{\bar{s}}, \mathscr{L}_{\bar{s}})$ does not contain any T-orbit.

Theorem 2.4.6 ([3, 2.10.10]) *The stack $\mathcal{TP}^{\mathrm{fr}}[Q]$ is a proper algebraic stack over $\mathbb{Z}$ with finite diagonal.*

Abelian Varieties

2.4.7. Following Alexeev [3, 1.1.3.2], a *stable semiabelic variety* over an algebraically closed field k is a scheme P/k together with an action of a semiabelian scheme G/k such that the following condition holds:

1. The dimension of G is equal to the dimension of each irreducible component of P.
2. P is seminormal.
3. There are only finitely many orbits for the G–action.
4. The stabilizer group scheme of every point of P is connected, reduced, and lies in the toric part T of G.

A *stable semiabelic pair* is a projective stable semiabelic variety P (with action of the semiabelian group scheme G) together with an ample line bundle $\mathscr{L}$ on P and a section $\theta \in H^0(P, \mathscr{L})$ whose zero locus does not contain any G–orbits.

2.4.8. If S is a scheme, then a *stable semiabelic pair* over S is a collection of data $(G, f : P \to S, \mathscr{L}, \theta \in f_*\mathscr{L})$ where:

1. G is a semiabelian scheme over S.
2. $f : P \to S$ is a projective flat morphism, and G acts on P over S.
3. $\mathscr{L}$ is a relatively ample invertible sheaf on P.

4. $\theta \in f_*\mathscr{L}$ is a section.
5. For every geometric point $\bar{s} \to S$ the fiber $(G_{\bar{s}}, P_{\bar{s}}, \mathscr{L}_{\bar{s}}, \theta_{\bar{s}} \in H^0(P_{\bar{s}}, \mathscr{L}_{\bar{s}}))$ is a stable semiabelic pair.

One can show that the sheaf $f_*\mathscr{L}$ is locally free and that its formation commutes with arbitrary base change on S. We define the *degree* of a stable semiabelic pair to be the rank of $f_*\mathscr{L}$.

2.4.9. Fix integers g and d, and let $\mathscr{M}_{g,d}$ denote the stack over $\mathbb{Z}$ which to any scheme S associates the groupoid of stable semiabelic pairs $(G, P, \mathscr{L}, \theta)$ with G of dimension g and $\mathscr{L}$ of degree d. Let $\mathscr{A}_{g,d} \subset \mathscr{M}_{g,d}$ denote the substack classifying pairs $(G, P, \mathscr{L}, \theta)$, where G is an abelian scheme.

By a standard Hilbert scheme argument one sees that the diagonal of $\mathscr{M}_{g,d}$ is representable. It therefore makes sense to talk about the connected components of $\mathscr{M}_{g,d}$ (we ignore the question of whether the stack $\mathscr{M}_{g,d}$ is algebraic; presumably this is the case). Furthermore the inclusion $\mathscr{A}_{g,d} \subset \mathscr{M}_{g,d}$ is representable by open immersions. Let $\mathscr{A}_{g,d}^{\mathrm{Alex}}$ denote the union of the connected components of $\mathscr{M}_{g,d}$ meeting $\mathscr{A}_{g,d}$.

Theorem 2.4.10 ([3, 5.10.1]) *The stack $\mathscr{A}_{g,d}^{\mathrm{Alex}}$ is a proper Artin stack over $\mathbb{Z}$ with finite diagonal.*

2.4.11. When $d = 1$, the stack $\mathscr{A}_{g,1}$ is canonically isomorphic to the moduli stack of principally polarized abelian varieties of dimension g. To see this denote temporarily by $\mathscr{A}_g'$ the moduli stack of principally polarized abelian varieties. Using 2.1.6 we get a functor

$$F : \mathscr{A}_{g,1} \to \mathscr{A}_g', \quad (A, P, \mathscr{L}, \theta) \mapsto (A, \lambda_{\mathscr{L}} : A \to A^t), \qquad (2.4.11.1)$$

which we claim is an isomorphism. Evidently every object of $\mathscr{A}_g'$ is étale locally in the essential image of F, so it suffices to show that F is fully faithful. This is the content of the following lemma:

Lemma 2.4.12 *Let S be a scheme and A/S an abelian scheme. Let $(P_i, \mathscr{L}_i, \theta_i)$ $(i = 1, 2)$ be two collections of data as follows:*

(i) $f_i : P_i \to S$ is an A-torsor;
(ii) $\mathscr{L}_i$ is a relatively ample invertible sheaf on P_i such that $f_{i*}\mathscr{L}_i$ is locally free of rank 1 on S;
(iii) $\theta_i \in f_{i*}\mathscr{L}_i$ is a section which is nonzero in every fiber.

Assume further that the two isomorphisms

$$\lambda_{\mathscr{L}_i} : A \to A^t \qquad (2.4.12.1)$$

are equal. Then there exists a unique pair (g, g^b) of isomorphisms

$$g : P_1 \to P_2, \quad g^b : g^*\mathscr{L}_2 \to \mathscr{L}_1, \qquad (2.4.12.2)$$

where g is an isomorphism of A-torsors and g^b is an isomorphism of line bundle on P_1 such that $g^{b*}(\theta_2) = \theta_1$.

Proof. Using the uniqueness we may by descent theory work locally on the fppf topology on S. We may therefore assume that the torsors P_i are trivial. Choose some trivializations $\iota_i : A \to P_i$.

Any isomorphism of torsors $g : A \to A$ is necessarily translation by a point. Indeed any isomorphism g can be written as $t_a \circ h$, where $h : A \to A$ is a homomorphism. Since g is also supposed to commute with the A-action we have for every $\alpha \in A$

$$g(\alpha) = g(e) + \alpha \tag{2.4.12.3}$$

which gives

$$h(\alpha) + a = a + \alpha. \tag{2.4.12.4}$$

Therefore h is the identity and g must be translation by a point.

Next note that since $\lambda_{\mathscr{L}_1} = \lambda_{\mathscr{L}_2}$ we have étale locally on S

$$t_a^*(\mathscr{L}_1 \otimes \mathscr{L}_2^{-1}) \simeq \mathscr{L}_1 \otimes \mathscr{L}_2^{-1} \tag{2.4.12.5}$$

for all $a \in A$. It follows that $\mathscr{L}_1 \otimes \mathscr{L}_2^{-1} \in \underline{\mathrm{Pic}}^0_{A/S}(S)$. In particular, there exists a unique point $a \in A$ such that $t_a^*\mathscr{L}_2 \simeq \mathscr{L}_1$. We take translation by this element a to be the map g. The choice of g^b is then uniquely determined by the condition $g^{b*}(\theta_2) = \theta_1$. $\quad\square$

Remark 2.4.13. The space $\mathscr{A}_{g,1}^{\mathrm{Alex}}$ therefore provides a compactification of $\mathscr{A}_{g,1}$. However, in general the stack $\mathscr{A}_{g,1}^{\mathrm{Alex}}$ has many irreducible componets. One of the main problems dealt with in this paper is how to "carve out" the component of $\mathscr{A}_{g,1}^{\mathrm{Alex}}$ and also to generalize the theory to higher degree polarizations.

3

Moduli of Broken Toric Varieties

In this chapter we discuss the logarithmic interpretation of the main component of the moduli spaces for "broken toric varieties" considered by Alexeev. This study contains many of the basic ideas used later for our study of abelian varieties. The reader only interested in abelian varieties can, however, skip directly to chapter 4.

3.1 The Basic Construction

3.1.1. Let X be a free abelian group of finite rank r, and let $X_{\mathbb{R}}$ denote $X \otimes_{\mathbb{Z}} \mathbb{R}$. A *polytope* in $X_{\mathbb{R}}$ is a subset equal to the convex hull of a finite set of points in $X_{\mathbb{R}}$. A polytope $Q \subset X_{\mathbb{R}}$ is *integral* if it is the convex hull of a finite set of points in $X \subset X_{\mathbb{R}}$. A *face* of a polytope $Q \subset X_{\mathbb{R}}$ is a subset F of the boundary ∂Q of the form $Q \cap H$ for some $r-1$–dimensional linear subspace $H \subset X_{\mathbb{R}}$. A face of a polytope is again a polytope. The *dimension* of a polytope Q is the dimension of the smallest affine subspace of $X_{\mathbb{R}}$ containing Q. A *vertex* of a polytope Q is a zero–dimensional face.

A *paving* of a polytope $Q \subset X_{\mathbb{R}}$ is a finite collection S of subpolytopes $\omega \subset Q$ such that the following hold:

(i) For any two elements $\omega, \eta \in S$, the intersection $\omega \cap \eta$ is also in S.
(ii) Any face of a polytope $\omega \in S$ is again in S.
(iii) $Q = \cup_{\omega \in S} \omega$, and for any two distinct elements $\omega, \eta \in S$ the interiors of ω and η are disjoint.

If Q is an integral polytope, then an *integral paving* of Q is a paving S such that each $\omega \in S$ is an integral polytope.

Let $Q \subset X_{\mathbb{R}}$ be an integral polytope, and let $Z \subset Q$ be the set of integral point. Let $\psi : Z \to \mathbb{R}$ be a function, and consider the convex hull in $\mathbb{R} \oplus X_{\mathbb{R}}$ of the set

$$G_{\psi} := \{(h, x) \in \mathbb{R} \oplus X_{\mathbb{R}} | x \in Z \text{ and } h \geq \psi(x)\}. \qquad (3.1.1.1)$$

M.C. Olsson, *Compactifying Moduli Spaces for Abelian Varieties.* Lecture Notes in Mathematics 1958.
© Springer-Verlag Berlin Heidelberg 2008

The lower boundary of G_ψ is then the graph of a piecewise linear function $g_\psi : Q \to \mathbb{R}$. A paving S of Q is called *regular* if it is equal to the domains of linearity of g_ψ for some $\psi : Z \to \mathbb{R}$.

3.1.2. Let X be a free abelian group of finite rank, and for an abelian group A (e.g. $\mathbb{Q}$ or $\mathbb{R}$) let X_A denote $X \otimes_\mathbb{Z} A$. Denote by T the torus $\mathrm{Spec}(\mathbb{Z}[X])$, and let $Q \subset X_\mathbb{R}$ denote an integral polytope of dimension ≥ 1. We denote by $\mathbb{X}$ the abelian group $\mathbb{Z} \oplus X$.

Let S be an integral regular paving of Q. We view S as a category with a unique morphism $\omega_i \to \omega_j$ between two polytopes ω_i and ω_j of S if ω_i is a face of ω_j and the empty set otherwise.

For a face ω_i of S, let N_i denote the integral points of the cone $\mathrm{Cone}(1, \omega_i) \subset \mathbb{X}_\mathbb{R}$, and define

$$N_S^{gp} := \varinjlim_{\omega_i \in S} N_i^{gp}, \tag{3.1.2.1}$$

where for an integral monoid M we write M^{gp} for its associated group. Also define P to be the integral points of $\mathrm{Cone}(1, Q) \subset \mathbb{X}_\mathbb{R}$. The $\mathbb{Z}$-grading on $\mathbb{X}$ makes P an $\mathbb{N}$-graded monoid.

Remark 3.1.3. Define $N_S := \varinjlim_{\omega_i \in S} N_i$ (limit in the category of integral monoids). Then by the universal property of the group associated to a monoid, the group associated to N_S is equal to N_S^{gp}.

3.1.4. For every $\omega_i \in S$, let $\rho_i : N_i \to N_S^{gp}$ be the natural map. If $p \in N_i \cap N_j$, then $p \in N_k$ for $\omega_k = \omega_i \cap \omega_j$. It follows that $\rho_i(p) = \rho_j(p)$. Consequently the ρ_i define a set map $\rho : P \to N_S^{gp}$ by sending $p \in N_i$ to $\rho_i(p)$.

For $p_1, p_2 \in P$ define $p_1 * p_2 \in N_S^{gp}$ by

$$p_1 * p_2 := \rho(p_1) + \rho(p_2) - \rho(p_1 + p_2). \tag{3.1.4.1}$$

Let $H_S \subset N_S^{gp}$ be the submonoid generated by the elements $p_1 * p_2$.

Lemma 3.1.5 (i) $p_1 * p_2 = p_2 * p_1$.
(ii) $p_1 * p_2 + p_3 * (p_1 + p_2) = \rho(p_1) + \rho(p_2) + \rho(p_3) - \rho(p_1 + p_2 + p_3)$.
(iii) *If* $p_1, p_2 \in N_i$, *then* $p_1 * p_2 = 0$.

Proof. This follows immediately from the definition. □

3.1.6. Following [3], let $\widetilde{SC}_1(\mathbb{X}_{\geq 0})$ denote the free monoid with generators symbols $(\bar{\chi}, \chi)$, where $\bar{\chi} \in \oplus_{\omega_i \in S} N_i$ and $\chi \in \cup N_i \subset \oplus_i N_i$ and $\lambda(\bar{\chi}) = \lambda(\chi)$, where $\lambda : \oplus N_i \to P$ denotes the natural projection. Define $SC_1(\mathbb{X}_{\geq 0})$ to be the quotient of $\widetilde{SC}_1(\mathbb{X}_{\geq 0})$ by the relations

$$(\bar{\chi}_1, \chi_1) + (\bar{\chi}_2, \chi_2) = (\bar{\chi}_1 + \bar{\chi}_2, \chi_1 + \chi_2) \tag{3.1.6.1}$$

when χ_1 and χ_2 lie in the same cone. Define $SC_1(\mathbb{X}_{\geq 0})/B_1$ to be the quotient defined by the relations

$$(\chi_1, \chi_2) + (\bar{\chi}_0, \chi_1) = (\bar{\chi}_0, \chi_2). \tag{3.1.6.2}$$

Let $\delta : \oplus_i N_i \to N_S^{gp}$ be the map sending $(\chi_i)_{\omega_i \in S}$ to $\sum_i \chi_i$, and let

$$\pi : \widetilde{SC}_1(\mathbb{X}_{\geq 0}) \to N_S^{gp} \tag{3.1.6.3}$$

be the map sending $(\bar{\chi}, \chi)$ to $\delta(\bar{\chi}) - \chi$.

Lemma 3.1.7 *The map π descends to $SC_1(\mathbb{X}_{\geq 0})/B_1$, and defines a surjection $SC_1(\mathbb{X}_{\geq 0})/B_1 \to H_S$.*

Proof. That the map descends to $SC_1(\mathbb{X}_{\geq 0})/B_1$ follows from the definition of the relations.

The image of π contains H_S since for $p \in N_i$ and $q \in N_j$ defining elements $\bar{p}, \bar{q} \in \oplus N_i$ the image of $(\bar{p} + \bar{q}, p + q)$ is equal to $p * q$.

To see that the image is contained in H_S, choose an ordering $\omega_1, \ldots \omega_n$ of the top–dimensional simplices of S, and consider an element

$$(\sum_{i=1}^n \bar{\chi}_i, \sum_i \chi_i), \tag{3.1.7.1}$$

where $\chi_i \in N_i$. Then the image of this element under π is equal to the image of

$$(\overline{\bar{\chi}_1 + \bar{\chi}_2}, \chi_1 + \chi_2) + (\overline{\chi_1 + \chi_2} + \bar{\chi}_3 + \cdots + \bar{\chi}_n, \sum_{i=1}^n \chi_i). \tag{3.1.7.2}$$

By induction on n the result follows. $\square$

Corollary 3.1.8 *The monoid H_S is finitely generated.*

Proof. This follows from the preceding lemma and [3, 2.9.5]. $\square$

3.1.9. Define $P \rtimes H_S$ to be the monoid with underlying set $P \times H_S$ and addition law given by

$$(p_1, q_1) + (p_2, q_2) := (p_1 + p_2, q_1 + q_2 + p_1 * p_2). \tag{3.1.9.1}$$

The formulas in 3.1.5 imply that this gives a well–defined commutative monoid structure. A straightforward verification shows that in fact the monoid $P \rtimes H_S$ is a commutative, integral, graded monoid with grading induced by the map

$$P \rtimes H_S \longrightarrow P \xrightarrow{\deg} \mathbb{N}. \tag{3.1.9.2}$$

Lemma 3.1.10 *The monoid $P \rtimes H_S$ is finitely generated.*

Proof. Let $p_1, \ldots, p_r \in P$ be elements such that for each $\omega \in S$ the monoid of integral points of $\mathrm{Cone}(1, \omega)$ is generated by the set of p_i's lying in $\mathrm{Cone}(1, \omega)$. Then $P \rtimes H_S$ is generated by H_S and the elements $(p_i, 0)$. $\square$

Lemma 3.1.11 *The morphism $H_S \to P \rtimes H_S$ is integral.*

Proof. By [24, 4.1], it suffices to show the following.

Consider elements $(p_i, h_i) \in P \rtimes H_S$ and $q_i \in H_S$ ($i = 1, 2$) such that

$$(p_1, h_1) + (0, q_1) = (p_2, h_2) + (0, q_2). \tag{3.1.11.1}$$

We must find $(p, h) \in P \rtimes H_S$, and elements $q_3, q_4 \in H_S$ such that

$$(p_1, h_1) = (p, h) + (0, q_3), \quad (p_2, h_2) = (p, h) + (0, q_4). \tag{3.1.11.2}$$

For this note that 3.1.11.1 can be rewritten as

$$(p_1, h_1 + q_1) = (p_2, h_2 + q_2), \tag{3.1.11.3}$$

and therefore $p_1 = p_2$ and $h_1 + q_1 = h_2 + q_2$. From this it follows that we can take $p = p_1$, $h = 0$, $q_3 = h_1$, and $q_4 = h_2$. □

3.1.12. Let $\mathscr{P}_S \to \mathrm{Spec}(\mathbb{Z}[H_S])$ be the projective scheme $\mathrm{Proj}(\mathbb{Z}[P \rtimes H_S])$. The natural map of monoids $P \rtimes H_S \to \mathbb{Z}[P \rtimes H_S]$ induces a log structure $M_{\mathscr{P}_S}$ on $\mathscr{P}_S$ so that there is a natural log smooth and integral morphism

$$(\mathscr{P}_S, M_{\mathscr{P}_S}) \to \mathrm{Spec}(H_S \to \mathbb{Z}[H_S]). \tag{3.1.12.1}$$

This log structure $M_{\mathscr{P}_S}$ is perhaps most easily constructed by noting that the $\mathbb{G}_m$-action on $\mathrm{Spec}(\mathbb{Z}[P \rtimes H_S])$ extends naturally to an action on the log scheme $\mathrm{Spec}(P \rtimes H_S \to \mathbb{Z}[P \rtimes H_S])$ over $\mathrm{Spec}(\mathbb{Z}[H_S])$ (see for example [42, paragraph preceding 5.14]). Since $\mathscr{P}_S$ is isomorphic to $[\mathrm{Spec}(\mathbb{Z}[P \rtimes H_S]) - \{0\}/\mathbb{G}_m]$ we obtain by descent theory also a log structure on $\mathscr{P}_S$. The morphism 3.1.12.1 is log smooth and integral since the morphism

$$\mathrm{Spec}(\mathbb{Z}[P \rtimes H_S]) \to \mathrm{Spec}(\mathbb{Z}[H_S]) \tag{3.1.12.2}$$

is log smooth and integral by 2.3.16 (note that $H_S^{gp} \to (P \rtimes H_S)^{gp}$ is injective with cokernel isomorphic $\mathbb{Z} \oplus X$).

Note that the projection $P \rtimes H_S \to P$ induces an action of the group scheme $D(\mathbb{X}) \simeq T \times \mathbb{G}_m$ (this isomorphism is the one induced by the decomposition $\mathbb{X} \simeq \mathbb{Z} \oplus X$) on $(\mathscr{P}_S, M_{\mathscr{P}_S}, \mathscr{O}_{\mathscr{P}_S}(1))$. This action is equivalent to an action of T on $(\mathscr{P}_S, M_{\mathscr{P}_S})$ together with a T–linearization of the sheaf $\mathscr{O}_{\mathscr{P}_S}(1)$. As in the introduction, we refer to the resulting polarized log scheme $(\mathscr{P}_S, M_{\mathscr{P}_S}, \mathscr{O}_{\mathscr{P}_S}(1))$ with its T–action over $\mathrm{Spec}(H_S \to \mathbb{Z}[H_S])$ as the *standard family associated to S*.

Remark 3.1.13. The log structure $M_{\mathscr{P}_S}$ has the following additional structure. Let $(\mathscr{C}, M_{\mathscr{C}})$ denote the cone $\mathrm{Spec}(P \rtimes H_S \to \mathbb{Z}[P \rtimes H_S])$, and let $\mathscr{C}^o$ denote the complement of the vertex. The global chart induces a surjection $\pi : P \rtimes H_S \to \overline{M}_{\mathscr{C}}$. This map is $\mathbb{G}_m$-invariant, and hence descends to a surjection $\gamma : P \rtimes H_S \to \overline{M}_{\mathscr{P}_S}$ which étale locally lifts to a chart.

3.1.14. Let $Q_\mathbb{Q}$ denote the rational points of Q, and let

$$\eta : P \to Q_\mathbb{Q} \tag{3.1.14.1}$$

by the map sending p to $(1/\deg(p)) \cdot p \in (1, Q_\mathbb{Q}) \subset \mathbb{X}_\mathbb{Q}$.

3.1.15. Let k be a ring, and $f : H_S \to k$ a morphism (where k is viewed as a monoid using the multiplicative structure). Define $F_f := f^{-1}(k^*) \subset H_S$. The submonoid $F_f \subset H_S$ is a face, so we can consider the localization H_{S,F_f} (the submonoid of H_S^{gp} generated by H_S and F_f^{gp}), and the quotient $H_{S,F_f}/F_f^{gp}$.

Lemma 3.1.16 *Let $p, q, r \in P$ be elements.*
(i) $(q * p) + (r * p) + (q + p) * (r + p) = (q + r) * p + (q + r + p) * p + q * r$.
(ii) $f(p * q) \in k^*$ if and only if both $f((2q) * p)$ and $f((2q + p) * p)$ are in k^*.
(iii) If $f(p * q) \in k^*$, then for every $m \geq 1$ the elements $f((2^m q) * p)$ and $f((2^m q + p) * p)$ are in k^*.
(iv) For every $n \in \mathbb{N}$ if $f(p * q) \in k^*$ then $f(p * (nq)) \in k^*$.
(v) For every $n, m \in \mathbb{N}$, if $f(p * q) \in k^*$ then $f((np) * (mq)) \in k^*$.

Proof. Statement (i) is a straightforward calculation. Statement (ii) follows from (i) by taking q and r in (i) both equal to q. Statement (ii) also implies (iii) by induction on m.

For (iv), choose m such that $2^m \geq n$. Then by (i)

$$(nq) * p + ((2^m - n)q) * p + (nq + p) * ((2^m - n)q + p) = (2^m q) * p + (2^m q + p) * p, \tag{3.1.16.1}$$

and by (iii) the image of the right side under f is in k^*. Therefore f applied to each of the terms on the left side must also be in k^*.

Statement (v) follows immediately from (iv). □

Lemma 3.1.17 *Fix $p \in P$ and let $\omega_i \in S$ be a simplex such that there exists $q \in N_i$ with $f(p*q) \in k^*$, and such that the line segment in Q connecting $\eta(p)$ and $\eta(q)$ meets ω_i in at least one other point. Then $f(q' * p) \in k^*$ for every $q' \in N_i$.*

Proof. The assumptions imply that there exists an integer $n \geq 1$ such that $nq + p \in N_i$. Taking $q = nq$, $r = p$, and $p = q'$ in (3.1.16 (i)) we obtain

$$nq * q' + p * q' + (nq + q') * (p + q') = (nq + p) * q' + (nq + p + q') * q' + nq * p. \tag{3.1.17.1}$$

Since $nq + p, nq + p + q' \in N_i$ and using (3.1.16 (iv)), this implies that

$$f((nq + p) * q') f((nq + p + q') * q') f(nq * p) \in k^*, \tag{3.1.17.2}$$

and this in turn implies that $f(p * q') \in k^*$. □

Lemma 3.1.18 *Let $p, q \in P$ be elements with $f(p*q) \in k^*$, and let $L \subset Q$ be the line segment connecting $\eta(p)$ and $\eta(q)$. Then for any $q' \in P$ with $\eta(q') \in L$ we have $f(p * q') \in k^*$. In particular, for any two numbers $n, m \in \mathbb{N}$ we have $f((np + mq) * p) \in k^*$.*

Proof. Since the polytope Q is integral, there exists an integer r and a parametrization $\rho : [0, r] \to L$ such that $\rho(0) = q$, $\rho(r) = p$, and for every integer $i \in [0, r-1]$ the segment $\rho([i, i+1])$ lies in a single simplex of S and $\rho(i) \in Q_{\mathbb{Q}}$. This last observation implies that there exists $q_i \in P$ with $\eta(q_i) = \rho(i)$ (we take $q_0 = q$). By the preceding lemma, if $f(q_i * p) \in k^*$, then $f(q * p) \in k^*$ for every $q \in P$ with $\eta(q) \in \rho([i, i+1])$. In particular, if $f(q_i * p) \in k^*$ then $f(q_{i+1} * p) \in k^*$ as well. By induction on i starting with $i = 0$ (this case is by assumption), the result follows. $\square$

Lemma 3.1.19 *Let $\omega_i \in S$ be a simplex, and let N_i denote the monoid of integral points in $\mathrm{Cone}(1, \omega_i)$. Let $p \in N_i$ be an interior point and $q, q' \in P$ elements such that $f(p * q)$ and $f(p * q')$ are in k^*. Fix $\alpha, \beta \in \mathbb{N}$.*
(i) $f((\alpha q + Np) * (\beta q' + Np)) \in k^*$ *for every $N \geq 1$.*
(ii) $f((\alpha q + \beta q') * p), f((\alpha q) * (\beta q')) \in k^*$.

Proof. For (i), note first that since p is an interior point the result holds for N sufficiently big. Therefore it suffices to prove that if the result holds for all integers greater than or equal to some $N \geq 2$ then it also holds for $N - 1$. For this take $q = \alpha q + (N-1)p$, $r = \beta q' + (N-1)p$, and $p = p$ in (3.1.16 (i)) so that

$$(\alpha q + (N-1)p) * p + (\beta q' + (N-1)p) * p + (\alpha q + Np) * (\beta q' + Np) \quad (3.1.19.1)$$

is equal to

$$(\alpha q + \beta q' + 2(N-1)p) * p + (\alpha q + \beta q' + (2N-1)p) * p + (\alpha q + (N-1)p) * (\beta q' + (N-1)p). \quad (3.1.19.2)$$

Since 3.1.19.1 maps to k^* under f by induction and 3.1.18, this implies (i) for $N - 1$.

For (ii), apply (3.1.16 (i)) with $q = \alpha q$, $r = \beta q'$, and $p = p$ to get

$$(\alpha q * p) + (\beta q' * p) + (\alpha q + p) * (\beta q' + p) = (\alpha q + \beta q') * p + (\alpha q + \beta q' + p) * p + (\alpha q) * (\beta q'). \quad (3.1.19.3)$$

By (i) and (3.1.16 (iv)) the left side of this expression maps to k^* under f, and therefore (ii) follows. $\square$

3.1.20. Define a new paving S' of Q, which is coarser than S, by declaring that two top–dimensional simplices $\omega_i, \omega_j \in S$ lie in the same simplex of S' if and only if there exists interior elements $p_i \in N_i$ and $p_j \in N_j$ with $f(p_i * p_j) \in k^*$. The preceding lemmas imply that this is well–defined. That is for three simplices $\omega_i, \omega_j, \omega_k \in S$ and interior elements $p_i \in N_i$, $p_j \in N_j$, and $p_k \in N_k$, if $f(p_i * p_k), f(p_j * p_k) \in k^*$ then $f(p_i * p_j) \in k^*$. Also the simplices of S' are convex by 3.1.18 so this really gives a paving. Since S is a refinement of S' there is a natural map $\pi : N_S^{gp} \to N_{S'}^{gp}$ which induces a surjection $H_S \to H_{S'}$.

Lemma 3.1.21 *The kernel of $\pi : N_S^{gp} \to N_{S'}^{gp}$ is equal to the subgroup of N_S^{gp} generated by $f^{-1}(k^*) \subset H_S$.*

Proof. For $\omega_j \in S'$, let $\Delta_j \subset S$ be the full subcategory whose objects are simplices $\omega_i \in S$ lying in ω_j. Then

$$N_S^{gp} = \varinjlim_{\omega_j \in S'} (\varinjlim_{\omega_i \in \Delta_j} N_i^{gp}). \tag{3.1.21.1}$$

To prove the lemma it therefore suffices to show that for any fixed $\omega_j \in S'$ the kernel of the map

$$\varinjlim_{\omega_i \in \Delta_j} N_i^{gp} \to \mathrm{Cone}(1, \omega_j)^{gp} \tag{3.1.21.2}$$

is generated by elements $p_1 * p_2$ with $p_1, p_2 \in \mathrm{Cone}(1, \omega_j)$. Let $K \subset \varinjlim_{\omega_i \in \Delta_j} N_i^{gp}$ be the subgroup generated by these elements.

For any abelian group A, we have by the universal property of direct limit and the group associated to a monoid

$$\mathrm{Hom}(\varinjlim_{\omega_i \in \Delta_j} N_i^{gp}, A) \simeq \varprojlim_{\omega_i \in \Delta_j} \mathrm{Hom}(N_i, A). \tag{3.1.21.3}$$

From this description it follows that to give an element

$$\lambda \in \mathrm{Hom}((\varinjlim_{\omega_i \in \Delta_j} N_i^{gp})/K, A) \tag{3.1.21.4}$$

is equivalent to giving a compatible collection of morphisms $\lambda_i : N_i \to A$ such that for any two elements $p_i \in N_i$ and $p_j \in N_j$ with $p_i + p_j \in N_k$ we have

$$\lambda_i(p_i) + \lambda_j(p_j) - \lambda_k(p_i + p_j) = 0. \tag{3.1.21.5}$$

This condition amounts to saying that the maps λ_i are induced by a homomorphism $\lambda : \mathrm{Cone}(1, \omega_j) \to A$. By Yoneda's lemma it follows that

$$\varinjlim_{\omega_i \in \Delta_j} N_i^{gp}/K \simeq \mathrm{Cone}(1, \omega_j)^{gp} \tag{3.1.21.6}$$

as desired. $\square$

Corollary 3.1.22 *The map π identifies $H_{S'}$ with the quotient of H_S by the face $f^{-1}(k^*)$.*

3.1.23. Let M_k be the log structure on $\mathrm{Spec}(k)$ induced by the map $f : H_S \to k$, and let

$$(\mathscr{P}_{S,k}, M_{\mathscr{P}_{S,k}}, \mathscr{O}_{\mathscr{P}_{S,k}}(1)) \to (\mathrm{Spec}(k), M_k) \tag{3.1.23.1}$$

be the polarized log scheme with action of T_k obtained by base change from $(\mathscr{P}_S, M_{\mathscr{P}_S}, \mathscr{O}_{\mathscr{P}_S}(1))$ by the map $\mathrm{Spec}(k) \to \mathrm{Spec}(\mathbb{Z}[H_S])$.

Lemma 3.1.24 *After replacing k by a finite flat ring extension, the family 3.1.23.1 is isomorphic to the base change of the standard family*

$$(\mathscr{P}_{S'}, M_{\mathscr{P}_{S'}}, \mathcal{O}_{\mathscr{P}_{S'}}) \to \mathrm{Spec}(H_{S'} \to \mathbb{Z}[H_{S'}]) \qquad (3.1.24.1)$$

by a map $\mathrm{Spec}(k) \to \mathrm{Spec}(\mathbb{Z}[H_{S'}])$ *induced by a map* $H_{S'} \to k$ *sending all nonzero elements to* $k - k^*$.

Proof. If $\rho : H_S \to k^*$ is a homomorphism, and $f' : H_S \to k$ is the map sending $h \in H_S$ to $\rho(h) \cdot f(h)$, then we obtain a second family

$$(\mathscr{P}'_{S,k}, M_{\mathscr{P}'_{S,k}}, \mathcal{O}_{\mathscr{P}'_{S,k}}(1)) \to (\mathrm{Spec}(k), M'_k). \qquad (3.1.24.2)$$

Since P^{gp} is a free abelian group we have $\mathrm{Ext}^1(P^{gp}, k^*) = 0$. It follows that the exact sequence

$$0 \to H_S^{gp} \to (P \rtimes H_S)^{gp} \to P^{gp} \to 0 \qquad (3.1.24.3)$$

induces a short exact sequence

$$0 \to \mathrm{Hom}(P^{gp}, k^*) \to \mathrm{Hom}((P \rtimes H_S)^{gp}, k^*) \to \mathrm{Hom}(H_S^{gp}, k^*) \to 0. \qquad (3.1.24.4)$$

Let $\tilde{\rho} : P \rtimes H_S \to k^*$ be a lifting of ρ. Then there is a commutative diagram

$$
\begin{array}{ccc}
P \rtimes H_S & \xrightarrow{(p,h) \mapsto (\tilde{\rho}(p)\rho(h),p,h)} & k^* \oplus P \rtimes H_S \\
\uparrow & & \uparrow \\
H_S & \xrightarrow{h \mapsto (\rho(h),h)} & k^* \oplus H_S
\end{array}
\qquad (3.1.24.5)
$$

which induces a commutative diagram of polarized log schemes

$$
\begin{array}{ccc}
(\mathscr{P}'_{S,k}, M_{\mathscr{P}'_{S,k}}, \mathcal{O}_{\mathscr{P}'_{S,k}}(1)) & \longrightarrow & (\mathscr{P}_{S,k}, M_{\mathscr{P}_{S,k}}, \mathcal{O}_{\mathscr{P}_{S,k}}(1)) \\
\downarrow & & \downarrow \\
(\mathrm{Spec}(k), M'_k) & \xrightarrow{\sigma} & (\mathrm{Spec}(k), M_k),
\end{array}
\qquad (3.1.24.6)
$$

where σ is an isomorphism of log schemes whose underlying morphism $\mathrm{Spec}(k) \to \mathrm{Spec}(k)$ is the identity.

Let $F \subset H_S$ be the face $f^{-1}(k^*)$, and let $H_{S,F}$ be the localization so that $H_{S'} = H_S/F$. Since $H_{S'}^{gp}$ is a finitely generated group, there exists after replacing k by a finite flat extension a homomorphism $\rho : H_S^{gp} \to k^*$ whose restriction to F^{gp} is the map induced by f. Replacing f by $\rho^{-1} \cdot f$ we may assume that f maps F^{gp} to 1. In this case the map $H_S \to k$ factors through a map $H_{S'} \to k$ sending all nonzero elements to 0. This defines the map $\mathrm{Spec}(k) \to \mathrm{Spec}(\mathbb{Z}[H_{S'}])$ and an isomorphism between M_k and the pullback of the log structure on $\mathrm{Spec}(\mathbb{Z}[H_{S'}])$. Furthermore, since F^{gp} is sent to $1 \in k$, there is a canonical isomorphism

$$\mathbb{Z}[P \rtimes H_S] \otimes_{\mathbb{Z}[H_S]} k \simeq \mathbb{Z}[(P \rtimes H_S) \oplus_{H_S} H_{S'}] \otimes_{\mathbb{Z}[H_{S'}]} k \simeq \mathbb{Z}[P \rtimes H_{S'}] \otimes_{\mathbb{Z}[H_{S'}]} k.$$
(3.1.24.7)

This defines the desired isomorphism between 3.1.23.1 and 3.1.24.1. □

Remark 3.1.25. Note that any two choices of the lifting ρ differ by a homomorphism $P^{gp} \to k^*$. Equivalently, the isomorphism in 3.1.24.1 is canonical up to the action of $D(\mathbb{X})$ on $(\mathscr{P}_S, M_{\mathscr{P}_S}, \mathscr{O}_{\mathscr{P}_S})$.

Corollary 3.1.26 *Let $g : \mathscr{P}_S \to \mathrm{Spec}(\mathbb{Z}[H_S])$ denote the structure morphism. Then the formation of the sheaves $R^i g_* \mathscr{O}_{\mathscr{P}_S}(d)$ ($d \geq 0$, $i \geq 0$) commutes with arbitrary base change $B' \to \mathrm{Spec}(\mathbb{Z}[H_S])$, and the sheaves $R^i g_* \mathscr{O}_{\mathscr{P}_S}(d)$ are zero for $i \geq 1$. Furthermore there is a canonical isomorphism of $\mathbb{Z}[H_S]$–algebras*

$$\oplus_{d \geq 0} H^0(\mathscr{P}_S, \mathscr{O}_{\mathscr{P}_S}(d)) \simeq \mathbb{Z}[P \rtimes H_S].$$
(3.1.26.1)

Proof. First note that by the construction there is a canonical map

$$\mathbb{Z}[P \rtimes H_S] \to \oplus_{d \geq 0} H^0(\mathscr{P}_S, \mathscr{O}_{\mathscr{P}_S}(d)).$$
(3.1.26.2)

By standard cohomology and base change results [20, III.12.11], it therefore suffices to show that for any geometric point $\bar{s} \to \mathrm{Spec}(\mathbb{Z}[H_S])$, we have $H^i(\mathscr{P}_{S,\bar{s}}, \mathscr{O}_{\mathscr{P}_{S,\bar{s}}}(d)) = 0$ for all $i \geq 1$ and $d \geq 0$ and that 3.1.26.2 induces an isomorphism

$$k(\bar{s}) \otimes_{\mathbb{Z}[H_S]} \mathbb{Z}[P \rtimes H_S] \to \oplus_{d \geq 0} H^0(\mathscr{P}_{S,\bar{s}}, \mathscr{O}_{\mathscr{P}_{S,\bar{s}}}(d)).$$
(3.1.26.3)

This follows from [3, 2.5.1 and 2.5.2]. □

Corollary 3.1.27 *Let (B, M_B) be a fine log scheme, and $g : (X, M_X) \to (B, M_B)$ a proper log smooth morphism of fine log schemes. Let L be an invertible sheaf on X such that for every geometric point $\bar{s} \to B$ the data $(M_B|_{k(\bar{s})}, X_{\bar{s}}, M_{X_{\bar{s}}}, L_{X_{\bar{s}}})$ over $\mathrm{Spec}(k(\bar{s}))$ is isomorphic to the base change of $(M_S, \mathscr{P}_S, M_{\mathscr{P}_S}, \mathscr{O}_{\mathscr{P}_S}(1))$ via some map $\mathbb{Z}[H_S] \to k(\bar{s})$. Then the natural map $\mathscr{O}_B \to g_* \mathscr{O}_X$ is an isomorphism, for any $d \geq 0$ the formation of $g_* L^d$ commutes with arbitrary base change, and the sheaves $R^i g_* L^d$ are zero for $i \geq 1$.*

Proof. Again cohomology and base change [20, III.12.11] implies that it suffices to verify these statement for each geometric fiber which is the previous corollary. □

Remark 3.1.28. The above discussion implies in particular that when considering the standard family it suffices to consider pavings S such that H_S is sharp (i.e. $H_S^* = \{0\}$). By the following proposition it therefore suffices to consider regular integral pavings of Q.

Proposition 3.1.29. *Let S be a paving of Q. Then S is regular if and only if the monoid H_S is sharp.*

Proof. For this note first the following characterization of sharp monoids:

Lemma 3.1.30 *Let M be a fine monoid. Then M is sharp if and only if there exists a homomorphism $h : M \to \mathbb{N}$ with $h^{-1}(0) = 0$.*

Proof. Since $\mathbb{N}$ is sharp, for any $h : M \to \mathbb{N}$ we have $h(M^*) = \{0\}$. From this the "if" direction follows.

For the "only if" direction, let M^{sat} denote the saturation of M and let $\overline{M}$ denote the quotient of M^{sat} by its group of unit. The inverse image of 0 under the composite $M \to M^{\text{sat}} \to \overline{M}$ is equal to $\{0\}$. Indeed, if an element $m \in M$ maps to a unit in M^{sat}, then there exists an element $y \in M^{\text{sat}}$ such that $x + y = 0$. Let n be an integer such that $ny \in M$. Then we have $nx + ny = 0$ in M so $x + ((n-1)x + ny) = 0$ and x is a unit in M. It follows that it suffices to prove the "only if" direction in the case when M is also saturated.

In this case, it follows from [25, 5.8 (1)] that there exists for any element $m \in M$ a morphism $h : M \to \mathbb{N}$ with $h(m) \neq 0$. Let $m_1, \ldots, m_r \in M$ be a set of generators and let $h_i : M \to \mathbb{N}$ be a map with $h_i(m_i) \neq 0$. Then $h := h_1 + \cdots + h_r : M \to \mathbb{N}$ is a map with $h^{-1}(0) = \{0\}$. $\square$

Thus 3.1.29 is equivalent to the statement that S is regular if and only if there exists a homomorphism $h : H_S^{gp} \to \mathbb{Z}$ sending the nonzero elements of H_S to the positive integers. From the exact sequence

$$0 \to \text{Hom}(P^{gp}, \mathbb{Z}) \to \text{Hom}(P \rtimes H_S, \mathbb{Z}) \to \text{Hom}(H_S, \mathbb{Z}) \to 0 \qquad (3.1.30.1)$$

we see that there exists such a homomorphism h if and only if there exists a function $g : P \to \mathbb{Z}$ which is linear on the monoid of integral points of each $\text{Cone}(1, \omega_i)$, and such that for any two elements $p, q \in P$ not lying in the same $\text{Cone}(1, \omega)$ we have $g(p) + g(q) - g(p+q) > 0$. Now giving such a function g is equivalent to giving a function $\psi : Z \to \mathbb{Z}$ (where Z denotes the integral points of Q) such that S is the regular paving associated to ψ by the construction in 3.1.1. $\square$

3.1.31. Let $H \to Q$ be an integral morphism of integral monoids such that $H^{gp} \to Q^{gp}$ is injective with torsion free cokernel. Let P denote the cokernel in the category of integral monoids of the morphism $H \to Q$. The monoid P is equal to the image of Q in the group $Q^{gp}/H^{gp} = P^{gp}$. Assume that H is sharp (i.e. $H^* = \{0\}$), and let $\pi : Q \to P$ be the projection.

For $p \in P$ define a partial order $\leq$ on $\pi^{-1}(p)$ by declaring that $f \leq g$ if there exists $h \in H$ such that $f + h = g$.

Lemma 3.1.32 *For any $p \in P$ there exists a unique element $\tilde{p} \in \pi^{-1}(p)$ such that $\tilde{p} \leq f$ for all $f \in \pi^{-1}(p)$.*

Proof. By [41, 2.1.5 (3)], there exists a minimal (for the partial order $\leq$) element $\tilde{p} \in \pi^{-1}(p)$. Let $f \in \pi^{-1}(p)$ be any other element. Since $\tilde{p}$ and f map to the same element in P, there exist elements $h_1, h_2 \in H$ such that

$$\tilde{p} + h_1 = f + h_2. \tag{3.1.32.1}$$

Since the morphism $H \to Q$ is integral, this implies by [24, 4.1] that there exists an element $g \in \pi^{-1}(p)$ and $h_3, h_4 \in H$ such that

$$\tilde{p} = g + h_3, \ f = g + h_4, \ h_1 + h_3 = h_2 + h_4. \tag{3.1.32.2}$$

Since $\tilde{p}$ is minimal, this implies that $h_3 = 0$, $\tilde{p} = g$, whence $\tilde{p} \le f$. This also proves that if f is also minimal then $f = \tilde{p}$. $\square$

Remark 3.1.33. It follows in particular that as a module over $\mathbb{Z}[H]$, there exists a canonical isomorphism

$$\mathbb{Z}[P \rtimes H_S] \simeq \oplus_{p \in P} \mathbb{Z}[H] \cdot \tilde{p}. \tag{3.1.33.1}$$

3.2 Automorphisms of the Standard Family over a Field

3.2.1. Let k be a field, S an integral regular paving of a polytope $Q \subset X_{\mathbb{R}}$, M_k the log structure on $\mathrm{Spec}(k)$ induced by the unique map $H_S \to k$ sending all nonzero elements to 0, and let

$$(\mathscr{P}_k, M_{\mathscr{P}_k}, \mathscr{O}_{\mathscr{P}_k}(1), T\text{–action}) \to (\mathrm{Spec}(k), M_k) \tag{3.2.1.1}$$

be the resulting standard family. Let $\mathscr{C}$ denote the cone

$$\mathrm{Spec}(\oplus_{d \ge 0} H^0(\mathscr{P}, \mathscr{O}_{\mathscr{P}_k}(d))) \simeq \mathrm{Spec}(k \otimes_{\mathbb{Z}[H_S]} \mathbb{Z}[P \rtimes H_S]), \tag{3.2.1.2}$$

and let $M_{\mathscr{C}}$ denote the log structure associated to the map $P \rtimes H_S \to \mathscr{O}_{\mathscr{C}}$.

Proposition 3.2.2. Let $\alpha : (\mathscr{P}_k, \mathscr{O}_{\mathscr{P}_k}(1)) \to (\mathscr{P}_k, \mathscr{O}_{\mathscr{P}_k}(1))$ be an automorphism of the underlying polarized scheme with T–action (recall $T = D(X)$). Then there exists a unique pair (α^b, β), where $\beta : M_k \to M_k$ is an automorphism of M_k and $\alpha^b : \alpha^* M_{\mathscr{P}_k} \to M_{\mathscr{P}_k}$ is an isomorphism compatible with the T–action such that the diagram

$$\begin{array}{ccc} \alpha^* M_{\mathscr{P}_k} & \xrightarrow{\alpha^b} & M_{\mathscr{P}_k} \\ \uparrow & & \uparrow \\ M_k & \xrightarrow{\beta} & M_k \end{array} \tag{3.2.2.1}$$

commutes.

Proof. Let $\alpha_{\mathscr{C}} : \mathscr{C} \to \mathscr{C}$ be the automorphism of the cone induced by α.

For an element $(p, h) \in P \rtimes H_S$, write $\zeta_{(p,h)}$ for the resulting element of $k \otimes_{\mathbb{Z}[H_S]} \mathbb{Z}[P \rtimes H_S]$.

The automorphism $\alpha_{\mathscr{C}} : \mathscr{C} \to \mathscr{C}$ is induced by a map

$$\alpha_{\mathscr{C}}^* : k \otimes_{\mathbb{Z}[H_S]} \mathbb{Z}[P \rtimes H_S] \to k \otimes_{\mathbb{Z}[H_S]} \mathbb{Z}[P \rtimes H_S] \qquad (3.2.2.2)$$

compatible with the $D(\mathbb{X})$–action. Looking at the character decomposition of $k \otimes_{\mathbb{Z}[H_S]} \mathbb{Z}[P \rtimes H_S]$, it follows that there exists a set map $\rho : P \to k^*$ such that $\alpha_{\mathscr{C}}^*$ is given by

$$\zeta_{(p,0)} \mapsto \rho(p)\zeta_{(p,0)}. \qquad (3.2.2.3)$$

Furthermore, for any simplex $\omega \in S$ the restriction of ρ to $\mathrm{Cone}(1, \omega) \subset P$ is a homomorphism. Consequently, ρ is an element of $\mathrm{Hom}(N_S^{gp}, k^*)$, where N_S^{gp} is defined as in 3.1.2.1.

Let $\beta : M_k \to M_k$ be the automorphism induced by the composite

$$\tilde{\beta} : H_S \hookrightarrow N_S^{gp} \to k^*. \qquad (3.2.2.4)$$

The map β is induced by the morphism of prelog structures $H_S \to k^* \oplus H_S$ sending $h \in H_S$ to $(\tilde{\beta}(h), h)$. Also define

$$\tilde{\rho} : P \rtimes H_S \to k^*, \quad (p, h) \mapsto \rho(p)\tilde{\beta}(h). \qquad (3.2.2.5)$$

The formula

$$\rho(p)\rho(q) = \rho(p+q)(\rho(p)\rho(q)\rho(p+q)^{-1}) = \rho(p+q)\tilde{\beta}(p*q) \qquad (3.2.2.6)$$

implies that in fact $\tilde{\rho}$ is a homomorphism. We then define α^b to be the isomorphism induced by the map

$$P \rtimes H_S \to k^* \oplus P \rtimes H_S, \quad (p, h) \mapsto (\tilde{\rho}(p, h), (p, h)). \qquad (3.2.2.7)$$

This completes the proof of the existence of (α^b, β).

For the uniqueness, it suffices to show that if $\sigma : M_{\mathscr{P}_k} \to M_{\mathscr{P}_k}$ and $\beta : M_k \to M_k$ are compatible automorphisms of the log structures which are also compatible with the T–action, then σ and β must be the identities. This can be verified after base change to an algebraic closure of k, and hence we may assume that k is algebraically closed.

For $\omega_i \in S$, let $\mathscr{P}_i \subset \mathscr{P}_k$ be the irreducible component $\mathrm{Proj}(k[N_i])$, and let $M_i \subset M_{\mathscr{P}_k}|_{\mathscr{P}_i}$ be the sub–log structure consisting of sections whose image in $\mathscr{O}_{\mathscr{P}_i}$ is nonzero. The log structure M_i is isomorphic to the standard log structure on $\mathrm{Proj}(k[N_i])$ induced by the map $N_i \to k[N_i]$. In particular, the map $M_i \to \mathscr{O}_{\mathscr{P}_i}$ is injective which implies that $\sigma|_{M_i}$ is the identity. Denote by $\mathscr{C}_i \subset \mathscr{C}_k$ the closed subscheme $\mathrm{Spec}(k[N_i])$, and let $\mathscr{C}_i^o \subset \mathscr{C}_i$ denote the complement of the vertex.

Lemma 3.2.3 *The natural map*

$$k^* \oplus (P \rtimes H_S)^{gp} \to H^0(\mathscr{C}_i^o, M_{\mathscr{P}_k}^{gp}|_{\mathscr{C}_i^o}) \qquad (3.2.3.1)$$

is an isomorphism.

Proof. Denote by Q_i the quotient of $M_{\mathscr{P}_k}^{gp}|_{\mathscr{C}_i^o}$ by M_i^{gp} so that there is an exact sequence

$$0 \to M_i^{gp} \to M_{\mathscr{P}_k}^{gp}|_{\mathscr{C}_i^o} \to Q_i \to 0. \tag{3.2.3.2}$$

The quotient Q_i is the constant sheaf associated to the abelian group $(P \rtimes H_S)^{gp}/N_i^{gp}$. Consideration of the commutative diagram (where the map c is an isomorphism since Q_i is a constant sheaf on a connected scheme)

$$
\begin{array}{ccccccccc}
0 & \longrightarrow & k^* \oplus N_i^{gp} & \longrightarrow & k^* \oplus (P \rtimes H_S)^{gp} & \longrightarrow & (P \rtimes H_S)^{gp}/N_i^{gp} & \longrightarrow & 0 \\
& & \downarrow & & \downarrow & & \downarrow{\scriptstyle c} & & \\
0 & \longrightarrow & H^0(\mathscr{C}_i^o, M_i^{gp}) & \longrightarrow & H^0(\mathscr{C}_i^o, M_{\mathscr{P}_k}^{gp}|_{\mathscr{C}_i^o}) & \longrightarrow & H^0(\mathscr{C}_i^o, Q_i) & &
\end{array}
$$

$$\tag{3.2.3.3}$$

shows that it suffices to prove that $H^0(\mathscr{C}_i^o, M_i^{gp})$ is equal to $k^* \oplus N_i^{gp}$. This follows from noting that there is an injection

$$H^0(\mathscr{C}_i^o, M_i^{gp}) \hookrightarrow H^0(\mathrm{Spec}(k[N_i^{gp}]), \mathscr{O}^*) = k^* \oplus N_i^{gp}. \tag{3.2.3.4}$$

□

Lemma 3.2.4 *The natural map*

$$k^* \oplus (P \rtimes H_S)^{gp} \to H^0(\mathscr{C}_k^o, M_{\mathscr{C}_k}^{gp}) \tag{3.2.4.1}$$

is an isomorphism.

Proof. Let W denote the set of top–dimensional simplices of S, and choose an ordering of W. For an ordered subset $i_1 < i_2 < \cdots < i_r$ of W let

$$j_{i_1 \cdots i_r} : \mathscr{C}_{i_1 \cdots i_r} \hookrightarrow \mathscr{C}_k \tag{3.2.4.2}$$

denote the intersection $\mathscr{C}_{i_1} \cap \cdots \cap \mathscr{C}_{i_r}$. There is then a commutative diagram

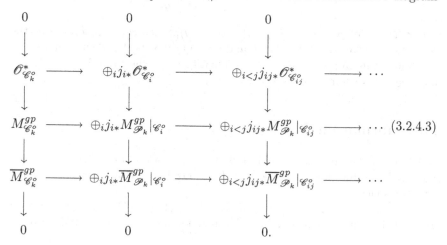

$$\tag{3.2.4.3}$$

By cohomological descent for proper morphisms the bottom row is exact [8, VIII.9.4], and by [3, 2.5.5] the top row is also exact in degrees ≤ 1. It follows that the middle row is also exact in degrees ≤ 1 so

$$H^0(\mathscr{C}_k^o, M_{\mathscr{C}_k^o}^{gp}) = \operatorname{Ker}(\oplus_i H^0(\mathscr{C}_i^o, M_{\mathscr{P}_k}^{gp}|_{\mathscr{C}_i^o}) \to \oplus_{i<j} H^0(\mathscr{C}_{ij}^o, M_{\mathscr{P}_k}^{gp}|_{\mathscr{C}_{ij}^o})),$$

(3.2.4.4)

which by 3.2.3 shows that

$$H^0(\mathscr{C}_k^o, M_{\mathscr{C}_k^o}^{gp}) = H^0(|Q|, k^* \oplus (P \rtimes H_S)^{gp}) = k^* \oplus (P \rtimes H_S)^{gp}. \quad (3.2.4.5)$$

Here $|Q|$ denotes the topological space associated to the polytope Q (i.e. the set Q with the topology induced by the real topology on $X_\mathbb{R}$). $\square$

It follows that the automorphism σ induces an automorphism $\delta : k^* \oplus (P \rtimes H_S)^{gp} \to k^* \oplus (P \rtimes H_S)^{gp}$. By the above, for every simplex ω_i, the map

$$H^0(\mathscr{C}_k^o, M_{\mathscr{C}_k^o}^{gp}) \to H^0(\mathscr{C}_i^o, M_{\mathscr{C}_k^o}^{gp}|_{\mathscr{C}_i^o})$$

(3.2.4.6)

is an isomorphism. Furthermore, since the map $M_i \to \mathcal{O}_{\mathscr{C}_i^o}$ is injective the automorphism δ induces the identity map on

$$H^0(\mathscr{C}_i^o, M_i^{gp}) \subset H^0(\mathscr{C}_i^o, M_{\mathscr{C}_k^o}^{gp}|_{\mathscr{C}_i^o}).$$

(3.2.4.7)

If p lies in $\operatorname{Cone}(1, \omega_i)$, then the image of p in $H^0(\mathscr{C}_i^o, M_{\mathscr{C}_k^o}^{gp}|_{\mathscr{C}_i^o})$ lies in $H^0(\mathscr{C}_i^o, M_i^{gp})$. Since 3.2.4.6 is injective it follows that $\delta(p) = p$. This implies that δ restricts to the identity map on P. Furthermore, for any two elements $p, q \in P$ we have

$$\begin{aligned}
\delta(p+q, 0) + \delta(0, p*q) &= \delta(p+q, p*q) \\
&= \delta(p, 0) + \delta(q, 0) \\
&= (p, 0) + (q, 0) \\
&= (p+q, p*q)
\end{aligned}$$

in $P \rtimes H_S$. It follows that $\delta(0, p*q) = (0, p*q)$. Since H_S is generated by the elements $p*q$ we conclude that δ, and hence also σ, is the identity map.

Finally this implies that β is the identity since the map $M_k \to M_{\mathscr{P}_k}$ is injective. This completes the proof of 3.2.2. $\square$

3.3 Deformation Theory

3.3.1. Let k be an infinite field, and $A' \to A$ a surjection of Artinian local rings with residue field k and kernel $J = \operatorname{Ker}(A' \to A)$ a finite dimensional k–vector space.

Let S be an integral regular paving of a polytope $Q \subset X_\mathbb{R}$, and let $c : H_S \to A'$ be a morphism sending all nonzero elements to the nilradical of A'

and inducing log structures $M_{A'}$, M_A, and M_k and $\mathrm{Spec}(A')$, $\mathrm{Spec}(A)$, and $\mathrm{Spec}(k)$ respectively. Let

$$(\mathscr{P}_A, M_{\mathscr{P}_A}, L_A, \rho) \to (\mathrm{Spec}(A), M_A) \qquad (3.3.1.1)$$

be the polarized log scheme with T–action ρ obtained by base change via the map $\mathbb{Z}[H_S] \to A$ from the standard family over $\mathbb{Z}[H_S]$.

Definition 3.3.2. A *deformation* $(\mathscr{P}_{A'}, M_{\mathscr{P}_{A'}}, L_{A'}, \rho')$ of $(\mathscr{P}_A, M_{\mathscr{P}_A}, L_A, \rho)$ to $(\mathrm{Spec}(A'), M_{A'})$ is a log smooth deformation $(\mathscr{P}_{A'}, M_{\mathscr{P}_{A'}})$ of $(\mathscr{P}_A, M_{\mathscr{P}_A})$ together with a deformation $L_{A'}$ of L_A and a lifting ρ' of the T–action ρ to $(\mathscr{P}_{A'}, M_{\mathscr{P}_{A'}}, L_{A'})$.

Proposition 3.3.3. *Any deformation*

$$(\mathscr{P}_{A'}, M_{\mathscr{P}_{A'}}, L_{A'}, \rho') \qquad (3.3.3.1)$$

of 3.3.1.1 to $(\mathrm{Spec}(A'), M_{A'})$ *is isomorphic to the base change of the standard family via the map* $\mathbb{Z}[H_S] \to A'$ *induced by c.*

The proof will be in several steps 3.3.4–3.3.9.

Fix a deformation 3.3.3.1. Let $\mathscr{C}_A = \mathrm{Spec}(A \otimes_{\mathbb{Z}[H_S]} \mathbb{Z}[P \rtimes H_S])$ denote the cone and $\mathscr{C}_A^o$ the complement of the vertex so that $\mathscr{P}_A = [\mathscr{C}_A^o/\mathbb{G}_m]$. Let $M_{\mathscr{C}_A}$ denote the canonical log structure on $\mathscr{C}_A$ induced by the map $P \rtimes H_S \to A \otimes_{\mathbb{Z}[H_S]} \mathbb{Z}[P \rtimes H_S]$. Also let $\mathscr{C}_{A'}^o$ denote the total space of the $\mathscr{O}_{\mathscr{P}_{A'}}^*$–torsor corresponding to $L_{A'}$ so that $\mathscr{P}_{A'} = [\mathscr{C}_{A'}^o/\mathbb{G}_m]$, and let $M_{\mathscr{C}_{A'}^o}$ denote the pullback of $M_{\mathscr{P}_{A'}}$ to $\mathscr{C}_{A'}^o$.

Lemma 3.3.4 *For any* $i \geq 1$, *the group* $H^i(\mathscr{C}_k^o, \mathscr{O}_{\mathscr{C}_k^o})^{\mathbb{G}_m}$ *of* $\mathbb{G}_m$–*invariants in* $H^i(\mathscr{C}_k^o, \mathscr{O}_{\mathscr{C}_k^o})$ *is zero.*

Proof. Since $\mathscr{P}_k = [\mathscr{C}_k^o/\mathbb{G}_m]$, there is a canonical isomorphism

$$H^i(\mathscr{C}_k^o, \mathscr{O}_{\mathscr{C}_k^o})^{\mathbb{G}_m} \simeq H^i(\mathscr{P}_k, \mathscr{O}_{\mathscr{P}_k}), \qquad (3.3.4.1)$$

which by 3.1.27 is zero. $\square$

Lemma 3.3.5 *The log tangent sheaf* $T_{(\mathscr{C}_k, M_{\mathscr{C}_k})/(k, M_k)}$ *is isomorphic as a sheaf with* $D(\mathbb{X})$–*action to* $\mathrm{Hom}(\mathbb{X}, \mathbb{Z}) \otimes_{\mathbb{Z}} \mathscr{O}_{\mathscr{C}_k}$ *(where* $D(\mathbb{X})$ *acts trivially on* $\mathrm{Hom}(\mathbb{X}, \mathbb{Z})$*).*

Proof. By [24, 1.8] the chart $P \rtimes H_S \to M_{\mathscr{C}_k}$ induces an isomorphism

$$d\log : \mathscr{O}_{\mathscr{C}_k^o} \otimes_{\mathbb{Z}} P^{gp} \simeq \Omega^1_{(\mathscr{C}_k^o, M_{\mathscr{C}_k^o})/(k, M_k)}, \qquad (3.3.5.1)$$

which by the construction in (loc. cit.) is compatible with the $D(\mathbb{X})$–action. $\square$

Lemma 3.3.6 *The log smooth deformation* $(\mathscr{C}_{A'}^o, M_{\mathscr{C}_{A'}^o})$ *of* $(\mathscr{C}_A^o, M_{\mathscr{C}_A^o})$ *(i.e. ignoring the action of* $D(\mathbb{X})$*) is isomorphic to the base change of the standard model.*

Proof. By [24, 3.14] the set of isomorphism classes of log smooth deformations of $(\mathscr{C}_A^o, M_{\mathscr{C}_A^o})$ form a torsor under the group

$$H^1(\mathscr{C}_k^o, T_{(\mathscr{C}_k^o, M_{\mathscr{C}_k^o})/(k, M_k)} \otimes J) \simeq H^1(\mathscr{C}_k^o, \mathscr{O}_{\mathscr{C}_k^o}) \otimes \operatorname{Hom}(X, J). \qquad (3.3.6.1)$$

Thus taking the difference of $(\mathscr{C}_{A'}^o, M_{\mathscr{C}_{A'}^o})$ and the standard model we obtain a class o in this group. Furthermore, since both the standard model and $(\mathscr{C}_{A'}^o, M_{\mathscr{C}_{A'}^o})$ come equipped with liftings of the $D(\mathbb{X})$–action the class o is invariant under the $\mathbb{G}_m$–action. By 3.3.4 we therefore have $o = 0$. □

This reduces the proof of 3.3.3 to the following. Let τ be an action of $D(\mathbb{X})$ on the log scheme

$$\operatorname{Spec}(P \rtimes H_S \to A' \otimes_{\mathbb{Z}[H_S]} \mathbb{Z}[P \rtimes H_S]) - \{\text{vertex}\} \qquad (3.3.6.2)$$

over $\operatorname{Spec}(H_S \to A')$ reducing to the standard action. Then we wish to show that there is an infinitesimal automorphism σ of $\operatorname{Spec}(P \rtimes H_S \to A' \otimes_{\mathbb{Z}[H_S]} \mathbb{Z}[P \rtimes H_S]) - \{\text{vertex}\}$ over $\operatorname{Spec}(H_S \to A')$ taking τ to the standard action. For such an isomorphism σ induces an isomorphism of log schemes

$$
\begin{aligned}
(\mathscr{P}_{A'}, M_{\mathscr{P}_{A'}}) &\simeq [(\mathscr{C}_{A'}^o, M_{\mathscr{C}_{A'}^o})/\mathbb{G}_m] \\
&\simeq [\operatorname{Spec}(P \rtimes H_S \to A' \otimes_{\mathbb{Z}[H_S]} \mathbb{Z}[P \rtimes H_S]) - \{\text{vertex}\}/\mathbb{G}_m],
\end{aligned}
$$

and the quotient of 3.3.6.2 by the $\mathbb{G}_m$-action is just the base change of the standard model. Furthermore the deformation $L_{A'}$ of the line bundle can be recovered as the invertible sheaf associated to the $\mathbb{G}_m$-torsor $\mathscr{C}_{A'}^o \to [\mathscr{C}_{A'}^o/\mathbb{G}_m]$. Let $(\mathscr{C}_{A'}, M_{\mathscr{C}_{A'}})$ denote the log scheme $\operatorname{Spec}(P \rtimes H_S \to A' \otimes_{\mathbb{Z}[H_S]} \mathbb{Z}[P \rtimes H_S])$.

Lemma 3.3.7 *The natural map*

$$H^0(\mathscr{C}_{A'}, \mathscr{O}_{\mathscr{C}_{A'}}) \to H^0(\mathscr{C}_{A'}^o, \mathscr{O}_{\mathscr{C}_{A'}^o}) \qquad (3.3.7.1)$$

is an isomorphism.

Proof. The short exact sequence of sheaves

$$0 \to J \otimes \mathscr{O}_{\mathscr{C}_k} \to \mathscr{O}_{\mathscr{C}_{A'}} \to \mathscr{O}_{\mathscr{C}_A} \to 0 \qquad (3.3.7.2)$$

induces a commutative diagram

$$
\begin{array}{ccccccccc}
0 & \longrightarrow & J \otimes H^0(\mathscr{C}_k, \mathscr{O}_{\mathscr{C}_k}) & \longrightarrow & H^0(\mathscr{C}_{A'}, \mathscr{O}_{\mathscr{C}_{A'}}) & \longrightarrow & H^0(\mathscr{C}_A, \mathscr{O}_{\mathscr{C}_A}) & \longrightarrow & 0 \\
 & & \downarrow & & \downarrow & & \downarrow & & \\
0 & \longrightarrow & J \otimes H^0(\mathscr{C}_k^o, \mathscr{O}_{\mathscr{C}_k^o}) & \longrightarrow & H^0(\mathscr{C}_{A'}^o, \mathscr{O}_{\mathscr{C}_{A'}^o}) & \longrightarrow & H^0(\mathscr{C}_A^o, \mathscr{O}_{\mathscr{C}_A^o}), & &
\end{array}
$$

$$(3.3.7.3)$$

where the top row is exact on the right since $\mathscr{C}_k$ is affine. From this it follows that it suffices to prove the result for $\mathscr{C}_k$ and $\mathscr{C}_A$. By induction on the integer n such that $\mathrm{rad}(A)^n = 0$ it follows that it suffices to show that the map

$$H^0(\mathscr{C}_k, \mathcal{O}_{\mathscr{C}_k}) \to H^0(\mathscr{C}_k^o, \mathcal{O}_{\mathscr{C}_k^o}) \tag{3.3.7.4}$$

is an isomorphism.

For simplicity write just $\mathscr{C}$ for $\mathscr{C}_k$. For $\omega_i \in S$ let $\mathscr{C}_i$ denote the cone $\mathrm{Spec}(k[N_i])$, and let $j_i : \mathscr{C}_i \hookrightarrow \mathscr{C}$ be the inclusion. Let $\mathscr{S}$ denote the set of top–dimensional simplices, and order these simplices in some way. Then by [3, 2.5.5] there is a natural isomorphism of sheaves

$$\mathcal{O}_{\mathscr{C}} \simeq \mathrm{Ker}(\oplus_{\omega_i \in \mathscr{S}} j_{i*} \mathcal{O}_{\mathscr{C}_i} \to \oplus_{\omega_i, \omega_j \in \mathscr{S}, \omega_i < \omega_j} j_{ij*} \mathcal{O}_{\mathscr{C}_{ij}}), \tag{3.3.7.5}$$

where $\mathscr{C}_{ij}$ denotes the intersection $\mathscr{C}_i \cap \mathscr{C}_j$ and $j_{ij} : \mathscr{C}_{ij} \hookrightarrow \mathscr{C}$ is the inclusion. We thus obtain a commutative diagram

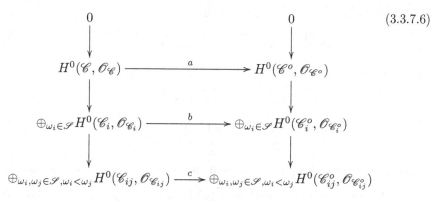

$$(3.3.7.6)$$

with exact columns. Since the dimension of Q is at least 1 by assumption, each $\mathscr{C}_i$ is a normal scheme of dimension ≥ 2. It follows that the map b is an isomorphism, and that the kernel of c is equal to the direct sum of $H^0(\mathscr{C}_{ij}, \mathcal{O}_{\mathscr{C}_{ij}})$ for $\omega_i, \omega_j \in \mathscr{S}$ with $\omega_i \cap \omega_j$ zero–dimensional. In particular the map a is injective. Furthermore, if $\alpha \in H^0(\mathscr{C}^o, \mathcal{O}_{\mathscr{C}^o})$ is a section mapping to zero in $\oplus_{\omega_i \in \mathscr{S}} H^0(\mathscr{C}_i^o, \mathcal{O}_{\mathscr{C}_i^o})$ then α induces a collection of sections $(\alpha_i) \in \oplus_{\omega_i} H^0(\mathscr{C}_i, \mathcal{O}_{\mathscr{C}_i})$ which are compatible on the overlaps, and hence are obtained from a section of $H^0(\mathscr{C}, \mathcal{O}_{\mathscr{C}})$. $\square$

Fix an element $u \in D(P^{gp})(A')$, and let τ_u be the resulting automorphism of $(\mathscr{C}_{A'}, M_{\mathscr{C}_{A'}})$. Since τ reduces to the standard action over A, the automorphism τ_u induces the identity automorphism on $\overline{M}_{\mathscr{C}_{A'}}$. It follows that for every element $p \in P$, there exists a unique element $\nu_p \in \Gamma(\mathscr{C}_{A'}^o, \mathcal{O}_{\mathscr{C}_{A'}^o}^*) = \Gamma(\mathscr{C}_{A'}, \mathcal{O}_{\mathscr{C}_{A'}}^*)$ reducing to $u(p)$ in $\Gamma(\mathscr{C}, \mathcal{O}_{\mathscr{C}}^*)$ such that

$$\tau_u(\beta(\tilde{p})) = \lambda(\nu_p) + \beta(\tilde{p}), \tag{3.3.7.7}$$

where $\tilde{p} \in P \rtimes H_S$ denotes the canonical lifting of p provided by 3.3.3, and β is the chart. Write

$$\nu_p = u(p)(1 + \sum_{\chi \in P} g_\chi^u(p)), \quad g_\chi^u(p) \in J \cdot \zeta_\chi \subset A' \otimes_{\mathbb{Z}[H]} \mathbb{Z}[P \rtimes H_S]. \quad (3.3.7.8)$$

Lemma 3.3.8 (i) *For any $\chi \in P$ with $u(\chi) \neq 1$, the element*

$$h_\chi(p) := g_\chi^u(p)(u(\chi) - 1)^{-1} \in J \cdot \zeta_\chi \quad (3.3.8.1)$$

is independent of the choice of u, and the map $p \mapsto 1 + \sum_{\chi \in P} h_\chi(p) \in \Gamma(\mathscr{C}_{A'}, \mathscr{O}^*_{\mathscr{C}_{A'}})$ is a homomorphism.
(ii) *For any section $u \in D(P^{gp})(A')$ we have*

$$(u(\chi) - 1)h_\chi(p) = g_\chi^u(p), \quad (3.3.8.2)$$

where $h_\chi(p)$ is defined as in (i) using any other section of $D(P^{pg})(A')$ not mapping to 1 in $D(P^{gp})(k)$.

Proof. Let $u' \in D(P^{gp})(A')$ be a second section. Since the action $\tau_{u'}$ agrees with the standard action modulo J, we have

$$\tau_{u'}(1 + \sum_\chi g_\chi^u(p)) = 1 + \sum_\chi g_\chi^u(p)u'(\chi). \quad (3.3.8.3)$$

Computing we find

$$\begin{aligned}
\tau_{u'} \circ \tau_u(\beta(\tilde{p})) = \ & \tau_{u'}(\lambda(u(p)) + \lambda(1 + \sum_\chi g_\chi^u(p)) + \beta(\tilde{p})) \\
= \ & \lambda(u'(p)) + \lambda(u(p)) + \lambda(1 + \sum_\chi g_\chi^u(p)u'(\chi)) \\
& + \lambda(1 + \sum_\chi g_\chi^{u'}(p)) + \beta(\tilde{p}).
\end{aligned} \quad (3.3.8.4)$$

From this and the corresponding formula for $\tau_u \circ \tau_{u'}(\beta(\tilde{p}))$ it follows that

$$\sum_\chi (g_\chi^u(p)u'(\chi) + g_\chi^{u'}(p)) = \sum_\chi (g_\chi^{u'}(p)u(\chi) + g_\chi^u(p)). \quad (3.3.8.5)$$

This implies that for every χ we have

$$g_\chi^u(p)u'(\chi) + g_\chi^{u'}(p) = g_\chi^{u'}(p)u(\chi) + g_\chi^u(p), \quad (3.3.8.6)$$

or equivalently

$$g_\chi^u(p)(u'(\chi) - 1) = g_\chi^{u'}(p)(u(\chi) - 1). \quad (3.3.8.7)$$

This proves the independence of u in the definition of h_χ in (i) and also proves (ii). That the map $p \mapsto 1 + \sum_\chi h_\chi(p)$ is a homomorphism follows from the observation that $g_\chi^u(p)$ is additive in p, whence $h_\chi(p)$ is also additive in p.   □

3.3.9. This lemma completes the proof of 3.3.3 because the infinitesimal automorphism defined by

$$P \rtimes H_S \to \Gamma(\mathscr{C}_{A'}, \mathscr{O}^*_{\mathscr{C}_{A'}}) \oplus P \rtimes H_S, \quad q \mapsto (1 + \sum_\chi h_\chi(\bar{q}), q), \quad (3.3.9.1)$$

where $\bar{q} \in P$ denotes the image of q in P, transforms the action τ into the standard action of $D(\mathbb{X})$.   □

3.4 Algebraization

Fix an integral regular paving S of a polytope $Q \subset X_{\mathbb{R}}$.

3.4.1. Let A be a complete noetherian local ring with maximal ideal $\mathfrak{m} \subset A$, and let A_n denote the reduction of A modulo $\mathfrak{m}^n$. Let k denote the residue field of A and assume $k \neq \mathbb{F}_2$. Fix a map $l : H_S \to A$ defining a log structure M_A on $\operatorname{Spec}(A)$, and let M_{A_n} denote the pullback of M_A to $\operatorname{Spec}(A_n)$. Assume that the map $l : H_S \to k$ sends all non–zero elements to 0.

Let

$$(X_A, M_{X_A}, L_A, \rho_A) \to (\operatorname{Spec}(A), M_A) \tag{3.4.1.1}$$

be a log smooth proper morphism with polarization and T–action, such that the reduction $(X_k, M_{X_k}, L_k, \rho_k)$ is isomorphic to the standard family over $(\operatorname{Spec}(k), M_k)$.

Proposition 3.4.2. *The family* $(X_A, M_{X_A}, L_A, \rho_A)$ *is isomorphic to the base change of the standard family over* $\mathbb{Z}[H_S]$ *via the map* $l : \mathbb{Z}[H_S] \to A$.

Proof. Let $(X_n, M_{X_n}, L_n, \rho_n)$ be the reduction of $(X_A, M_{X_A}, L_A, \rho_A)$ modulo $\mathfrak{m}^n$.

By the deformation theory, there is an isomorphism of compatible systems

$$\{(X_n, M_{X_n}, L_n, \rho_n)\}_n \simeq \{(\mathscr{P}_{S,A_n}, M_{\mathscr{P}_{S,A_n}}, \mathscr{O}_{\mathscr{P}_{S,A_n}}(1), \text{standard action})\}_n. \tag{3.4.2.1}$$

By the Grothendieck existence theorem [17, 5.1.4] it follows that (X_n, L_n) is isomorphic to the base change of the standard family $(\mathscr{P}_{S,A}, \mathscr{O}_{\mathscr{P}_{S,A}}(1))$.

To see that this isomorphism is also compatible with the T–actions, for any integer d define $R_d = H^0(X_A, L_A^{\otimes d})$. The module R_d is a finite A–module, and the two actions correspond to two decompositions $R_d = \oplus_\chi M_\chi$ and $R_d = \oplus_\chi N_\chi$. These two decompositions agree modulo $\mathfrak{m}^n$ for every integer n, and hence they must agree. Since X_A can be recovered as $\operatorname{Proj}(\oplus_d R_d)$ it follows that the two actions agree.

It remains only to see that the log structure M_{X_A} agrees with the log structure obtained from the standard family. Let M' denote the log structure on X_A obtained from the standard family.

Since the sheaves $\overline{M}_{X_A}$ and $\overline{M}'$ are constructible, the proper base change theorem [8, XII.5.1] implies that

$$H^0(X_A, \overline{M}_{X_A}) = H^0(X_k, \overline{M}_{X_k}) \simeq H^0(X_A, \overline{M}'). \tag{3.4.2.2}$$

Let $\gamma' : P \times H_S \to \overline{M}'$ be the natural map described in 3.1.13, and let $\gamma : P \rtimes H_S \to \overline{M}_{X_A}$ be the map induced by the isomorphism 3.4.2.2.

Lemma 3.4.3 *The map* γ *lifts étale locally on* X_A *to a chart for* M_{X_A}.

Proof. Since X_A is proper, it suffices to consider étale neighborhoods of geometric points $\bar{x} \to X_A$ in the closed fiber. Let $F_{\bar{x}} \subset P \rtimes H_S$ be the inverse image under the map $P \rtimes H_S \to \overline{M}_{X_A, \bar{x}}$. Since $(P \rtimes H_S)^{gp}$ is torsion free, the map $\gamma : (P \rtimes H_S)^{gp} \to \overline{M}^{gp}_{X_A, \bar{x}}$ admits a lifting to a map $\tilde{\gamma} : (P \rtimes H_S)^{gp} \to M^{gp}_{X_A, \bar{x}}$ compatible with the given map $H_S \to M_{X_A, \bar{x}}$ (see for example [23, proof of 4.1]). By [24, 2.10] it follows that it suffices to show that the map

$$(P \rtimes H_S)/F_{\bar{x}} \to \overline{M}_{X_A, \bar{x}} \tag{3.4.3.1}$$

is an isomorphism. This can be verified on the closed fiber in which case it follows from the fact that γ' locally lifts to a chart.   □

Lemma 3.4.4 *Let I denote the functor on the category of X_A–schemes which to any scheme T associates the set of isomorphisms $M_{X_A}|_T \simeq M'|_T$ over $M_A|_T$ compatible with the maps γ and γ'. Then I is a separated algebraic space.*

Proof. The assertion is étale local on X_A. We may therefore assume that there exists charts $\beta : P \rtimes H_S \to M_{X_A}$ and $\beta' : P \rtimes H_S \to M'$ lifting γ and γ' respectively. Define

$$\mathscr{S}_{P \rtimes H_S/H_S} := [\mathrm{Spec}(\mathbb{Z}[P \rtimes H_S])/D(P^g)]. \tag{3.4.4.1}$$

We view $\mathscr{S}_{P \rtimes H_S/H_S}$ as a moduli stack for log structure as in [42, 5.20]. Let $g : X_A \to \mathscr{S}_{P \rtimes H_S/H_S} \times \mathscr{S}_{P \rtimes H_S/H_S}$ be the map defined by the log structures M_{X_A} and M'. Then

$$I \simeq \mathscr{S}_{P \rtimes H_S/H_S} \times_{\Delta, \mathscr{S}_{P \rtimes H_S/H_S} \times \mathscr{S}_{P \rtimes H_S/H_S}, g} X_A. \tag{3.4.4.2}$$

Since the diagonal of $\mathscr{S}_{P \rtimes H_S/H_S}$ is separated the result follows.   □

Remark 3.4.5. In fact I is representable by a scheme, but this is not needed.

To construct an isomorphism between M_{X_A} and M', note that the isomorphisms over the reductions define a compatible family of maps $X_n \to I$. Since I is separated and X_A is proper, this compatible collection is uniquely algebraizable to a map $X_A \to I$ which gives the desired isomorphism. This completes the proof of 3.4.2.   □

3.5 Approximation

3.5.1. Let (B, M_B) be a fine log scheme, $\bar{b} \to B$ a geometric point, S an integral regular paving of Q, and $H_S \to \Gamma(B, M_B)$ a chart such that the map $H_S \to k(\bar{b})$ sends all non–zero elements to 0. Let

$$(X_B, M_{X_B}, L_B, \rho_B) \to (B, M_B) \tag{3.5.1.1}$$

be a proper log smooth family with polarization and T–action such that the fiber $(X_{\bar{b}}, M_{X_{\bar{b}}}, L_{\bar{b}}, \rho_{\bar{b}})$ over $(\bar{b}, M_{\bar{b}})$ is isomorphic to the base change of the standard family via the map $\mathbb{Z}[H_S] \to k(\bar{b})$ induced by the chart.

Proposition 3.5.2. *In an étale neighborhood of $\bar{b}$, the family 3.5.1.1 is isomorphic to the base change via $B \to \mathrm{Spec}(\mathbb{Z}[H_S])$ of the standard family.*

Proof. Let $(\mathscr{P}, M_{\mathscr{P}}, L_{\mathscr{P}}, \rho')$ denote the standard family.

By the preceding section, we know the result over the completion $\widehat{\mathscr{O}}_{B,\bar{b}}$. Let I_1 denote the functor over B

$$\underline{\mathrm{Isom}}(X_B, \mathscr{P}). \tag{3.5.2.1}$$

By [45, 4.1 and 4.2] it is a separated algebraic space. Over I_1 there is a tautological isomorphism $\sigma : X_B \times_B I_1 \to \mathscr{P} \times_B I_1$. Let I_2 be the functor over I_1 classifying isomorphisms between the polarizations. The space I_2 can be described as follows. Let L_1 be the sheaf $\mathrm{pr}_1^* L_B$ on $X_B \times_B I_1$, and let L_2 be the pullback of $L_{\mathscr{P}}$ along the tautological isomorphism

$$X_B \times_B I_1 \to \mathscr{P} \times_B I_1 \tag{3.5.2.2}$$

over I_1 (so L_2 is also an invertible sheaf on $X_B \times_B I_1$. Let $\widetilde{I}_2 \to X_B \times_B I_1$ be the $\mathbb{G}_m$–torsor of isomorphisms between L_1 and L_2. Then I_2 is equal to the fiber product of the diagram

$$\underline{\mathrm{Hom}}(X_B, \widetilde{I}_2)$$
$$\downarrow \tag{3.5.2.3}$$
$$B \xrightarrow{\ \mathrm{id}\ } \underline{\mathrm{Hom}}(X_B, X_B),$$

where for two B–schemes U and V we write $\underline{\mathrm{Hom}}(U, V)$ for the functor on the category of B–schemes which to any T/B associates the set of T–morphisms $U \times_B T \to V \times_B T$. It is shown in [45, 4.1] that if U is proper and flat over B and V is separated over B then $\underline{\mathrm{Hom}}(U, V)$ is a separated algebraic space over B. In particular, I_2 is an algebraic space separated over I_1 and so I_2 is also separated over B. Now the tautological isomorphism $\sigma : (X_B, L_B) \to (\mathscr{P}, L_{\mathscr{P}})$ over I_2 may not be compatible with the T–actions. We claim that the condition that it is compatible with the T–action is represented by a closed subscheme, denoted I_3, of I_2. This can be seen by noting that the two actions define two decompositions of

$$\mathcal{R} = \oplus_{d \geq 0} f_* L^d \tag{3.5.2.4}$$

into a direct sum of locally free sheaves of rank 1. Since this algebra is finitely generated, this implies that I_3 is defined as a subfunctor of I_2 by the condition that a finite set of maps between locally free sheaves of rank 1 on I_2 vanish. This is evidently representable by a closed subscheme.

By [6, 2.5], the map $\mathrm{Spec}(\widehat{\mathscr{O}}_{B,\bar{b}}) \to I_3$ defined by the isomorphisms over the completion can be approximated by a morphism in some étale neighborhood of $\bar{b}$. It remains only to approximate the isomorphism between the log structures M_{X_B} and $M_{\mathscr{P}}$ (which we view as a log structure also on X_B). By the proper

base change theorem, we may assume as in 3.4.3 that there is a map γ' : $P \rtimes H_S \to \overline{M}_{X_B}$ which locally lifts to a chart. Let $\gamma : P \rtimes H_S \to \overline{M}_{\mathscr{P}}$ be the map defined in 3.1.13. Then as in 3.4.4 we can define a functor Q over X_B which to any X_B–scheme associates the set of isomorphisms between the pullbacks of M_{X_B} and $M_{\mathscr{P}}$ and are also compatible with the maps γ and γ'. By 3.4.4 (note that the proof made no use of the fact that the base was the spectrum of a complete local ring) the functor Q is represented by a separated algebraic space over X_B. Let I_5 denote the functor over B which to any scheme $W \to B$ associated the set of sections of the projection $Q \times_B W \to X_B \times_B W$. If $f : X_B \to B$ denotes the structure morphism, then in the notation of [45, 1.5] we have $I_5 = f_*Q$. In particular, I_5 is represented by a quasi–separated algebraic space over B. From this and [6, 2.5] it follows that we can also approximate the isomorphisms over the completion between the log structures.
$\square$

3.6 Automorphisms over a General Base

3.6.1. Fix an integral regular paving S of a polytope $Q \subset X_{\mathbb{R}}$.

Let $B = \mathrm{Spec}(A)$ be an affine scheme, M_B the log structure on B defined by a map $H_S \to A$, and

$$(X_B, M_{X_B}, L_B, \rho_B) \to (B, M_B) \tag{3.6.1.1}$$

the standard family over (B, M_B).

Let Aut be the group of automorphisms of the data

$$(M_B, X_B, M_{X_B}, L_B, \rho_B), \tag{3.6.1.2}$$

and let Aut$'$ denote the group of automorphisms of the underlying polarized scheme with T–action.

Proposition 3.6.2. *The forgetful map* $h : \mathrm{Aut} \to \mathrm{Aut}'$ *is a bijection.*

Proof. To see that the map h is injective, let $a : M_{X_B} \to M_{X_B}$ and $b : M_B \to M_B$ be automorphisms defining an element of the kernel. By the case when B is a field already considered, the induced maps $\bar{a} : \overline{M}_{X_B} \to \overline{M}_{X_B}$ and $\bar{b} : \overline{M}_B \to \overline{M}_B$ are the identity morphisms.

Let $\gamma : P \rtimes H_S \to \overline{M}_{X_B}$ denote the surjection defined in 3.1.13. Then since $\bar{a}$ is the identity, there exists a map of sheaves $\tilde{k} : \overline{M}_{X_B} \to \mathscr{O}^*_{X_B}$ such that for any local section $m \in M_{X_B}$ we have $a(m) = \lambda(\tilde{k}(m)) + m$. Let $k : P \rtimes H_S \to A^*$ be the map obtained by composing with γ and taking global sections. Now for every $p \in P$ there exists a point $x \in X_B$ such that the image of p in $\mathscr{O}_{X,x}$ is non–zero. From this it follows that $k(p,0) = 1$ for every p. This in turn implies that $k(0, p * q) = 1$ for every $p, q \in P$ since

$$1 = k(p,0) \cdot k(q,0) = k(p+q, p*q) = k(p+q, 0) \cdot k(0, p*q). \tag{3.6.2.1}$$

It follows that k is the constant map sending all elements to 1 and hence a and b are trivial. This proves the injectivity of h.

For the surjectivity consider an automorphism

$$\sigma : (X_B, L_B, \rho_B) \to (X_B, L_B, \rho_B). \qquad (3.6.2.2)$$

Let $\mathcal{R}$ denote the algebra $\oplus_{d \geq 0} H^0(X_B, L_B^{\otimes d})$. By 3.1.13 there is a canonical isomorphism of graded algebras

$$\mathcal{R} \simeq A \otimes_{\mathbb{Z}[H_S]} \mathbb{Z}[P \rtimes H_S]. \qquad (3.6.2.3)$$

The automorphism σ therefore induces an automorphism of the algebra $A \otimes_{\mathbb{Z}[H_S]} \mathbb{Z}[P \rtimes H_S]$ compatible with the $D(P^{gp})$–action. It follows that there exists a set map

$$\kappa : P \to A^* \qquad (3.6.2.4)$$

such that σ is induced by the map $\zeta_{(p,0)} \mapsto \kappa(p) \cdot \zeta_{(p,0)}$ in $A \otimes_{\mathbb{Z}[H_S]} \mathbb{Z}[P \rtimes H_S]$, where $\zeta_{(p,0)} \in A \otimes_{\mathbb{Z}[H_S]} \mathbb{Z}[P \rtimes H_S]$ denotes the element defined by $(p,0) \in P \rtimes H_S$. The map κ need not be a homomorphism, but for any $\omega_i \in S$ the restriction of κ to the cone N_i is linear. The map κ therefore corresponds to a map

$$\delta : N_S^{gp} = \varinjlim N_i^g \to A^*. \qquad (3.6.2.5)$$

Note that for $p, q \in P$ we have

$$\delta(p * q) = \kappa(p)\kappa(q)\kappa(p + q)^{-1}. \qquad (3.6.2.6)$$

Also if $\beta : P \rtimes H_S \to \mathcal{R}$ denotes the natural map, then

$$\beta(p, 0) \cdot \beta(q, 0) = \beta(p + q, 0)\beta(0, p * q), \qquad (3.6.2.7)$$

and σ acts trivially on $\beta(0, p * q)$. It follows that

$$\kappa(p) \cdot \kappa(q) \cdot \beta(0, p * q) = \kappa(p + q) \cdot \beta(0, p * q). \qquad (3.6.2.8)$$

In other words, the diagram

$$
\begin{array}{ccc}
H_S & \longrightarrow & A \\
{\scriptstyle h \mapsto (\delta(h), h)} \downarrow & & \downarrow {\scriptstyle \mathrm{id}} \\
A^* \oplus H_S & \longrightarrow & A
\end{array}
\qquad (3.6.2.9)
$$

commutes.

Let $a : M_{X_B} \to M_{X_B}$ and $b : M_B \to M_B$ be the automorphisms induced by the maps

$$P \rtimes H_S \to A^*, \quad (p, h) \mapsto \kappa(p) \cdot \delta(h), \qquad (3.6.2.10)$$

and

$$H_S \to A^*, \quad h \mapsto \delta(h). \qquad (3.6.2.11)$$

Then these maps induce an automorphism of $(M_B, X_B, M_{X_B}, L_B, \rho_B)$ inducing σ. This completes the proof of 3.6.2. $\quad \square$

3.7 The Stack $\mathscr{K}_Q$

3.7.1. Fix a finitely generated free abelian group X, and let $Q \subset X_{\mathbb{R}}$ be an integral polytope.

Let $\mathscr{K}_Q$ denote the fibered category over $\mathrm{Spec}(\mathbb{Z})$ which to any scheme B associates the groupoid of data

$$(M_B, f : (X, M_X) \to (B, M_B), L, \theta, \rho) \tag{3.7.1.1}$$

where

(i) M_B is a fine log structure on B,
(ii) $f : (X, M_X) \to (B, M_B)$ is a log smooth morphism whose underlying morphism of schemes $X \to B$ is proper.
(iii) L is a relatively ample invertible sheaf on X/B.
(iv) ρ is an action of $D(X)$ on the triple $((X, M_X), L)$ over (B, M_B).
(v) $\theta \in f_*L$ is a section.

We require that the following conditions hold for every geometric point $\bar{s} \to B$:

(vi) The zero–locus of $\theta_{\bar{s}}$ in $\mathscr{P}_{\bar{s}}$ does not contain any T–orbits.
(vii) There exists a paving S of Q such that the data

$$(X_{\bar{s}}, M_{X_{\bar{s}}}, L_{\bar{s}}, \rho_{\bar{s}}) \to (\bar{s}, M_B|_{\bar{s}}) \tag{3.7.1.2}$$

is isomorphic to the base change of the standard family

$$(\mathscr{P}_S, M_{\mathscr{P}_S}, \mathscr{O}_{\mathscr{P}_S}(1), D(X)\text{–action}) \tag{3.7.1.3}$$

via the map $\mathbb{Z}[H_S] \to k(\bar{s})$ sending all non–zero elements of H_S to 0 (note that implicit here is the assumption that H_S has no nonzero invertible elements; see 3.1.28 for more discussion about this).

3.7.2. Note that there is a natural log structure $M_{\mathscr{K}_Q}$ on the fibered category $\mathscr{K}_Q$. To give such a log structure is equivalent to giving for any 1–morphism $t : B \to \mathscr{K}_Q$ a log structure on B which is functorial in the morphism t (see [46, Appendix B] for more details). If t corresponds to data 3.7.1.1 then we simply take the log structure M_B.

The main theorem of this chapter is the following:

Theorem 3.7.3 (i) *The fibered category $\mathscr{K}_Q$ is a proper irreducible algebraic stack with finite diagonal.*
(ii) *The log stack $(\mathscr{K}_Q, M_{\mathscr{K}_Q})$ is log smooth over $\mathrm{Spec}(\mathbb{Z})$.*
(iii) *The stack $\mathscr{K}_Q$ is canonically isomorphic to the closure (with the reduced structure) of the irreducible open substack classifying smooth polarized toric varieties of the stack $\mathcal{TP}^{\mathrm{fr}}[Q]$ defined in [3, 2.10.10].*

Remark 3.7.4. Statement (3.7.3 (ii)) implies that locally in the smooth topology the stack $\mathcal{K}_Q$ is isomorphic to $\mathrm{Spec}(\mathbb{Z}[M])$ for some fine monoid M (this follows from [24, 3.5]). In particular, the normalization of $\mathcal{K}_Q$ has at most toroidal singularities.

3.7.5. The proof of 3.7.3 occupies the remainder of this section. Note that clearly $\mathcal{K}_Q$ is a stack with respect to the étale topology.

Lemma 3.7.6 *Let* $\mathcal{TP}^{\mathrm{fr}}[Q]$ *be the stack defined in [3, 2.10.10]. Then the functor* $\mathcal{K}_Q \to \mathcal{TP}^{\mathrm{fr}}[Q]$ *which forgets the log structures is fully faithful. In particular, by (loc. cit.) the diagonal*

$$\Delta : \mathcal{K}_Q \to \mathcal{K}_Q \times \mathcal{K}_Q \qquad (3.7.6.1)$$

is representable and finite.

Proof. Let B be a scheme and

$$(M_B, X_B, M_{X_B}, L_B, \theta, \rho), (M'_B, X'_B, M_{X'_B}, L'_B, \theta', \rho') \in \mathcal{K}_Q(B) \qquad (3.7.6.2)$$

two objects. Set

$$I := \mathrm{Spec}(B) \times_{\mathcal{K}_Q \times \mathcal{K}_Q} \mathcal{K}_Q, \quad I' := \mathrm{Spec}(B) \times_{\mathcal{TP}^{\mathrm{fr}}[Q] \times \mathcal{TP}^{\mathrm{fr}}[Q]} \mathcal{TP}^{\mathrm{fr}}[Q].$$
$$(3.7.6.3)$$

A priori, I is just a sheaf. Let $\bar{b} \to B$ be a geometric point over which the two families are isomorphic. By the deformation theory, algebraization, and approximation we can after replacing B by an étale neighborhood of b assume that there exists an integral regular paving S of Q, maps $\rho, \rho' : H_S \to \Gamma(B, \mathcal{O}_B)$ inducing the same map to $k(\bar{b})$, such that the families (that is, the given data without the sections) are isomorphic to the base changes of the standard family over $\mathbb{Z}[H_S]$ by the maps $B \to \mathrm{Spec}(H_S)$ induced by ρ and ρ'.

In this case, the argument used to prove the surjectivity part of 3.6.2 shows that the map $I(B) \to I'(B)$ is surjective. From 3.6.2 it therefore follows that $I(B) \to I'(B)$ is bijective. This implies the lemma. □

3.7.7. To see that $\mathcal{K}_Q$ admits a smooth cover, proceed as follows. For any integral regular paving S of Q, let $\widetilde{\mathcal{K}_{Q,S}}$ be the functor over $\mathbb{Z}[H_S]$ which to any $B \to \mathrm{Spec}(\mathbb{Z}[H_S])$ associates the set of sections of the base change of $\mathcal{O}_{\mathscr{P}_S}(1)$ giving the base change of the standard family to B the structure of an object of $\mathcal{K}_Q(B)$. The functor $\widetilde{\mathcal{K}_{Q,S}}$ is representable by an open subspace of the total space of the bundle $\mathcal{O}_{\mathscr{P}_S}(1)$. Furthermore, 3.1.26 shows that the induced map $\widetilde{\mathcal{K}_{Q,S}} \to \mathcal{K}_Q$ is smooth. Indeed it suffices to show that the map is formally smooth which follows from the vanishing of H^1 with coefficients in L.

Since the collection of $\widetilde{\mathcal{K}_{Q,S}}$ as S varies covers $\mathcal{K}_Q$ it follows that $\mathcal{K}_Q$ is an algebraic stack. In fact this also shows that $\mathcal{K}_Q$ is quasi–compact.

Lemma 3.7.8 *The stack $\mathcal{K}_Q$ is proper.*

Proof. It suffices to verify the valuative criterion for properness since we already know that the stack $\mathcal{K}_Q$ is a quasi–compact algebraic stack with finite diagonal [43, paragraph after 1.3]. In other words, given a discrete valuation ring V with fraction field K, and a morphism $\eta : \mathrm{Spec}(K) \to \mathcal{K}_Q$, we wish to show that there exists an extension $\mathrm{Spec}(V) \to \mathcal{K}_Q$ after possible making an extension $V \to V'$. Furthermore, it suffices to prove this for morphisms η whose image is contained in some dense open $\mathcal{U} \subset \mathcal{K}_Q$ (this also follows from Chow's lemma for Artin stacks [43] and the corresponding result for schemes).

We take for $\mathcal{U}$ the open substack of $\mathcal{K}_Q$ classifying families all of whose geometric fibers are irreducible. This is the same as the locus where the log structure $M_{\mathcal{K}_Q}$ is trivial, and by 3.4.2 the substack $\mathcal{U} \subset \mathcal{K}_Q$ is dense.

In this case, the valuative criterion essentially follows from [3, proof of 2.8.1]. We proceed with the notation of (loc. cit., proof of the case when the generic fiber is geometrically irreducible). Let S be the paving of Q defined in [3, 2.8.2] corresponding to the function g_ψ. We assume that the function g_ψ is integer valued on $\mathbb{X} \cap Q$ (this always holds after a finite extension of V). Then the extension of the family of the generic fiber is defined by the V–subalgebra $R \subset K[\mathbb{X}]$ with generators $\pi^{g_\psi(\chi)} \cdot 1_\chi$, where $\pi \in V$ is a uniformizer, 1_χ denotes the basis element of $K[\mathbb{X}]$ defined by χ.

Recall that $g_\psi : P \to \mathbb{Z}$ is obtained from a function $\tilde{g}_\psi : Q \to \mathbb{R}$ by

$$g_\psi(p) = \deg(p)\tilde{g}_\psi(p/\deg(p)). \tag{3.7.8.1}$$

Furthermore, the function $\tilde{g}_\psi$ is convex so that

$$\begin{aligned}
g_\psi(p+q)/\deg(p+q) = {} & \tilde{g}_\psi((\deg(p)/\deg(p+q)) \cdot (p/\deg(p)) \\
& + (\deg(q)/\deg(p+q)) \cdot (q/\deg(q))) \\
\leq {} & ((\deg(p)/\deg(p+q)) \cdot \tilde{g}_\psi(p/\deg(p))) \\
& + ((\deg(q)/\deg(p+q)) \cdot \tilde{g}_\psi(q/\deg(q))) \\
= {} & (g_\psi(p)/\deg(p+q)) + (g_\psi(q)/\deg(p+q)).
\end{aligned}$$

Therefore

$$g_\psi(p) + g_\psi(q) - g_\psi(p+q) \in \mathbb{N}. \tag{3.7.8.2}$$

Define $\rho : H_S \to V$ by

$$p * q \mapsto \pi^{g_\psi(p) + g_\psi(q) - g_\psi(p+q)}, \tag{3.7.8.3}$$

and let $\lambda : P \rtimes H_S \to R$ be the map sending

$$(p, h) \mapsto \rho(h) \cdot \pi^{g_\psi(p)} \cdot 1_p. \tag{3.7.8.4}$$

Then this defines an isomorphism

$$R \simeq \mathbb{Z}[P \rtimes H_S] \otimes_{\mathbb{Z}[H_S]} V \tag{3.7.8.5}$$

of graded algebras compatible with the torus actions. Therefore the valuative criterion holds and the lemma is shown. $\square$

3.7.9. The fully faithful functor $\mathscr{K}_Q \to \mathcal{TP}^{\mathrm{fr}}[Q]$ is therefore also proper, and hence a closed immersion. Since it contains the open substack of $\mathcal{TP}^{\mathrm{fr}}[Q]$ classifying objects with irreducible underlying space, the image of $\mathscr{K}_Q$ in $\mathcal{TP}^{\mathrm{fr}}[Q]$ is contained in the main irreducible component of $\mathcal{TP}^{\mathrm{fr}}[Q]$ (with the reduced structure). Since $\mathscr{K}_Q$ is also reduced this completes the proof of 3.7.3. $\square$

4

Moduli of Principally Polarized Abelian Varieties

In this chapter we give a modular interpretation of the normalization of the main component of Alexeev's stack $\mathscr{A}_{g,1}^{\text{Alex}}$. The results of this chapter will be subsumed by the results of chapter 5, but we treat this case first as it is technically simpler and essentially follows the outline of the preceding chapter (and in particular at various points we make use of results of Alexeev).

4.1 The Standard Construction

4.1.1. Fix a lattice X isomorphic to $\mathbb{Z}^r$ for some integer r, and let $X_{\mathbb{R}}$ denote $X \otimes_{\mathbb{Z}} \mathbb{R}$.

A *paving* of $X_{\mathbb{R}}$ is a set S of polytopes $\omega \in X_{\mathbb{R}}$ satisfying

(i) For any two elements $\omega, \eta \in S$, the intersection $\omega \cap \eta$ is also in S.
(ii) Any face of a polytope $\omega \in S$ is again in S.
(iii) $X_{\mathbb{R}} = \cup_{\omega \in S}\omega$ and for any two distinct $\omega, \eta \in S$ the interiors of ω and η are disjoint.
(iv) For any bounded subset $W \subset X_{\mathbb{R}}$ there exists only finitely many $\omega \in S$ with $W \cap \omega \neq \emptyset$.

A paving S of $X_{\mathbb{R}}$ is called *integral* if each $\omega \in S$ is an integral polytope in $X_{\mathbb{R}}$. Following [3, 5.7.3], an integral paving S of $X_{\mathbb{R}}$ is called *regular* if there exists a non–homogeneous $\mathbb{R}$–valued quadratic function a on X with positive definite homogeneous part such that S is the set of domains of linearity of the function defined by the lower envelope of the convex hull of the set

$$G_a := \{(x, a(x))|x \in X\}. \tag{4.1.1.1}$$

A paving S of $X_{\mathbb{R}}$ is called X–*invariant* if it is invariant under the translation action of X.

M.C. Olsson, *Compactifying Moduli Spaces for Abelian Varieties*. Lecture Notes in Mathematics 1958.
© Springer-Verlag Berlin Heidelberg 2008

Lemma 4.1.2 *A regular paving S of $X_{\mathbb{R}}$ is X-invariant.*

Proof. If $y \in X$, then the function

$$L(x) := a(x + y) - a(x) \tag{4.1.2.1}$$

is a linear function of x since a is a quadratic function. Therefore if a' denotes the quadratic function $a(- + y)$ and if $g' : X_{\mathbb{R}} \to \mathbb{R}$ is the corresponding piecewise linear function (whose graph is the lower envelope of $G_{a'}$) then $g'(x) - L(x)$ is a piecewise linear function on $X_{\mathbb{R}}$ whose values on $x \in X$ are equal to $a(x)$. It follows that if g denotes the piecewise linear function defined by a (the lower envelope of G_a) then $g = g' - L$. Since L is linear this implies that the domains of linearity of g and g' are the same. $\square$

4.1.3. Let S be an integral regular paving of $X_{\mathbb{R}}$. Let P denote the integral points of $\mathrm{Cone}(1, X) \subset \mathbb{X} := \mathbb{Z} \oplus X$. The group X acts on $\mathbb{X}$ by $y \in X$ sending (d, x) to $(d, x + dy)$. Define

$$\rho : P \to \varinjlim_{\omega \in S} \mathrm{Cone}(1, \omega)^{gp} \tag{4.1.3.1}$$

to be the set map which sends $p \in \mathrm{Cone}(1, \omega)$ to the corresponding element of $\mathrm{Cone}(1, \omega)^{gp}$, and let $\widetilde{H}_S \subset \varinjlim_{\omega \in S} \mathrm{Cone}(1, \omega)^{gp}$ be the submonoid generated by elements $p * q := \rho(p) + \rho(q) - \rho(p + q)$ $(p, q \in P)$. The action of X on $\mathbb{X}$ induces an action of X on $\widetilde{H}_S$, and we define H_S to be the quotient of $\widetilde{H}_S$ by this action. For $\omega_i \in S$, let $N_i \subset P$ denote the integral points of $\mathrm{Cone}(1, \omega_i)$, and for $p, q \in P$ let $p * q \in H_S$ also denote the image of the element $p * q \in \widetilde{H}_S$. As in 3.1.5 one sees by elementary calculations that the following formulas hold in $\widetilde{H}_S$:

Lemma 4.1.4 (i) $p_1 * p_2 = p_2 * p_1$.
(ii) $p_1 * p_2 + p_3 * (p_1 + p_2) = \rho(p_1) + \rho(p_2) + \rho(p_3) - \rho(p_1 + p_2 + p_3)$.
(iii) *If* $p_1, p_2 \in N_i$, *then* $p_1 * p_2 = 0$.

4.1.5. As in 3.1.9 define $P \rtimes H_S$ to be the monoid with underlying set $P \times H_S$ and addition law given by

$$(p_1, q_1) + (p_2, q_2) := (p_1 + p_2, q_1 + q_2 + p_1 * p_2). \tag{4.1.5.1}$$

The formulas in 4.1.4 imply that this gives a well–defined commutative monoid structure. A straightforward verification shows that in fact the monoid $P \rtimes H_S$ is a commutative, integral, graded monoid (but not finitely generated) with grading induced by the map

$$P \rtimes H_S \xrightarrow{} P \xrightarrow{\deg} \mathbb{N}. \tag{4.1.5.2}$$

Lemma 4.1.6 *The monoid H_S is finitely generated.*

Proof. Define $\widetilde{SC}_1(\mathbb{X}_{\geq 0})$ to be the free monoid with generators pairs $(\sum \bar{\chi}_i, \chi)$, where $\sum \bar{\chi}_i \in \oplus_{w_i \in S} N_i$ and $\chi \in P$ such that the image of $\sum \bar{\chi}_i$ under the natural map

$$\oplus N_i \to P \tag{4.1.6.1}$$

is equal to χ. Let $SC_1(\mathbb{X}_{\geq 0})'$ denote the quotient of $\widetilde{SC}_1(\mathbb{X}_{\geq 0})$ by the relations

$$(\bar{\chi}_1, \chi_1) + (\bar{\chi}_2, \chi_2) = (\bar{\chi}_1 + \bar{\chi}_2, \chi_1 + \chi_2) \tag{4.1.6.2}$$

when χ_1 and χ_2 lie in the same cone, and let $SC_1(\mathbb{X}_{\geq 0})'/B_1$ denote the quotient of $SC_1(\mathbb{X}_{\geq 0})'$ by the relations

$$(\chi_1, \chi_2) + (\bar{\chi}_0, \chi_1) = (\bar{\chi}_0, \chi_2). \tag{4.1.6.3}$$

Finally define $SC_1(\mathbb{X}_{\geq 0})/B_1$ to be the quotient $SC_1(\mathbb{X}_{\geq 0})'/B_1$ by the natural X–action. By [3, 5.8.2], the monoid $SC_1(\mathbb{X}_{\geq 0})/B_1$ is finitely generated.

For every $w_i \in S$, let $\rho_i : \mathrm{Cone}(1, w_i)^{gp} \to \varinjlim_{w_i \in S} \mathrm{Cone}(1, w_i)^{gp}$ be the natural map. There is a map

$$\widetilde{SC}_1(\mathbb{X}) \to \varinjlim_{w_i \in S} \mathrm{Cone}(1, w_i)^{gp} \tag{4.1.6.4}$$

sending $(\sum_i \bar{\chi}_i, \chi)$ to $(\sum_i \rho_i(\bar{\chi}_i)) - \rho_k(\chi)$, where $\chi \in N_k$. By the definition of the relations 4.1.6.2 and 4.1.6.3 this map descends to a map

$$\pi : SC_1(\mathbb{X}_{\geq 0})'/B_1 \to \varinjlim_{w_i \in S} \mathrm{Cone}(1, w_i)^{gp} \tag{4.1.6.5}$$

Since any element $p * q \in \widetilde{H}_S$ is equal to the image of $(\bar{p} + \bar{q}, p + q)$ the image of π contains $\widetilde{H}_S$. In fact the image of π is equal to $\widetilde{H}_S$. This is shown by induction as in the proof of 3.1.7 by noting that the image of an element

$$(\sum_{i=1}^{n} \bar{\chi}_i, \sum_i \chi_i), \tag{4.1.6.6}$$

under π is equal to the image of

$$(\bar{\chi}_1 + \bar{\chi}_2, \chi_1 + \chi_2) + (\overline{\chi_1 + \chi_2} + \bar{\chi}_3 + \cdots + \bar{\chi}_n, \sum_{i=1}^{n} \chi_i). \tag{4.1.6.7}$$

Here we have ordered the submonoids $N_i = \mathrm{Cone}(1, w_i) \cap (\mathbb{Z} \oplus X)$ in some way and $\chi_i \in N_i$. Taking the quotient by the X–action we therefore obtain a surjection $SC_1(\mathbb{X}_{\geq 0})/B_1 \to H_S$ which implies the lemma since $SC_1(\mathbb{X}_{\geq 0})/B_1$ is finitely generated. $\square$

Remark 4.1.7. In what follows we will only need the weaker result that H_S^{sat} is finitely generated. This can be seen more directly as follows. Since H_S^{sat} is saturated, it suffices to show that H_S^{gp} is a finitely generated group [41, 2.2.11].

Since H_S^{gp} is equal to $\mathbb{Z} \otimes_{\mathbb{Z}[X]} \widetilde{H}_S^{gp}$ (where the map $\mathbb{Z}[X] \to \mathbb{Z}$ sends all elements of X to 1), it suffices to show that $\widetilde{H}_S^{gp}$ is finitely generated as a $\mathbb{Z}[X]$–module. This follows from noting that there is a short exact sequence

$$0 \to \widetilde{H}_S^{gp} \to \varinjlim N_\omega^{gp} \to P^{gp} \to 0, \tag{4.1.7.1}$$

and $\varinjlim N_\omega^{gp}$ and P^{gp} are immediately seen to be finitely generated $\mathbb{Z}[X]$–modules.

Lemma 4.1.8 *The monoid H_S is sharp (i.e. $H_S^* = \{0\}$).*

Proof. Let $a : X \to \mathbb{R}$ be a quadratic function with positive definite quadratic part defining S, and let $\tilde{g} : X_\mathbb{R} \to \mathbb{R}$ be the piecewise linear function given by the lower envelope of the convex hull of the set

$$G_a := \{(x, a(x)) | x \in X\} \subset X_\mathbb{R} \times \mathbb{R}. \tag{4.1.8.1}$$

Extend $\tilde{g}$ to a function $g : P \to \mathbb{R}$ by the formula

$$g(p) := \deg(p)\tilde{g}(p/\deg(p)). \tag{4.1.8.2}$$

By the definition of the paving associated to a, the restriction of g to each $P_\omega \subset P$ (where $\omega \in S$) is a linear function, and for two elements $p, q \in P$ which do not lie in the same P_ω we have

$$g(p) + g(q) - g(p + q) > 0. \tag{4.1.8.3}$$

The function g therefore defines a map

$$g : H_S \to \mathbb{R}_{>0} \tag{4.1.8.4}$$

sending the generators of H_S to nonzero elements of $\mathbb{R}_{>0}$. Since the additive monoid $\mathbb{R}_{>0}$ is torsion free this implies that H_S is sharp. $\square$

Lemma 4.1.9 *The morphism $H_S \to P \rtimes H_S$ is integral. In particular $H_S^{gp} \to (P \rtimes H_S)^{gp}$ is injective with cokernel isomorphic to $P^{gp} \simeq X \otimes \mathbb{Z}$.*

Proof. This follows from the same argument proving 3.1.11. $\square$

4.1.10. Let B be a scheme and fix data as follows:

(i) An abelian scheme A/B;
(ii) An ample line bundle $\mathcal{M}$ on A defining a principal polarization $A \to A^t$;
(iii) A rigidification of $\mathcal{M}$ at $0 \in A$;
(iv) A morphism of monoids $\beta : H_S \to \mathcal{O}_B$;
(v) A homomorphism $c : X \to A^t \simeq A$ defining a semiabelian scheme G/B as in 2.1.8;

(vi) A trivialization $\psi : X \times B \to c^* \mathcal{M}^{-1}$ inducing a trivialization $\tau : B \times X \times X \to (c \times c)^* \mathcal{B}^{-1}$ compatible with the symmetric biextension structure (where $\mathcal{B}$ denotes the Poincare bundle which we view as a bundle over $A \times A$ using the principal polarization).

Define $P \rtimes H_S$ to be the monoid whose elements are (p, h) with $p \in P$ and $h \in H_S$, and whose addition law is given by $(p, h) + (p', h') = (p + p', h + h' + p * p')$ (where we also abusively write $p * p'$ for the image of $p * p'$ in H_S). Let $\mathcal{R}$ denote the quasi–coherent sheaf on A given by

$$\mathcal{R} := \oplus_{(d,x) \in P} \mathcal{M}^d \otimes L_x. \qquad (4.1.10.1)$$

Since c is a homomorphism, for any $x, y \in X$ there exists a unique isomorphism of rigidified line bundles can : $L_x \otimes L_y \simeq L_{x+y}$. Define an algebra structure on $\mathcal{R}$ by defining multiplication using the maps

$$\rho((d, x) * (d', y)) \cdot \text{can} : (\mathcal{M}^d \otimes L_x) \otimes (\mathcal{M}^{d'} \otimes L_y) \to \mathcal{M}^{d+d'} \otimes L_{x+y}. \quad (4.1.10.2)$$

Note that locally on A, if we choose a trivialization of $\mathcal{M}$ and compatible trivializations of the L_x, then $\mathcal{R}$ is isomorphic $\mathbb{Z}[P \rtimes H_S] \otimes_{\mathbb{Z}[H_S]} \mathcal{O}_A$ (where $\mathbb{Z}[H_S] \to \mathcal{O}_A$ is induced by the composite $H_S \to \mathcal{O}_B \to \mathcal{O}_A$). The scheme $\widetilde{\mathscr{P}} := \text{Proj}_A(\mathcal{R})$ has a natural log structure $M_{\widetilde{\mathscr{P}}}$ defined locally by the morphism $P \rtimes H_S \to \mathbb{Z}[P \rtimes H_S] \otimes_{\mathbb{Z}[H_S]} \mathcal{O}_A$, and there is a canonical log smooth morphism

$$(\widetilde{\mathscr{P}}, M_{\widetilde{\mathscr{P}}}) \to (B, M_B), \qquad (4.1.10.3)$$

where M_B is the log structure on B defined by the map $H_S \to \mathcal{O}_B$.

Lemma 4.1.11 *The morphism 4.1.10.3 is log smooth an integral.*

Proof. By the construction there exists after replacing $\widetilde{\mathscr{P}}$ by a smooth covering a smooth morphism

$$\pi : \widetilde{\mathscr{P}} \to B \times_{\text{Spec}(\mathbb{Z}[H_S])} \text{Spec}(\mathbb{Z}[P \rtimes H_S]) \qquad (4.1.11.1)$$

such that the log structure $M_{\widetilde{\mathscr{P}}}$ is isomorphic to the pullback of the log structure on $\text{Spec}(\mathbb{Z}[P \rtimes H_S])$ induced by the natural map $P \rtimes H_S \to \mathbb{Z}[P \rtimes H_S]$. Combining 4.1.9, 2.3.16, and 2.3.14 we obtain the lemma. $\square$

4.1.12. It is sometimes useful to have an alternate description of the log structure $M_\mathscr{C}$ on the cone $\mathscr{C} = \underline{\text{Spec}}_A(\mathcal{R})$. Let $s : \mathscr{C} \to A$ denote the projection. Then the log structure $M_\mathscr{C}$ is the log structures associated to the prelog structure which to any $U \to \mathscr{C}$ associates the set of pairs $((d, x), v)$, where $(d, x) \in P$ and v is a trivialization of $s^*(\mathcal{M}^d \otimes L_x)$ over U. The map $M_\mathscr{C} \to \mathcal{O}_\mathscr{C}$ sends $((d, x), v)$ to the image of v under the canonical map $s^*(\mathcal{M}^d \otimes L_x) \to \mathcal{O}_\mathscr{C}$.

We will often use this description as follows. Let $\lambda : A \to A$ denote an automorphism of A (not necessarily compatible with the group structure) and $\tilde{\lambda} : \lambda^* \mathcal{R} \to \mathcal{R}$ an isomorphism induced by isomorphisms $\lambda_{(d,x)} : \lambda^*(\mathcal{M}^d \otimes$

L_x) $\to$ $(\mathcal{M}^d \otimes L_x)$. Then the induced isomorphism $\tilde{\lambda} : \mathscr{C} \to \mathscr{C}$ extends canonically to an isomorphism of log schemes $(\mathscr{C}, M_{\mathscr{C}}) \to (\mathscr{C}, M_{\mathscr{C}})$. Indeed the log structure $\tilde{\lambda}^* M_{\mathscr{C}}$ is associated to the prelog structure which to any $U \to \mathscr{C}$ associates the set of pairs $((d,x), v')$, where $(d,x) \in P$ and $v' \in s^* \lambda^*(\mathcal{M}^d \otimes L_x)$, so the isomorphisms $\lambda_{(d,x)}$ induce isomorphisms of log structures.

Lemma 4.1.13 *The log structure $M_{\widetilde{\mathscr{P}}}$ is a fine log structure.*

Proof. By construction, $M_{\widetilde{\mathscr{P}}}$ is a quasi–coherent log structure in the sense of [24, 2.1]. Hence it suffices to show that for any geometric point $\bar{p} \to \widetilde{\mathscr{P}}$ the monoid $\overline{M}_{\widetilde{\mathscr{P}}, \bar{p}}$ is a finitely generated monoid. For this note that there exists a surjection $P \rtimes H_S \to \overline{M}_{\widetilde{\mathscr{P}}, \bar{p}}$ identifying $\overline{M}_{\widetilde{\mathscr{P}}, \bar{p}}$ with $P \rtimes H_S / F$ for some face $F \subset P \rtimes H_S$ containing at least one element f of the form $((l, \eta), 0)$ with $(l, \eta) \in P$. Let $Q \subset P$ denote the submonoid generated by the cones over simplices ω containing $\eta / l \in X_{\mathbb{R}}$. Then for any $(d, x) \in P$ and $h \in H_S$ we have

$$((d,x), h) + Nf = ((d + Nl, x + N\eta), h + (d,x) * (Nl, N\eta)). \qquad (4.1.13.1)$$

Now for N sufficiently big, $(d + Nl, x + N\eta)$ is in Q, and therefore the map $Q \rtimes H_S \to P \rtimes H_S / F$ is surjective. Since Q is a finitely generated monoid, and hence also $Q \rtimes H_S$, it follows that $P \rtimes H_S / F$ is finitely generated. $\square$

4.1.14. There is also a natural G–action on $(\widetilde{\mathscr{P}}, M_{\widetilde{\mathscr{P}}})$ over (B, M_B). The action $g : (\widetilde{\mathscr{P}}, M_{\widetilde{\mathscr{P}}}) \to (\widetilde{\mathscr{P}}, M_{\widetilde{\mathscr{P}}})$ of a point $g \in G(B)$ corresponding to a point $a \in A(B)$ and compatible isomorphisms $\iota_g : t_a^* L_x \to L_x$ is obtained by locally on A choosing a trivialization of $t_a^* \mathcal{M} \otimes \mathcal{M}^{-1}$ and defining the resulting map $g^* : t_a^* \mathcal{R} \to \mathcal{R}$ to be the one induced by the isomorphisms ι_g. The resulting morphism

$$A \times_{a, A} \widetilde{\mathscr{P}} \to \widetilde{\mathscr{P}} \qquad (4.1.14.1)$$

is then independent of the choice of trivialization of $t_a^* \mathcal{M} \otimes \mathcal{M}^{-1}$ and hence defined globally. Furthermore, the local description shows that it extends naturally to a morphism of log schemes over (B, M_B).

Let us verify that this really defines an action of G on $(\widetilde{\mathscr{P}}, M_{\widetilde{\mathscr{P}}})$ over (B, M_B). This amounts to the following:

Lemma 4.1.15 *Let $g' \in G(B)$ be a second point. Then $(g + g')^* = g'^* \circ g^*$.*

Proof. The point g (resp. g') is given by a point $a \in A(B)$ (resp. $a' \in A(B)$) together with compatible isomorphisms $\epsilon_x : t_a^* L_x \to L_x$ (resp. $\epsilon_x' : t_{a'}^* L_x \to L_x$). The point $g + g'$ then corresponds to $a + a'$ and the isomorphisms

$$\epsilon'' : t_{a+a'}^* L_x \simeq t_{a'}^* t_a^* L_x \xrightarrow{t_{a'}^* \epsilon_x} t_{a'}^* L_x \xrightarrow{\epsilon'} L_x. \qquad (4.1.15.1)$$

The composite $g'^* \circ g^*$ is given locally on A by choosing trivializations of $\mathcal{M}$ and $t_{a+a'}^* \mathcal{M}$ and then using the map of algebras over A

$$t_{a'}^* t_a^* (\oplus_{(d,x)} L_x) \xrightarrow{\ t_{a',\epsilon}^*\ } t_{a'}^* (\oplus_{(d,x)} L_x) \xrightarrow{\ \epsilon'\ } \oplus_{(d,x)} L_x. \qquad (4.1.15.2)$$

By the above description of $g + g'$ this agrees with $(g + g')^*$. This shows the result on the level of the underlying schemes. In fact, combining this with 4.1.12 this also shows that the two morphisms of log schemes are equal. $\square$

4.1.16. There is also an X–action on $(\widetilde{\mathscr{P}}, M_{\widetilde{\mathscr{P}}}, \mathscr{O}_{\widetilde{\mathscr{P}}}(1))$. Recall from 2.2.12 that ψ and τ define isomorphisms

$$\psi(y)^d \tau(y,x) : t_y^* (\mathscr{M}^d \otimes L_x) \to \mathscr{M}^d \otimes L_{x+dy}. \qquad (4.1.16.1)$$

For $y \in X$ define

$$T_y^* : t_{c(y)}^* \mathscr{R} \to \mathscr{R} \qquad (4.1.16.2)$$

to be the map obtained by taking the direct sum of the isomorphisms $\psi(y)^d \tau(y,x)$.

Lemma 4.1.17 *The map 4.1.16.2 is an algebra homomorphism.*

Proof. We have to show that the diagram

$$
\begin{array}{ccc}
t_{c(y)}^*(\mathscr{M}^d \otimes L_x) \otimes t_{c(y)}^*(\mathscr{M}^{d'} \otimes L_{x'}) & \xrightarrow{\ \rho((d,x)*(d',x'))\mathrm{can}\ } & t_{c(y)}^*(\mathscr{M}^{d+d'} \otimes L_{x+x'}) \\
{\scriptstyle \psi(y)^d \tau(y,x) \otimes \psi(y)^{d'} \tau(y,x')}\downarrow & & \downarrow{\scriptstyle \psi(y)^{d+d'} \tau(y,x+x')} \\
(\mathscr{M}^d \otimes L_{x+dy}) \otimes (\mathscr{M}^{d'} \otimes L_{x'+d'y}) & \xrightarrow{\ \rho((d,x+dy)*(d',x'+d'y))\mathrm{can}\ } & \mathscr{M}^{d+d'} \otimes L_{x+x'+(d+d')y}
\end{array}
$$

$$(4.1.17.1)$$

commutes. Since $\rho((d, x + dy) * (d', x' + d'y))$ is equal to $\rho((d, x) * (d', x'))$ (since H_S is the quotient of $\widetilde{H}_S$ by the X–action) this follows from (2.2.13 (iii)'). $\square$

4.1.18. Note that if locally on A we choose a trivialization of $\mathscr{M}$ and compatible trivializations of the L_x, then T_y^* is induced by the map

$$P \rtimes H_S \to (P \rtimes H_S) \oplus \mathscr{O}_A^*, \quad ((d,x),h) \mapsto (((d, x + dy), h), \psi(y)^d \tau(y,x)). \qquad (4.1.18.1)$$

It follows that the morphism $T_y : \widetilde{\mathscr{P}} \to \widetilde{\mathscr{P}}$ extends naturally to a morphism of polarized log schemes $T_y : (\widetilde{\mathscr{P}}, M_{\widetilde{\mathscr{P}}}, \mathscr{O}_{\widetilde{\mathscr{P}}}(1)) \to (\widetilde{\mathscr{P}}, M_{\widetilde{\mathscr{P}}}, \mathscr{O}_{\widetilde{\mathscr{P}}}(1))$ over (B, M_B). Observe also that the map $T_y : (\widetilde{\mathscr{P}}, M_{\widetilde{\mathscr{P}}}) \to (\widetilde{\mathscr{P}}, M_{\widetilde{\mathscr{P}}})$ depends only on τ and not on the choice of ψ inducing τ. Indeed if ψ' is a second trivialization of $c^* \mathscr{M}^{-1}$ inducing τ, then ψ and ψ' differ by a homomorphism $X \to \mathbb{G}_m$. To see this last statement, note that by the definition of τ for any two element $x, y \in X$ we have

$$\psi(x+y)\psi(x)^{-1}\psi(y)^{-1} = \tau(x,y) = \psi'(x+y)\psi'(x)^{-1}\psi'(y)^{-1} \qquad (4.1.18.2)$$

as sections of $\mathscr{M}(x+y) \otimes \mathscr{M}(x)^{-1} \otimes \mathscr{M}(y)^{-1}$. This formula implies that ψ/ψ' is a homomorphism $X \to \mathbb{G}_m$.

Lemma 4.1.19 *For any $y, y' \in X$ the two morphisms*

$$T_{y+y'}, T_{y'} \circ T_y : (\widetilde{\mathscr{P}}, M_{\widetilde{\mathscr{P}}}, \mathscr{O}_{\widetilde{\mathscr{P}}}(1)) \to (\widetilde{\mathscr{P}}, M_{\widetilde{\mathscr{P}}}, \mathscr{O}_{\widetilde{\mathscr{P}}}(1)) \qquad (4.1.19.1)$$

are equal.

Proof. Chasing through the definitions one finds that this is equivalent to (2.2.13 (ii)'). □

Lemma 4.1.20 *The actions of G and X on the fine log scheme $(\widetilde{\mathscr{P}}, M_{\widetilde{\mathscr{P}}})$ commute.*

Proof. For a point $g \in G(B)$ mapping to $a \in A(B)$, the translation action $S_g : (\widetilde{\mathscr{P}}, M_{\widetilde{\mathscr{P}}}) \to (\widetilde{\mathscr{P}}, M_{\widetilde{\mathscr{P}}})$ is obtained from maps $i_g^x : t_a^* L_x \to L_x$ such that $i_g^{x+y} = i_g^x \otimes i_g^y$. It follows that the two maps

$$T_y^* \circ S_g^*, \quad S_g^* \circ T_y^* : \mathcal{R} \to \mathcal{R} \qquad (4.1.20.1)$$

locally differ by a scalar which implies the lemma (here we also use 4.1.12 to see that the maps of log structures agree). □

In fact we can say a bit more. On the polarized log scheme $(\widetilde{\mathscr{P}}, M_{\widetilde{\mathscr{P}}}, \mathscr{O}_{\widetilde{\mathscr{P}}}(1))$ there is both an action of X and an action of $T = D(X)$.

Lemma 4.1.21 *Let $y \in X$ and $u \in T(B)$ be sections, and T_y and S_u the resulting automorphisms of $(\widetilde{\mathscr{P}}, M_{\widetilde{\mathscr{P}}}, \mathscr{O}_{\widetilde{\mathscr{P}}}(1))$. Then*

$$u(y) \circ S_u \circ T_y = T_y \circ S_u, \qquad (4.1.21.1)$$

where $\mathbb{G}_m$ acts on $(\widetilde{\mathscr{P}}, M_{\widetilde{\mathscr{P}}}, \mathscr{O}_{\widetilde{\mathscr{P}}}(1))$ via multiplication on $\mathscr{O}_{\widetilde{\mathscr{P}}}(1)$.

Proof. The morphism $S_u \circ T_y$ is induced by the maps

$$\mathscr{M}^d \otimes L_x \xrightarrow{u(x)} \mathscr{M}^d \otimes L_x \xrightarrow{\psi(y)^d \tau(y,x)} \mathscr{M}^d \otimes L_{x+dy}, \qquad (4.1.21.2)$$

and $T_y \circ S_u$ is given by

$$\mathscr{M}^d \otimes L_x \xrightarrow{\psi(y)^d \tau(y,x)} \mathscr{M}^d \otimes L_{x+dy} \xrightarrow{u(x+y)} \mathscr{M}^d \otimes L_{x+dy}. \qquad (4.1.21.3)$$

Since $u(x + y) = u(x)u(y)$ this implies the lemma. □

4.1.22. Now assume that the image of H_S in $\mathscr{O}_B$ is contained in the nilradical of $\mathscr{O}_B$. Then as explained in [4, 3.17] the action of X on $\widetilde{\mathscr{P}}$ is properly discontinuous, and hence we can form the quotient $\mathscr{P} := \widetilde{\mathscr{P}}/X$. Furthermore, by descent theory the log structure $M_{\widetilde{\mathscr{P}}}$ and the invertible sheaf $\mathscr{O}_{\widetilde{\mathscr{P}}}(1)$ descend to give a fine log structure $M_{\mathscr{P}}$ on $\mathscr{P}$ and a polarization $L_{\mathscr{P}}$.

The result is a polarized fine log scheme with G–action

$$(\mathscr{P}, M_{\mathscr{P}}, L_{\mathscr{P}}, G\text{–action}) \to (B, M_B). \qquad (4.1.22.1)$$

By 4.1.11 the morphism

$$(\mathscr{P}, M_{\mathscr{P}}) \to (B, M_B) \qquad (4.1.22.2)$$

is log smooth and integral. We refer to the data $(G, M_B, \mathscr{P}, M_{\mathscr{P}}, L_{\mathscr{P}}, G\text{–action})$ as a *standard family* over B.

Though many of the proofs involve this standard family, for technical reasons it will be useful to consider a slight modification of the standard family. Namely, let $(B^{\mathrm{sat}}, M_B^{\mathrm{sat}})$ denote the saturation of the log scheme (B, M_B) ([23], 3.0.3), and let $(\mathscr{P}_{B^{\mathrm{sat}}}, M_{\mathscr{P}}^{\mathrm{sat}}, L_{\mathscr{P}}) \to (\mathrm{Spec}(B^{\mathrm{sat}}), M_B^{\mathrm{sat}})$ denote the polarized log scheme obtained by base change $(\mathrm{Spec}(B^{\mathrm{sat}}), M_B^{\mathrm{sat}}) \to (\mathrm{Spec}(B), M_B)$. The group G still acts on $(\mathscr{P}_{B^{\mathrm{sat}}}, M_{\mathscr{P}}^{\mathrm{sat}}, L_{\mathscr{P}})$. We refer to the data

$$(G, M_B^{\mathrm{sat}}, \mathscr{P}_{B^{\mathrm{sat}}}, M_{\mathscr{P}}^{\mathrm{sat}}, L_{\mathscr{P}}, G\text{–action}) \qquad (4.1.22.3)$$

as the *saturation of a standard family*.

Remark 4.1.23. There is a canonical section $\theta \in H^0(\mathscr{P}, L_{\mathscr{P}})$. To give such a section is equivalent to giving a section $\tilde{\theta} \in H^0(\widetilde{\mathscr{P}}, \mathscr{O}_{\widetilde{\mathscr{P}}}(1))^X$. Such a section can be written down explicitly [4, 3.22].

4.2 Automorphisms over a Field

4.2.1. Let k be a field, and A/k an abelian variety with a rigidified line bundle $\mathscr{M}$ defining a principal polarization on A. Let G/k be the semi–abelian variety defined by a homomorphism $c : X \to A^t$, for some lattice X. Let S be a regular paving of $X_{\mathbb{R}}$, and let M_k be the log structure on $\mathrm{Spec}(k)$ associated to the map $H_S \to k$ sending all nonzero elements to 0. Assume given ψ and τ as in 4.1.10 giving rise to the standard family

$$(\mathscr{P}, M_{\mathscr{P}}, L_{\mathscr{P}}, G\text{–action}) \to (\mathrm{Spec}(k), M_k). \qquad (4.2.1.1)$$

Proposition 4.2.2. *Let* $\alpha : (\mathscr{P}, L_{\mathscr{P}}) \to (\mathscr{P}, L_{\mathscr{P}})$ *be an automorphism of the underlying polarized scheme with G–action. Then there exists a unique pair* (α^b, β), *where* $\beta : M_k \to M_k$ *is an automorphism and* $\alpha^b : \alpha^* M_{\mathscr{P}} \to M_{\mathscr{P}}$ *is an isomorphism compatible with the G–action such that the diagram*

$$
\begin{array}{ccc}
\alpha^* M_{\mathscr{P}} & \xrightarrow{\alpha^b} & M_{\mathscr{P}} \\
\uparrow & & \uparrow \\
M_k & \xrightarrow{\beta} & M_k
\end{array}
\qquad (4.2.2.1)
$$

commutes.

Proof. We first consider the problem of finding (α^b, β) compatible with the T-action.

Let $(\widetilde{\mathscr{P}}, M_{\widetilde{\mathscr{P}}}, \mathscr{O}_{\widetilde{\mathscr{P}}}(1))$ be as in the construction of the standard model. By [3, 4.3.1] there exists a unique automorphism $\tilde{\alpha}$ of $(\widetilde{\mathscr{P}}, \mathscr{O}_{\widetilde{\mathscr{P}}}(1))$ compatible with the T and X–actions inducing α after taking the quotient by X (see also section 5.3). Let $\pi : \widetilde{\mathscr{P}} \to A$ denote the projection.

Lemma 4.2.3 *There exists a unique automorphism $\bar{\alpha} : A \to A$ such that the diagram*

$$
\begin{array}{ccc}
\widetilde{\mathscr{P}} & \xrightarrow{\ \tilde{\alpha}\ } & \widetilde{\mathscr{P}} \\
{\scriptstyle \pi}\downarrow & & \downarrow{\scriptstyle \pi} \\
A & \xrightarrow{\ \bar{\alpha}\ } & A
\end{array}
\tag{4.2.3.1}
$$

commutes.

Proof. The uniqueness is immediate since π is faithfully flat.

To prove existence, note first that the irreducible components $\mathscr{P}_i$ of $\widetilde{\mathscr{P}}$ are given by the schemes $\mathrm{Proj}_A(\oplus_{(d,x)\in P_i} \mathcal{M}^d \otimes L_x)$, where P_i denotes the integral points of $\mathrm{Cone}(1, \omega_i)$ for some top–dimensional simplex $\omega_i \in S$. Furthermore, if the intersection of two components $\mathscr{P}_i$ and $\mathscr{P}_j$ is nonempty, then their intersection is given by $\mathrm{Proj}_A(\oplus_{(d,x)\in P_{ij}} \mathcal{M}^d \otimes L_x)$, where P_{ij} denotes the integral points of the cone over $\omega_{ij} := \omega_i \cap \omega_j$. In particular, for any two irreducible components $\mathscr{P}_i$ and $\mathscr{P}_j$ of $\widetilde{\mathscr{P}}$ with $\mathscr{P}_i \cap \mathscr{P}_j \neq \emptyset$, the map $\mathscr{P}_i \cap \mathscr{P}_j \to A$ is surjective. It therefore suffices to show that for any irreducible component $\mathscr{P}_i$ of $\widetilde{\mathscr{P}}$ there exists a unique morphism $\bar{\alpha} : A \to A$ such that the diagram

$$
\begin{array}{ccc}
\mathscr{P}_i & \xrightarrow{\ \tilde{\alpha}\ } & \tilde{\alpha}(\mathscr{P}_i) \\
{\scriptstyle \pi}\downarrow & & \downarrow{\scriptstyle \pi} \\
A & \xrightarrow{\ \bar{\alpha}\ } & A
\end{array}
\tag{4.2.3.2}
$$

commutes. For this note that if $\mathscr{U}_i \subset \mathscr{P}_i$ is the maximal open subset on which T acts faithfully, then $A = [\mathscr{U}_i/T]$. In fact locally over A, we have $\mathscr{P}_i \simeq \mathrm{Spec}(\mathscr{O}_A[P_i])$ with P_i equal to the integral points of $\mathrm{Cone}(1, \omega_i) \subset \mathbb{X}_{\mathbb{R}}$ for some simplex $\omega_i \in S$, and $\mathscr{U}_i = \mathrm{Spec}_A(\mathscr{O}_A[P_i^{gp}])$. The open subset $\mathscr{U}_i$ is also $\tilde{\alpha}$–invariant since the automorphism $\tilde{\alpha}$ commutes with the T–action. The restriction of $\tilde{\alpha}$ to $\mathscr{U}_i$ therefore defines a map $\bar{\alpha}$, and the diagram 4.2.3.2 commutes since $\mathscr{U}_i$ is dense in $\mathscr{P}_i$ and $\mathscr{P}_i$ is reduced. $\square$

By [34, 6.4], we can write $\bar{\alpha}$ uniquely as $\bar{\alpha} = h + t_a$, where $h : A \to A$ is a homomorphism and $a \in A(k)$ is a point. The obstruction to lifting a to a point of G lies in $H^1(\mathrm{Spec}(k)_{\mathrm{et}}, T)$ which since k is a field is zero. Choose a lifting $\tilde{a} \in G$ of a. Replacing α by $\rho_{-\tilde{a}} \circ \alpha$ we may assume that $a = 0$ (here ρ denotes the action of G on $\widetilde{\mathscr{P}}$). In this case $\bar{\alpha}$ is a homomorphism and α is induced by a map

$$\bar{\alpha}^*(\oplus_{(d,x)}\mathcal{M}^d \otimes L_x) \to \oplus_{(d,x)}\mathcal{M}^d \otimes L_x. \qquad (4.2.3.3)$$

The compatibility with the T–action further implies that this map is obtained from the direct sum of isomorphisms $\alpha^*_{(d,x)} : \bar{\alpha}^*(\mathcal{M}^d \otimes L_x) \to \mathcal{M}^d \otimes L_x$ (look at the eigenspaces in $\pi_*(\mathcal{O}_{\widetilde{\mathscr{P}}}(1)))$. Now since $\bar{\alpha}^*(\mathcal{M}^d \otimes L_x)$ and $\mathcal{M}^d \otimes L_x$ are both rigidified at 0, giving such maps $\alpha^*_{(d,x)}$ is equivalent to giving a set map $\rho : P \to k^*$. Furthermore, the compatibility with the algebra structure implies that for every simplex $\omega_i \in S$ the restriction of ρ to the integral points of $\mathrm{Cone}(1,\omega_i)$ is linear. In other words, ρ is obtained from a homomorphism

$$\rho : \varinjlim_{\omega \in S} \mathrm{Cone}(1,\omega)^{gp} \to k^*. \qquad (4.2.3.4)$$

Let $\tilde{\gamma} : \tilde{H}_S \to k^*$ be the map induced by restriction. The compatibility with the X–action implies that $\rho(d,x) = \rho(d,x+dy)$ for every $x, y \in X$ and $d \geq 1$. This equation implies that $\tilde{\gamma}$ descends to a morphism $\gamma : H_S \to k^*$. Let β be the automorphism of M_k induced by the map

$$H_S \to H_S \oplus k^*, \quad h \mapsto (h,\gamma(h)), \qquad (4.2.3.5)$$

and let $\alpha^b : \alpha^* M_{\widetilde{\mathscr{P}}} \to M_{\widetilde{\mathscr{P}}}$ be the isomorphism of log structures obtained from the maps $\alpha^*_{(d,x)}$ and the description in 4.1.12 of the log structure $M_{\widetilde{\mathscr{P}}}$. This proves existence of the pair (α^b, β), except we still have to prove that α^b is compatible with the G–action.

So consider a point $g \in G(B)$ with image $a \in A(B)$ and the resulting two isomorphisms

$$t^*_g \circ \alpha^b, \alpha^b \circ t^*_g : \alpha^* t^*_g M_{\widetilde{\mathscr{P}}} \to M_{\widetilde{\mathscr{P}}}. \qquad (4.2.3.6)$$

Let σ_g be the automorphism $(\alpha^b \circ t^*_g) \circ (t^*_g \circ \alpha^b)^{-1}$ of $M_{\widetilde{\mathscr{P}}}$. Observe that since α^b is compatible with the T–action the automorphism σ_g depends only on $a \in A(B)$ and not on the lifting $g \in G(B)$ of a.

Let $\mathscr{C}$ denote the cone $\underline{\mathrm{Spec}}_A(\mathcal{R})$, and let $\mathscr{C}^o$ denote the complement of the vertex so that $\widetilde{\mathscr{P}} = [\mathscr{C}^o/\mathbb{G}_m]$. Let $M_{\mathscr{C}}$ denote the natural log structure on $\mathscr{C}$ giving locally over A by the map $P \to \mathcal{O}_A \otimes_{\mathbb{Z}[H_S]} \mathbb{Z}[P \rtimes H_S]$. The restriction to $\mathscr{C}^o$ is canonically isomorphic to the pullback of $M_{\widetilde{\mathscr{P}}}$. The global chart defines a surjection $\beta : P \rtimes H_S \to \overline{M}_{\mathscr{C}}$, and it follows from the construction of σ_g that the automorphism $\tilde{\sigma}_g : M_{\mathscr{C}^o} \to M_{\mathscr{C}^o}$ is compatible with this map β. From this and the compatibility with the $\mathbb{G}_m$–action on $\mathscr{C}^o$, it follows that the automorphism σ_g is induced by a homomorphism

$$\rho_g : P^{gp} \to \Gamma(\mathscr{C}^o, \mathcal{O}^*_{\mathscr{C}^o})^{\mathbb{G}_m} = \mathbb{G}_m(B). \qquad (4.2.3.7)$$

The association $g \mapsto \rho_g$ is clearly compatible with base change $B' \to B$, and hence induces a map of schemes $G \to \mathrm{Hom}(P^{gp}, \mathbb{G}_m)$. Furthermore, as mentioned earlier the compatibility with the T–action implies that this morphism descends to a morphism $A \to \mathrm{Hom}(P^{gp}, \mathbb{G}_m)$. Note that P^{gp} is a direct limit of finitely generated free abelian groups, and hence $\mathrm{Hom}(P^{gp}, \mathbb{G}_m)$ is a projective limit of tori. Since A is proper this implies that the map $A \to \mathrm{Hom}(P^{gp}, \mathbb{G}_m)$ is constant. Since it sends 0 to the constant map 1 this implies that σ_g is the identity for all g. This completes the proof of the existence of (α^b, β).

To prove the uniqueness, it suffices to prove that there are no non–trivial automorphisms of the log structure $M_{\widetilde{\mathscr{P}}}$ compatible with the G–action. This is because the map $M_k \to M_{\widetilde{\mathscr{P}}}$ is injective.

Let $\sigma : M_{\widetilde{\mathscr{P}}} \to M_{\widetilde{\mathscr{P}}}$ be an automorphism of the log structure compatible with the G–action. We first show that $\bar{\sigma} : \overline{M}_{\widetilde{\mathscr{P}}} \to \overline{M}_{\widetilde{\mathscr{P}}}$ is the identity. Since σ is compatible with the G–action, to prove that $\bar{\sigma}$ is the identity it suffices to show that $\bar{\sigma}$ induces the identity over the fiber $\widetilde{\mathscr{P}}_0$ of $\widetilde{\mathscr{P}}$ over $0 \in A$. Note that $\widetilde{\mathscr{P}}_0 \simeq \mathrm{Proj}(k \otimes_{\mathbb{Z}[H_S]} \mathbb{Z}[P \rtimes H_S])$. We may also assume that k is algebraically closed.

For $\omega_i \in S$, let $\mathscr{P}_i \subset \widetilde{\mathscr{P}}_0$ be the irreducible component $\mathrm{Proj}(k[N_i])$ (where N_i is the integral points of $\mathrm{Cone}(1, \omega_i)$), and let $M_i \subset M_{\widetilde{\mathscr{P}}_0}|_{\mathscr{P}_i}$ be the sub–log structure consisting of sections whose image in $\mathscr{O}_{\mathscr{P}_i}$ is nonzero. The log structure M_i is isomorphic to the standard log structure on $\mathrm{Proj}(k[N_i])$ induced by the map $N_i \to k[N_i]$. In particular, the map $M_i \to \mathscr{O}_{\mathscr{P}_i}$ is injective which implies that $\sigma|_{M_i}$ is the identity. Denote by $\mathscr{C}_k$ the cone $\mathrm{Spec}(k \otimes_{\mathbb{Z}[H_S]} \mathbb{Z}[P \rtimes H_S])$, let $\mathscr{C}_i \subset \mathscr{C}_k$ be the closed subscheme $\mathrm{Spec}(k[N_i])$, and let $\mathscr{C}_i^o \subset \mathscr{C}_i$ denote the complement of the vertex.

Lemma 4.2.4 *The natural map*

$$k^* \oplus (P \rtimes H_S)^{gp} \to H^0(\mathscr{C}_i^o, M^{gp}_{\mathscr{P}_i}|_{\mathscr{C}_i^o}) \tag{4.2.4.1}$$

is an isomorphism.

Proof. Denote by Q_i the quotient $M^{gp}_{\mathscr{P}_k}|_{\mathscr{C}_i^o}/M^{gp}_i$ so that there is an exact sequence

$$0 \to M^{gp}_i \to M^{gp}_{\mathscr{P}_k}|_{\mathscr{C}_i^o} \to Q_i \to 0. \tag{4.2.4.2}$$

The quotient Q_i is the constant sheaf associated to the abelian group $(P \rtimes H_S)^{gp}/N^{gp}_i$. Consideration of the commutative diagram

$$\begin{array}{ccccccccc}
0 & \longrightarrow & k^* \oplus N^{gp}_i & \longrightarrow & k^* \oplus (P \rtimes H_S)^{gp} & \longrightarrow & (P \rtimes H_S)^{gp}/N^{gp}_i & \longrightarrow & 0 \\
& & \downarrow & & \downarrow & & \downarrow{\scriptstyle y} & & \\
0 & \longrightarrow & H^0(\mathscr{C}_i^o, M^{gp}_i) & \longrightarrow & H^0(\mathscr{C}_i^o, M^{gp}_{\mathscr{P}_k}|_{\mathscr{C}_i^o}) & \longrightarrow & H^0(\mathscr{C}_i^o, Q_i) & &
\end{array}$$
$$\tag{4.2.4.3}$$

shows that it suffices to prove that $H^0(\mathscr{C}_i^o, M^{gp}_i)$ is equal to $k^* \oplus N^{gp}_i$ (note that the map labelled y is an isomorphism since $\mathscr{C}_i^o$ is connected). This follows from noting that there is an injection

$$H^0(\mathscr{C}_i^o, M^{gp}_i) \hookrightarrow H^0(\mathrm{Spec}(k[N^{gp}_i]), \mathscr{O}^*) = k^* \oplus N^{gp}_i. \tag{4.2.4.4}$$

$\square$

Lemma 4.2.5 *The natural map*

$$k^* \oplus (P \rtimes H_S)^{gp} \to H^0(\mathscr{C}_k^o, M_{\mathscr{C}_k^o}^{gp}) \qquad (4.2.5.1)$$

is an isomorphism.

Proof. This is the same as in 3.2.4. Let W denote the set of top–dimensional simplices of S, and choose an ordering of W. For an ordered subset $i_1 < i_2 < \cdots < i_r$ of W let

$$j_{i_1 \cdots i_r} : \mathscr{C}_{i_1 \cdots i_r} \hookrightarrow \mathscr{C}_k \qquad (4.2.5.2)$$

denote the intersection $\mathscr{C}_{i_1} \cap \cdots \cap \mathscr{C}_{i_r}$. There is then a commutative diagram

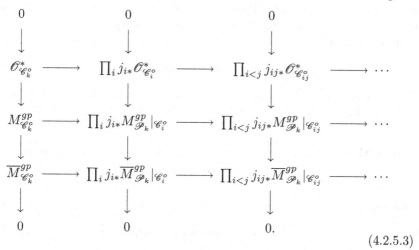

$$(4.2.5.3)$$

By cohomological descent for proper morphisms the bottom row is exact [8, VIII.9.4], and by [4, proof of 4.3] the top row is also exact in degrees ≤ 1. It follows that the middle row is also exact in degrees ≤ 1 so

$$H^0(\mathscr{C}_k^o, M_{\mathscr{C}_k^o}^{gp}) = \mathrm{Ker}(\prod_i H^0(\mathscr{C}_i^o, M_{\mathscr{P}_k}^{gp}|_{\mathscr{C}_i^o}) \to \prod_{i<j} H^0(\mathscr{C}_{ij}^o, M_{\mathscr{P}_k}^{gp}|_{\mathscr{C}_{ij}^o})),$$
$$(4.2.5.4)$$

which by 4.2.4 shows that

$$H^0(\mathscr{C}_k^o, M_{\mathscr{C}_k^o}^{gp}) = H^0(X_{\mathbb{R}}, k^* \oplus (P \rtimes H_S)^{gp}) = k^* \oplus (P \rtimes H_S)^{gp}, \qquad (4.2.5.5)$$

where $X_{\mathbb{R}}$ is viewed as a topological space with the topology induced by that on $\mathbb{R}$. $\square$

4.2.6. It follows that for every simplex $\omega_i \in S$ with corresponding closed subscheme $\mathscr{C}_i \subset \mathscr{C}_k$

$$H^0(\mathscr{C}_k^o, M_{\mathscr{C}_k^o}^{gp}) \to H^0(\mathscr{C}_i^o, M_{\mathscr{C}_k}|_{\mathscr{C}_i^o}) \qquad (4.2.6.1)$$

is an isomorphism. Since σ_0 necessarily acts trivially on the sublog structure $M_{\mathscr{C}_i} \subset M_{\mathscr{C}_k}|_{\mathscr{C}_i^o}$ it follows that the automorphism of $(P \rtimes H_S)^{gp}$ induced by σ_0 by taking global sections acts trivially on the submonoid $N_i \subset P$ for every $\omega_i \in S$. Therefore for any $p \in P$ we have $\sigma_0(p,0) = (p,0)$. Since H_S is generated by elements of the form $p * q$ it follows that σ_0 defines the identity automorphism of $P \rtimes H_S$. Therefore σ_0 is the identity, and hence $\bar{\sigma}$ is also the identity.

It follows that for every $m \in P \rtimes H_S$ there exists a unique element $\nu_m \in \Gamma(\mathscr{C}_k^o, \mathscr{O}_{\mathscr{C}_k^o}^*)$ such that $\sigma(\beta(m)) = \lambda(\nu_m) + \beta(m)$, where $\beta : P \rtimes H_S \to M_{\mathscr{C}_k^o}$ is the chart. Furthermore, ν_m is $\mathbb{G}_m$–equivariant and hence lies in

$$\Gamma(\mathscr{C}_k^o, \mathscr{O}_{\mathscr{C}_k^o}^*)^{\mathbb{G}_m} = \Gamma(\widetilde{\mathscr{P}}, \mathscr{O}_{\widetilde{\mathscr{P}}}^*) = k^*. \tag{4.2.6.2}$$

It follows that σ is induced by a homomorphism $\nu : P \rtimes H_S \to k^*$. Since σ induces the identity on $M_i \subset M_{\mathscr{P}_0}$ we also have $\nu(p,0) = 1$ for all $p \in P$. From the formula

$$\nu(p,0)\nu(q,0) = \nu(p+q,0)\nu(0,p*q) \tag{4.2.6.3}$$

we conclude that ν also restricts to the identity on H_S. From this it follows that σ is the identity. This completes the proof of 4.2.2. $\square$

Proposition 4.2.7. *The statement of 4.2.2 remains true if we replace*

$$(\mathscr{P}, M_{\mathscr{P}}, L_{\mathscr{P}}, G\text{--action}) \to (\text{Spec}(k), M_k) \tag{4.2.7.1}$$

by the saturation

$$(\mathscr{P}, M_{\mathscr{P}}^{\text{sat}}, L_{\mathscr{P}}, G\text{--action}) \to (\text{Spec}(k), M_k^{\text{sat}}) \tag{4.2.7.2}$$

in the statement of the proposition.

Proof. By the universal property of the saturation of a log structure, the pair (α^b, β) extends uniquely to compatible isomorphisms $\alpha^* M_{\mathscr{P}}^{\text{sat}} \to M_{\mathscr{P}}^{\text{sat}}$ and $M_k^{\text{sat}} \to M_k^{\text{sat}}$ extending α to an automorphism of the data

$$(M_k^{\text{sat}}, \mathscr{P}, M_{\mathscr{P}}^{\text{sat}}, L_{\mathscr{P}}, G\text{--action}). \tag{4.2.7.3}$$

Thus the only issue is the uniqueness. In other words, let $\iota^{\text{sat}} : M_k^{\text{sat}} \to M_k^{\text{sat}}$ and $\sigma^{\text{sat}} : M_{\mathscr{P}}^{\text{sat}} \to M_{\mathscr{P}}^{\text{sat}}$ denote compatible automorphisms of the log structures compatible with the G–action. We claim that ι^{sat} and σ^{sat} are the identity automorphisms. This follows from the same argument used in the proof of 4.2.2. $\square$

4.3 Deformation Theory

4.3.1. Let k be an infinite field, and $B' \to B$ a surjection of artinian local rings with residue field k such that the kernel $J := \text{Ker}(B' \to B)$ is a finite dimensional k–vector space.

Let X be a free abelian group of rank r, and S a regular paving of $X_{\mathbb{R}}$. Fix a homomorphism $\rho : H_S \to B'$ sending all nonzero elements of H_S to the nilradical of B, and let $M_{B'}$ (resp. M_B, M_k) be the induced log structure on $\mathrm{Spec}(B')$ (resp. $\mathrm{Spec}(B)$, $\mathrm{Spec}(k)$). Let A/B be an abelian scheme over B with a rigidified line bundle $\mathscr{M}$ defining a principal polarization, and let $c : X \to A^t$ be a homomorphism defining a semiabelian scheme G/B. Also fix a trivialization $\psi : 1_X \to c^* \mathscr{M}^{-1}$ as in 4.1.10, and let

$$(\mathscr{P}, M_{\mathscr{P}}, L_{\mathscr{P}}, G\text{–action}) \to (\mathrm{Spec}(B), M_B) \tag{4.3.1.1}$$

be the resulting standard family.

Let A'/B' be a lifting of A to an abelian scheme over B', and let $c' : X \to (A')^t$ be a morphism lifting c defining a semiabelian scheme G'/B'.

Definition 4.3.2. A *log smooth deformation* $(\mathscr{P}', M_{\mathscr{P}'}, L_{\mathscr{P}'}, G'\text{–action})$ of $(\mathscr{P}, M_{\mathscr{P}}, L_{\mathscr{P}}, G\text{–action})$ to $(\mathrm{Spec}(B'), M_{B'})$ consists of the following data:

(i) A log smooth deformation $(\mathscr{P}', M_{\mathscr{P}'}) \to (\mathrm{Spec}(B'), M_{B'})$ of $(\mathscr{P}, M_{\mathscr{P}})$ in the sense of [23, 8.1].

(ii) A lifting $L_{\mathscr{P}'}$ of the invertible sheaf $L_{\mathscr{P}}$ to $\mathscr{P}'$.

(iii) An action of G' on $(\mathscr{P}', M_{\mathscr{P}'})$ over $(\mathrm{Spec}(B'), M_{B'})$ reducing to the given action of G on $(\mathscr{P}, M_{\mathscr{P}})$ over B.

Proposition 4.3.3. *Let*

$$(\mathscr{P}', M_{\mathscr{P}'}, L_{\mathscr{P}'}, G'\text{–action})/(B', M_{B'}) \tag{4.3.3.1}$$

be a log smooth deformation of $(\mathscr{P}, M_{\mathscr{P}}, L_{\mathscr{P}}, G\text{–action})$ *to* $(\mathrm{Spec}(B'), M_{B'})$. *Then there exists a unique lifting* $\mathscr{M}'$ *of the polarization* $\mathscr{M}$ *to* A', *and a unique lifting* $\psi' : 1_X \to c'^* \mathscr{M}'^{-1}$ *of* ψ *such that* $(\mathscr{P}', M_{\mathscr{P}'}, L_{\mathscr{P}'}, G'\text{–action})$ *is isomorphic as a log smooth deformation to the standard family constructed using* $(A', \mathscr{M}')$ *and* ψ'.

Proof. Let T denote the torus corresponding to X. The deformation 4.3.3 defines a deformation $(\mathscr{P}', M_{\mathscr{P}'}, L_{\mathscr{P}'}, T\text{–action})$ of $(\mathscr{P}, M_{\mathscr{P}}, L_{\mathscr{P}}, T\text{–action})$ simply by restricting the action to the torus. By [3, 4.3] giving this deformation $(\mathscr{P}', M_{\mathscr{P}'}, L_{\mathscr{P}'}, T\text{–action})$ is equivalent to giving a deformation $(\widetilde{\mathscr{P}}', M_{\widetilde{\mathscr{P}}'}, L_{\widetilde{\mathscr{P}}'}, T \text{ and } X \text{ action})$ of $(\widetilde{\mathscr{P}}, M_{\widetilde{\mathscr{P}}}, L_{\widetilde{\mathscr{P}}}, T \text{ and } X \text{ action})$ to the log scheme $(\mathrm{Spec}(B'), M_{B'})$, where now T also acts on the line bundle $L_{\widetilde{\mathscr{P}}'}$ such that for $u \in T$ and $x \in X$ the actions satisfy the "Heisenberg relation"

$$T_u^* S_x^* = u(x) S_x^* T_u^*. \tag{4.3.3.2}$$

In fact the action of G' on $(\mathscr{P}', M_{\mathscr{P}'})$ lifts canonically to an action on $(\widetilde{\mathscr{P}}', M_{\widetilde{\mathscr{P}}'})$. This is because the morphism $(\widetilde{\mathscr{P}}', M_{\widetilde{\mathscr{P}}'}) \to (\mathscr{P}', M_{\mathscr{P}'})$ is étale. For then by the invariance of the étale site under infinitessimal thickenings to lift the action it suffices to do so after base change to B. But by the construction of the standard model there is a lifting of the G–action on $(\mathscr{P}, M_{\mathscr{P}})$ to $(\widetilde{\mathscr{P}}, M_{\widetilde{\mathscr{P}}})$.

Lemma 4.3.4 *Let* $f : \widetilde{\mathscr{P}} \to A$ *denote the projection. Then the natural map* $\mathscr{O}_A \to Rf_*\mathscr{O}_{\widetilde{\mathscr{P}}}$ *is an isomorphism in the derived category* $D(\mathscr{O}_A)$.

Proof. Since A is flat over B, and f is a flat morphism, a standard induction on the length of B reduces the proof to the case of $B = k$.

If $U \subset A$ is an open set over which $\mathscr{M}$ is trivial and there exists compatible trivializations of the L_x, then

$$\widetilde{\mathscr{P}} \times_A U \simeq \mathrm{Proj}(k \otimes_{\mathbb{Z}[H_S]} \mathbb{Z}[P \rtimes H_S]) \times_{\mathrm{Spec}(B)} U. \qquad (4.3.4.1)$$

In this case the cohomology can be computed as follows. For a top–dimensional simplex $\omega_i \in S$ let $j_i : \mathscr{P}_i \subset \widetilde{\mathscr{P}}$ be the irreducible component corresponding to ω_i. Choose an ordering of the top–dimensional simplices, and for $i_1 < \cdots < i_r$ let $\mathscr{P}_{i_1 i_2 \cdots i_r}$ denote the intersection $\mathscr{P}_{i_1} \cap \cdots \cap \mathscr{P}_{i_r}$. Then, as shown in [4, proof of 4.3], the natural map of complexes

$$\mathscr{O}_{\widetilde{\mathscr{P}}} \to (\prod_i j_{i*}\mathscr{O}_{\mathscr{P}_i} \to \prod_{i,j} j_{ij*}\mathscr{O}_{\mathscr{P}_{ij}} \to \cdots) \qquad (4.3.4.2)$$

is a quasi–isomorphism. Since each $\mathscr{P}_{i_1 \cdots i_r}$ is a projective toric variety over U the sheaves $R^k f_* \mathscr{O}_{\mathscr{P}_{i_1 \cdots i_r}}$ are zero for $k > 0$. This implies that $Rf_*\mathscr{O}_{\widetilde{\mathscr{P}}}$ is computed by the complex

$$(\prod_i j_{i*}f_*\mathscr{O}_{\mathscr{P}_i} \to \prod_{i,j} j_{ij*}f_*\mathscr{O}_{\mathscr{P}_{ij}} \to \cdots) = (\prod_i \mathscr{O}_A \to \prod_{i,j} \mathscr{O}_A \to \cdots). \qquad (4.3.4.3)$$

This computation shows that $R^i f_*\mathscr{O}_{\widetilde{P}}$ is locally on A isomorphic to the singular cohomology with coefficients in $\Gamma(A, \mathscr{O}_A)$) $H^i(X_{\mathbb{R}}, \Gamma(A, \mathscr{O}_A))$. Since $X_{\mathbb{R}}$ is contractible this proves the lemma. $\square$

In fact the proof shows the following:

Corollary 4.3.5 *For any* $i \geq 0$, *the formation of* $R^i f_*\mathscr{O}_{\widetilde{\mathscr{P}}}$ *commutes with arbitrary base change on* A, *and these sheaves are zero for* $i > 0$.

Using the method of the lemma we also obtain:

Corollary 4.3.6 *There is a canonical isomorphism* $\mathscr{M} \to f_*\mathscr{O}_{\widetilde{\mathscr{P}}}(1)^T$ *(invariants under the torus action).*

Proof. There is a natural map of sheaves $\mathscr{M} \to \oplus_{(d,x)\in P}\mathscr{M}^d \otimes L_x$ obtained from the isomorphism $\mathscr{M} \simeq \mathscr{M} \otimes L_0$ and this map induces a morphism $\epsilon : \mathscr{M} \to f_*\mathscr{O}_{\widetilde{\mathscr{P}}}(1)^T$. To check that this morphism is an isomorphism we may work locally on A where we choose a trivialization of $\mathscr{M}$ and compatible trivializations of the L_x. For a simplex $\omega_i \in S$, let $N_i^{(1)}$ denote the elements of degree 1 in the monoid of integral elements in $\mathrm{Cone}(1, \omega_i)$. Then as in the proof of 4.3.4 one sees that $Rf_*\mathscr{O}_{\widetilde{\mathscr{P}}}(1)$ can be represented by the complex

$$\left(\prod_i (\oplus_{n\in N_i^{(1)}} \mathscr{O}_A \cdot n) \to \prod_{i,j} (\oplus_{n\in N_{ij}^{(1)}} \mathscr{O}_A \cdot n) \to \cdots \right). \tag{4.3.6.1}$$

An element $\chi \in T$ acts on $\prod_i (\oplus_{n\in N_i^{(1)}} \mathscr{O}_A \cdot n)$ by $n \mapsto \chi(n) \cdot n$. From this it follows that the only nonzero components of an invariant section of $\prod_i (\oplus_{n\in N_i^{(1)}} \mathscr{O}_A \cdot n)$ are the components with $n = (1,0)$. From this the corollary follows. $\square$

Corollary 4.3.7 *For any $i > 0$ the sheaf $R^i f_* \mathscr{O}_{\widetilde{\mathscr{P}}}(1)$ is zero.*

Proof. We have to show that 4.3.6.1 has no higher cohomology. For this note that the entire complex decomposes into eigenspaces according to the T-action. Let $(1,x) \in P$ be an element and let σ_0 be the minimal cell containing x. Then the complex obtain from the x-eigenspaces in 4.3.6.1 computes the cellular cohomology of the set

$$\hat{\sigma}_0 := \cup_{\sigma_0 \subset \omega \in S} \omega \subset X_{\mathbb{R}}. \tag{4.3.7.1}$$

The corollary then follows from [4, 1.5 (v)] which implies that the topological space $\hat{\sigma}_0$ is contractible (in fact $\hat{\sigma}_0$ is the so-called dual Voronoi cell). $\square$

4.3.8. The vanishing of $R^1 f_* \mathscr{O}_{\widetilde{\mathscr{P}}}$ implies that the sequence

$$0 \to J \otimes_B f_* \mathscr{O}_{\widetilde{\mathscr{P}}} \to f_* \mathscr{O}_{\widetilde{\mathscr{P}}'} \to f_* \mathscr{O}_{\widetilde{\mathscr{P}}} \to 0 \tag{4.3.8.1}$$

is exact. Therefore $Z' := (|A|, f_* \mathscr{O}_{\widetilde{\mathscr{P}}'})$ is a flat deformation of A to B', where $|A|$ denotes the underlying topological space and $\mathscr{O}_{\widetilde{\mathscr{P}}'}$ is viewed as a sheaf of rings on $|\widetilde{\mathscr{P}}|$. There is also a natural action of G' on Z' reducing to the translation action on A. For a section $g \in G'(B')$ we have a commutative diagram

$$
\begin{array}{ccc}
|\widetilde{\mathscr{P}'}| & \xrightarrow{t_g} & |\widetilde{\mathscr{P}'}| \\
\downarrow{\scriptstyle f} & & \downarrow{\scriptstyle f} \\
|A| & \xrightarrow{t_g} & |A|,
\end{array}
\tag{4.3.8.2}
$$

and an isomorphism $t_g^{-1} \mathscr{O}_{\widetilde{\mathscr{P}}'} \to \mathscr{O}_{\widetilde{\mathscr{P}}'}$. This defines the action of g on Z'. Furthermore, the same analysis applies after making any flat extension of B'. Since G' is smooth over B' and Z' is flat over B', a morphism $G' \times Z' \to Z'$ is determined by its scheme valued points on flat B'-schemes (Yoneda's lemma applied to the category of flat B'-schemes). Hence the above construction defines an action of G' on Z'.

Lemma 4.3.9 *The action of G' on Z' factors through A'.*

Proof. It suffices to show that T acts trivially on $f_* \mathscr{O}_{\widetilde{\mathscr{P}}'}$. This follows from observing that the action of T on $J \otimes_B f_* \mathscr{O}_{\widetilde{\mathscr{P}}}$ and $f_* \mathscr{O}_{\widetilde{\mathscr{P}}}$ is trivial, and hence since T is reductive the torus T must also act trivially on the middle term of 4.3.8.1. $\square$

Since Z' with its A'–action is a flat deformation of the trivial A–torsor, the action of A' on Z' must also be torsorial. Chosing a lifting of the origin in A to Z' we can therefore identity Z' with A'. We therefore obtain a deformation $f' : \widetilde{\mathscr{P}}' \to A'$ of f compatible with the G'–actions.

For $x \in X$, let L'_x denote the rigidified invertible sheaf on A' corresponding to $c'(x)$. The same formula used in 4.1.10.2 defines an algebra structure on $\oplus_{(d,x)} L'_x$, and we let $\widetilde{\mathscr{P}}''$ denote $\mathrm{Proj}_{A'}(\oplus_{(d,x)} L'_x)$. The arguments of 4.1.16–4.1.20 define a G'–action on $\widetilde{\mathscr{P}}''$. Note that for any invertible sheaf $\mathscr{M}'$ on A' we have $\widetilde{\mathscr{P}}'' \simeq \mathrm{Proj}_{A'}(\oplus_{(d,x)} \mathscr{M}'^d \otimes L_x)$ (though the tautological invertible sheaves are different).

Let $\widetilde{\mathscr{P}}'_0$ and $\widetilde{\mathscr{P}}''_0$ denote the fibers over $0 \in A'$.

Lemma 4.3.10 *There exists an isomorphism of log smooth deformations of* $(\widetilde{\mathscr{P}}_0, M_{\widetilde{\mathscr{P}}_0})$

$$\sigma : (\widetilde{\mathscr{P}}'_0, M_{\widetilde{\mathscr{P}}'_0}) \to (\widetilde{\mathscr{P}}''_0, M_{\widetilde{\mathscr{P}}''_0}) \tag{4.3.10.1}$$

compatible with the torus actions.

Proof. First note that the log tangent sheaf $T_{(\widetilde{\mathscr{P}}, M_{\widetilde{\mathscr{P}}})/(B, M_B)}$ is isomorphic to $\mathcal{O}_{\widetilde{\mathscr{P}}} \otimes_{\mathbb{Z}} \mathrm{Hom}(X, \mathbb{Z})$. This can be seen as follows. Let $\mathscr{C}_B = \mathrm{Spec}(B \otimes_{\mathbb{Z}[H_S]} \mathbb{Z}[P \rtimes H_S])$ be the cone so that $\widetilde{\mathscr{P}} = [\mathscr{C}_B^o / \mathbb{G}_m]$, where $\mathscr{C}_B^o$ is the complement of the vertex. By [24, 1.8], the chart $P \rtimes H_S \to M_{\mathscr{C}_B}$ induces an isomorphism

$$d \log : \mathcal{O}_{\mathscr{C}_B^o} \otimes_{\mathbb{Z}} P^{gp} \simeq \Omega^1_{(\mathscr{C}_B^o, M_{\mathscr{C}_B^o})/(B, M_B)}. \tag{4.3.10.2}$$

Furthermore, the action of $\mathbb{G}_m$ on $\Omega^1_{(\mathscr{C}_B^o, M_{\mathscr{C}_B^o})/(B, M_B)}$ is induced from the standard action on $\mathcal{O}_{\mathscr{C}_B^o}$ and the trivial action on P^{gp}. The differentials $\Omega^1_{\mathscr{C}_B^o / \widetilde{\mathscr{P}}}$ admit a similar description. For any scheme W and $\mathbb{G}_m$–torsor $Q \to W$, there is a canonical generator $d \log(u)$ of $\Omega^1_{Q/W}$. This can be seen as follows. Write $\mathbb{G}_m$ as $\mathrm{Spec}(\mathbb{Z}[u^{\pm}])$. Then the section $d \log(u) = du/u \in \Omega^1_{\mathbb{G}_m/\mathbb{Z}}$ is invariant under the translation action. Hence it induces a generator of $\Omega^1_{Q/W}$ by locally on W choosing a trivialization of Q and pulling back $d \log(u)$. The translation invariance then implies that this is independent of the choice of trivialization and hence defined globally.

It follows from the construction of the isomorphism 4.3.10.2 that the map

$$\Omega^1_{(\mathscr{C}_B^o, M_{\mathscr{C}_B^o})/(B, M_B)} \to \Omega^1_{\mathscr{C}_B^o / \widetilde{\mathscr{P}}} \tag{4.3.10.3}$$

sends $d \log(1 \otimes p)$ to $\deg(p) \cdot d \log(u)$. In other words, the morphism 4.3.10.3 can be identified with the morphism obtained by tensoring the map $\deg : P^{gp} \to \mathbb{Z}$ with $\mathcal{O}_{\mathscr{C}_B^o}$. Since the kernel of this degree map is isomorphic to X, this yields an isomorphism between $\Omega^1_{(\widetilde{\mathscr{P}}, M_{\widetilde{\mathscr{P}}})/(B, M_B)}|_{\mathscr{C}_B^o}$ and $\mathcal{O}_{\mathscr{C}_B^o} \otimes_{\mathbb{Z}} X$. Furthermore, the $\mathbb{G}_m$–action is simply that induced by the action on $\mathcal{O}_{\mathscr{C}_B^o}$. By descent theory we therefore obtain an isomorphism $\Omega^1_{(\widetilde{\mathscr{P}}, M_{\widetilde{\mathscr{P}}})/(B, M_B)} \simeq \mathcal{O}_{\widetilde{\mathscr{P}}} \otimes_{\mathbb{Z}} X$.

Combining this with 4.3.5 we see that

$$H^1(\widetilde{\mathscr{P}}_0, T_{(\widetilde{\mathscr{P}}_0, M_{\widetilde{\mathscr{P}}_0})/(B, M_B)}) = 0. \tag{4.3.10.4}$$

By [24, 3.14] it therefore follows that there exists an isomorphism 4.3.10.1 of log smooth deformations of $(\widetilde{\mathscr{P}}_0, M_{\widetilde{\mathscr{P}}_0})$ which may or may not respect the torus actions. Furthermore, standard deformation theory [5, p. 69] shows that the deformations of the line bundle $\mathscr{O}_{\widetilde{\mathscr{P}}}(1)$ to $\widetilde{\mathscr{P}}_0'$ are naturally a torsor under $H^1(\widetilde{\mathscr{P}}_0, k \otimes_B \mathscr{O}_{\widetilde{\mathscr{P}}_0})$ which is zero by 4.3.5. Thus the isomorphism σ extends to an isomorphism of polarized log schemes.

The only issue that remains is the torus action. In what follows we therefore assume that $(\widetilde{\mathscr{P}}_0'', M_{\widetilde{\mathscr{P}}_0''}, \mathscr{O}_{\widetilde{\mathscr{P}}_0''}(1))$ and $(\widetilde{\mathscr{P}}_0', M_{\widetilde{\mathscr{P}}_0'}, L_{\widetilde{\mathscr{P}}_0'})$ have been identified. Let $\mathscr{C}_{B'}$ denote the cone over $\widetilde{\mathscr{P}}_0'$ which comes equipped with an action of the torus $D(\mathbb{X})$, and let $j : \mathscr{C}_{B'}^o \hookrightarrow \mathscr{C}_{B'}$ denote the complement of the vertex.

Sub-Lemma 4.3.11 *The natural map $H^0(\mathscr{C}_{B'}, \mathscr{O}_{\mathscr{C}_{B'}}) \to H^0(\mathscr{C}_{B'}^o, \mathscr{O}_{\mathscr{C}_{B'}^o})$ is an isomorphism.*

Proof. Let $\mathscr{C}_k$ denote the reduction to k of $\mathscr{C}_{B'}$. The short exact sequence of sheaves

$$0 \to J \otimes \mathscr{O}_{\mathscr{C}_k} \to \mathscr{O}_{\mathscr{C}_{B'}} \to \mathscr{O}_{\mathscr{C}_B} \to 0 \tag{4.3.11.1}$$

induces a commutative diagram

$$
\begin{array}{ccccccccc}
0 & \longrightarrow & J \otimes H^0(\mathscr{C}_k, \mathscr{O}_{\mathscr{C}_k}) & \longrightarrow & H^0(\mathscr{C}_{B'}, \mathscr{O}_{\mathscr{C}_{B'}}) & \longrightarrow & H^0(\mathscr{C}_B, \mathscr{O}_{\mathscr{C}_B}) & \longrightarrow & 0 \\
& & \downarrow & & \downarrow & & \downarrow & & \\
0 & \longrightarrow & J \otimes H^0(\mathscr{C}_k^o, \mathscr{O}_{\mathscr{C}_k^o}) & \longrightarrow & H^0(\mathscr{C}_{B'}^o, \mathscr{O}_{\mathscr{C}_{B'}^o}) & \longrightarrow & H^0(\mathscr{C}_B^o, \mathscr{O}_{\mathscr{C}_B^o}) & &
\end{array}
\tag{4.3.11.2}
$$

where the top row is exact since $\mathscr{C}_k$ is affine. It follows that it suffices to prove the result for $\mathscr{C}_k$ and $\mathscr{C}_B$. By induction on the integer n such that $\mathrm{rad}(B)^n = 0$ we are thus reduced to the case when $B' = k$.

For simplicity write simply $\mathscr{C}$ for $\mathscr{C}_k$ etc. For each top–dimensional simplex $\omega_i \in S$ write $\mathscr{C}_i$ for $\mathrm{Spec}(k[N_i]) \subset \mathscr{C}$, where N_i denotes the integral points of $\mathrm{Cone}(1, \omega_i)$. For two simplices $\omega_i, \omega_j \in S$ let $\mathscr{C}_{ij}$ denote the intersection $\mathscr{C}_i \cap \mathscr{C}_j$, and write $j_i : \mathscr{C}_i \hookrightarrow \mathscr{C}$ and $j_{ij} : \mathscr{C}_{ij} \hookrightarrow \mathscr{C}$ for the inclusions. Choose any ordering of the top–dimensional simplices of S. Then the same argument used in the proof of [3, 2.5.5] shows that

$$\mathscr{O}_{\mathscr{C}} \simeq \mathrm{Ker}(\prod_i j_{i*}\mathscr{O}_{\mathscr{C}_i} \to \prod_{i<j} j_{ij*}\mathscr{O}_{\mathscr{C}_{ij}}), \tag{4.3.11.3}$$

where the map is obtained by taking the difference of the restriction maps. From this we obtain a commutative diagram

$$\begin{array}{ccccc}
H^0(\mathscr{C}, \mathscr{O}_{\mathscr{C}}) & \longrightarrow & \prod_i H^0(\mathscr{C}_i, \mathscr{O}_{\mathscr{C}_i}) & \longrightarrow & \prod_{i<j} H^0(\mathscr{C}_{ij}, \mathscr{O}_{\mathscr{C}_{ij}}) \\
\downarrow{a} & & \downarrow{b} & & \downarrow{c} \\
H^0(\mathscr{C}^o, \mathscr{O}_{\mathscr{C}^o}) & \longrightarrow & \prod_i H^0(\mathscr{C}_i^o, \mathscr{O}_{\mathscr{C}_i^o}) & \longrightarrow & \prod_{i<j} H^0(\mathscr{C}_{ij}^o, \mathscr{O}_{\mathscr{C}_{ij}^o}).
\end{array}$$

$$(4.3.11.4)$$

Since each $\mathscr{C}_i$ is a normal scheme of dimension ≥ 2 the morphisms b and c are isomorphisms. This implies that a is also an isomorphism. □

Returning to the proof of 4.3.10, we assume given an action τ of $D(\mathbb{X})$ on the log scheme $(\mathscr{C}_{B'}^o, M_{\mathscr{C}_{B'}^o})$ reducing to the standard action over B and with $\mathbb{G}_m \subset D(\mathbb{X})$ acting in the usual way. We have to show that there exists an infinitesimal automorphism of $(\mathscr{C}_{B'}^o, M_{\mathscr{C}_{B'}^o})$ compatible with the $\mathbb{G}_m$–action taking τ to the standard action.

This is essentially the same as in 3.3. Let $u \in D(\mathbb{X})(B')$ be a section, and let τ_u be the automorphism of $(\mathscr{C}_{B'}^o, M_{\mathscr{C}_{B'}^o})$ defined by τ. Since τ_u reduces to the standard action modulo J, the automorphism τ_u is compatible with the map $P \rtimes H_S \to \overline{M}_{\mathscr{C}_{B'}^o}$ defined by the chart $\beta : P \rtimes H_S \to M_{\mathscr{C}_{B'}^o}$. Therefore, there exists a homomorphism

$$\nu : P^{gp} \to H^0(\mathscr{C}_{B'}^o, \mathscr{O}_{\mathscr{C}_{B'}^o}^*) = (B' \otimes_{\mathbb{Z}[H_S]} \mathbb{Z}[P \rtimes H_S])^*, \qquad (4.3.11.5)$$

where for the right equality we use 4.3.11, such that

$$\tau_u(\beta(p, h)) = \lambda(\nu_p) + \beta(p, h) \qquad (4.3.11.6)$$

for all $(p, h) \in P \rtimes H_S$. Viewing $B' \otimes_{\mathbb{Z}[H_S]} \mathbb{Z}[P \rtimes H_S]$ as a free B'–module on P as in 3.1.33, write

$$\nu_p = u(p)(1 + \sum_{\chi \in P} g_\chi^u(p)) \qquad (4.3.11.7)$$

with $g_\chi^u(p) \in J \cdot \chi$.

Sub-Lemma 4.3.12 (i) *For any $\chi \in P$ with $u(\chi) \neq 1$, the element*

$$h_\chi(p) := g_\chi^u(p)(u(\chi) - 1)^{-1} \in J \cdot \tilde\chi \qquad (4.3.12.1)$$

is independent of the choice of u, and the map $p \mapsto 1 + \sum_{\chi \in P} h_\chi(p) \in \Gamma(\mathscr{C}_{B'}, \mathscr{O}_{\mathscr{C}_{B'}}^*)$ is a homomorphism.
(ii) *For any section $u \in D(P^{gp})(B')$ we have*

$$(u(\chi) - 1)h_\chi(p) = g_\chi^u(p), \qquad (4.3.12.2)$$

where $h_\chi(p)$ is defined as in (i) using any other section of $D(P^{pg})(B')$ not mapping to 1 in $D(P^{gp})(k)$.

Proof. The proof is the same as the proof of 3.3.8. □

This sub-lemma completes the proof of 4.3.10 because the infinitesimal automorphism of $(\mathscr{C}_{B'}, M_{\mathscr{C}_{B'}})$ defined by

$$P \rtimes H_S \to \Gamma(\mathscr{C}_{B'}, \mathscr{O}^*_{\mathscr{C}_{B'}}) \oplus (P \rtimes H_S), \quad (p,h) \mapsto (1 + \sum_\chi h_\chi(p), (p,h)),$$

$$(4.3.12.3)$$

transforms the action τ into the standard action of $D(\mathbb{X})$. □

Lemma 4.3.13 *There exists an isomorphism of log smooth deformations of* $(\widetilde{\mathscr{P}}, M_{\widetilde{\mathscr{P}}})$

$$\sigma : (\widetilde{\mathscr{P}'}, M_{\widetilde{\mathscr{P}'}}) \to (\widetilde{\mathscr{P}''}, M_{\widetilde{\mathscr{P}''}}) \qquad (4.3.13.1)$$

compatible with the G–actions.

Proof. It suffices to construct for every scheme–valued point $a : W \to A$ an isomorphism

$$\sigma_a : (\widetilde{\mathscr{P}'}, M_{\widetilde{\mathscr{P}'}}) \times_{A,a} W \to (\widetilde{\mathscr{P}''}, M_{\widetilde{\mathscr{P}''}}) \times_{A,a} W. \qquad (4.3.13.2)$$

For this choose (after possibly replacing W by an étale cover) a lifting $\tilde{a} \in G(W)$ of a, and let

$$\sigma_0 : (\widetilde{\mathscr{P}'}, M_{\widetilde{\mathscr{P}'}}) \times_{A,0} W \to (\widetilde{\mathscr{P}''}, M_{\widetilde{\mathscr{P}''}}) \times_{A,0} W \qquad (4.3.13.3)$$

be an isomorphism as in 4.3.10 (in particular σ_0 is compatible with the torus action). Define σ_a to be the unique isomorphism making the following diagram commute:

$$(4.3.13.4)$$

where $t'_{\tilde{a}}$ and $t''_{\tilde{a}}$ denote the actions of G. Note that since σ_0 is compatible with the T–actions this does not depend on the choice of $\tilde{a}$. We thus obtain the isomorphism σ. □

Fix such an isomorphism σ.

Lemma 4.3.14 *The sequence*

$$0 \to J \otimes (f_* L_{\widetilde{\mathscr{P}_k}})^T \to (f'_* L_{\widetilde{\mathscr{P}'}})^T \to (f_* L_{\widetilde{\mathscr{P}}})^T \to 0 \qquad (4.3.14.1)$$

is exact. In particular, by 4.3.6 the sheaf $\mathscr{M}' := (f'_* L_{\widetilde{\mathscr{P}'}})^T$ is a deformation of $\mathscr{M}$ to A'.

Proof. By 4.3.7 we have an exact sequence

$$0 \to J \otimes (f_*L_{\widetilde{\mathscr{P}}_k}) \to (f'_*L_{\widetilde{\mathscr{P}}'}) \to (f_*L_{\widetilde{\mathscr{P}}}) \to 0 \tag{4.3.14.2}$$

of sheaves with T–action on A. Let $\mathcal{D}$ denote this category of sheaves of $\mathcal{O}_A$–modules on A_{et} with action of T. The category $\mathcal{D}$ is an abelian category with enough injectives and there is a left exact functor

$$I : \mathcal{D} \to (\mathcal{O}_A\text{-modules}), \quad F \mapsto F^T \tag{4.3.14.3}$$

which takes the T–invariants. To prove the lemma it suffices to show that $R^1I(f_*L_{\widetilde{\mathscr{P}}_k}) = 0$. Since T is a diagonalizable group, it is well-known that the restriction of I to the full subcategory $\mathcal{D}^{\mathrm{qcoh}}$ of quasi-coherent $\mathcal{O}_A$-modules is an exact functor, and that in fact the category $\mathcal{D}^{\mathrm{qcoh}}$ is semisimple. For a finite subset $\Delta \subset S$ let $\widetilde{\mathscr{P}}_\Delta \subset \widetilde{\mathscr{P}}_k$ denote the closed subscheme whose components correspond to the simplices $\omega \in \Delta$. Then each $\widetilde{\mathscr{P}}_\Delta$ is of finite type and T–invariant. We then have an isomorphism of sheaves with T–action

$$f_*L_{\widetilde{\mathscr{P}}_k} = \varprojlim_{\Delta \subset S} f_*L_{\widetilde{\mathscr{P}}_\Delta}, \tag{4.3.14.4}$$

and each $f_*L_{\widetilde{\mathscr{P}}_\Delta} \in \mathcal{D}^{\mathrm{qcoh}}$. The character decompositions of the $f_*L_{\widetilde{\mathscr{P}}_\Delta}$ then give a decomposition $f_*L_{\widetilde{\mathscr{P}}_k} = \prod_\chi F_\chi$ where F_χ is of rank 1. To prove the lemma it then suffices to show that $R^1I(F_\chi) = 0$ for all χ which is immediate. $\square$

4.3.15. It follows that $L_{\widetilde{\mathscr{P}}''} \otimes \mathcal{M}'$ and $L^\dagger_{\widetilde{\mathscr{P}}''} := \sigma_*L_{\widetilde{\mathscr{P}}'}$ are two deformations of $\mathcal{O}_{\widetilde{\mathscr{P}}}(1)$.

By 4.3.4, for an integer i the natural map

$$H^i(A_k, \mathcal{O}_{A_k}) \to H^i(\widetilde{\mathscr{P}}_k, \mathcal{O}_{\widetilde{\mathscr{P}}_k}) \tag{4.3.15.1}$$

is an isomorphism. Furthermore, the set of liftings of $\mathcal{M}$ form a torsor under $H^1(A_k, \mathcal{O}_{A_k}) \otimes J$. Let $o \in H^1(\widetilde{\mathscr{P}}_k, \mathcal{O}_{\widetilde{\mathscr{P}}_k}) \otimes J$ denote the class corresponding to $[L_{\widetilde{\mathscr{P}}''} \otimes \mathcal{M}'] - [L^\dagger_{\widetilde{\mathscr{P}}''}]$. Viewing this class as an element of $H^1(A_k, \mathcal{O}_{A_k}) \otimes J$ we see that after replacing $\mathcal{M}'$ by another lifting we can force $o = 0$ (and the choice of lifting $\mathcal{M}'$ is unique with this property). It follows that then $(\widetilde{\mathscr{P}}', M_{\widetilde{\mathscr{P}}'}, L_{\widetilde{\mathscr{P}}'})$ is isomorphic to the standard construction using $(A', \mathcal{M}')$.

It remains to show that the X–action on $(\widetilde{\mathscr{P}}', M_{\widetilde{\mathscr{P}}'}, L_{\widetilde{\mathscr{P}}'})$ is obtained from a deformation of ψ. Let $y \in X$ be an element, and let $S_y : (\widetilde{\mathscr{P}}', M_{\widetilde{\mathscr{P}}'}, L_{\widetilde{\mathscr{P}}'}) \to (\widetilde{\mathscr{P}}', M_{\widetilde{\mathscr{P}}'}, L_{\widetilde{\mathscr{P}}'})$ be the action. Let $c'^t : X \to A'$ denote the composite

$$X \xrightarrow{\ c'\ } A'^t \xrightarrow{\ \lambda_{\mathcal{M}'}^{-1}\ } A'. \tag{4.3.15.2}$$

Lemma 4.3.16 *The diagram*

$$
\begin{array}{ccc}
\widetilde{\mathscr{P}}' & \xrightarrow{\ S_y\ } & \widetilde{\mathscr{P}}' \\
{\scriptstyle f'}\downarrow & & \downarrow{\scriptstyle f'} \\
A' & \xrightarrow{\ t_{c'{}^t(y)}\ } & A'
\end{array}
\tag{4.3.16.1}
$$

commutes.

Proof. First note that there exists a morphism $\rho_y : A' \to A'$ such that the diagram

$$
\begin{array}{ccc}
\widetilde{\mathscr{P}}' & \xrightarrow{\ S_y\ } & \widetilde{\mathscr{P}}' \\
{\scriptstyle f'}\downarrow & & \downarrow{\scriptstyle f'} \\
A' & \xrightarrow{\ \rho_y\ } & A'
\end{array}
\tag{4.3.16.2}
$$

commutes. Indeed $f' \circ S_y$ is given by a map (of sheaves on the topological space underlying A)

$$
\mathscr{O}_{A'} \to f'_* \circ S_{y*}\mathscr{O}_{\widetilde{\mathscr{P}}'} = t_{c^t(y)*} \circ f'_*\mathscr{O}_{\widetilde{\mathscr{P}}'} = t_{c^t(y)*}\mathscr{O}_{A'},
\tag{4.3.16.3}
$$

so this defines ρ_y. By [34, 6.4] we can write ρ_y uniquely as $h + t_{\alpha(y)}$, where $h : A' \to A'$ is a homomorphism and $\alpha(y) \in A'(B')$. Since h reduces to the identity over B this implies that h is the identity [34, 6.1]. Therefore $\rho_y = t_{\alpha(y)}$ for some $\alpha(y) \in A'(B')$. We thus obtain a homomorphism

$$
\alpha : X \to A'(B')
\tag{4.3.16.4}
$$

by sending α to $\alpha(y)$. We claim that $\alpha = (c^t)'$. Unwinding the definitions this amounts to saying that $L_{c'(y)}$ is isomorphic to $(\rho_y^*\mathscr{M}') \otimes \mathscr{M}'^{-1}$, or equivalently $\rho_y^*\mathscr{M}' \simeq L_{c'(y)} \otimes \mathscr{M}'$. This follows from noting that the relation (see 4.1.21)

$$
u(y)T_u^* S_y^* = S_y^* T_u^*
\tag{4.3.16.5}
$$

for a section $u \in T$ implies that

$$
\rho_y^*\mathscr{M}' = \rho_y^{-1}(f'_* L_{\widetilde{\mathscr{P}}'})_0 \simeq (f'_* L_{\widetilde{\mathscr{P}}'})_y \simeq L_{c'(y)} \otimes \mathscr{M}',
\tag{4.3.16.6}
$$

where for $\chi : T \to \mathbb{G}_m$ we write $(f'_* L_{\widetilde{\mathscr{P}}'})_\chi$ for the χ-eigenspace for the T-action. $\square$

4.3.17. It follows that the X-action on the log scheme $(\widetilde{\mathscr{P}}', M_{\widetilde{\mathscr{P}}'})$ is induced by isomorphisms

$$
\iota_y^{(d,x)} : t_{c^t(y)}^* (\mathscr{M}'^d \otimes L_{c'(x)}) \to \mathscr{M}'^d \otimes L_{c'(x+dy)}.
\tag{4.3.17.1}
$$

As explained in 2.2.12 giving such isomorphisms is equivalent to giving for every $d \geq 1$ and $x, y \in X$ a trivialization of $\mathscr{M}'(y)^d \otimes \mathscr{B}'_{(c'(y),c'(x))}$, where $\mathscr{B}'$

denotes the Poincare bundle over $A' \times A'^t$. Define ψ' to be the trivialization of $c^* \mathcal{M}^{-1}$ obtained by taking $d = 1$ and $x = 0$ (note that for any $a' \in A$ the line $\mathcal{B}'_{(a',0)}$ is canonically trivialized). The fact that the isomorphisms $\iota_y^{(d,x)}$ are compatible with the monoid law on the log structure then amounts precisely to the commutativity of the diagram (2.2.13 (iii)'), and the fact that it is an action is equivalent to (2.2.13 (ii)'). Therefore the trivialization of $(c \times c)^* \mathcal{B}$ defined by ψ' is compatible with the symmetric biextension structure. This completes the proof of 4.3.3.

$\square$

Remark 4.3.18. The above computations can also be used to compute the cohomology groups $H^i(\mathcal{P}, L^d_{\mathcal{P}})$ (see [4, 4.4]). The conclusion is that for any integer $d \geq 1$ the groups $H^i(\mathcal{P}, \mathcal{L}^d_{\mathcal{P}})$ are zero for $i > 0$ and $H^0(\mathcal{P}, L^d_{\mathcal{P}})$ has rank d^g (where g is the dimension of G).

4.3.19. From the data

$$(G, \mathcal{P}, L_{\mathcal{P}}, G - \text{action}) \qquad (4.3.19.1)$$

we can recover the paving S, the map $c : X \to A^t$, as well as the map $\lambda_{\mathcal{M}} : A \to A^t$.

Indeed by the equivalence of categories in [3, 4.3.1] (see also the discussion in 5.3) we obtain the X-torsor $\pi : \widetilde{\mathcal{P}} \to \mathcal{P}$ and the line bundle $\mathcal{O}_{\widetilde{\mathcal{P}}}(1) = \pi^* L_{\mathcal{P}}$ together with the lifting of the T_X-action to $(\widetilde{\mathcal{P}}, \mathcal{O}_{\widetilde{\mathcal{P}}}(1))$. In fact we can also recover the G-action on $\widetilde{\mathcal{P}}$. Indeed a lifting of the G-action on $\mathcal{P}$ to an action on $\widetilde{\mathcal{P}}$ is unique. Indeed if $\tilde{\rho}_1, \tilde{\rho}_2$ are two liftings of the G-action on ρ then for any $g \in G$ the two maps

$$\rho_1(g), \rho_2(g) : \widetilde{\mathcal{P}} \to \widetilde{\mathcal{P}} \qquad (4.3.19.2)$$

define an automorphism of $\widetilde{\mathcal{P}}$ over $\mathcal{P}$ which commutes with the X-action. In this way we obtain a morphism of schemes

$$G \to X \qquad (4.3.19.3)$$

which must be constant since G is connected. Therefore $\rho_1 = \rho_2$.

In this way we recover $(\widetilde{\mathcal{P}}, \mathcal{O}_{\widetilde{\mathcal{P}}}(1), G - \text{action})$ from 4.3.19.1. The projection

$$f : \widetilde{\mathcal{P}} \to A \qquad (4.3.19.4)$$

can also be recovered from this, up to translation by a point of A. Indeed suppose

$$f_1, f_2 : \widetilde{\mathcal{P}} \to A \qquad (4.3.19.5)$$

are two surjections which commute with the G-actions (where G acts on A by translation through the projection $G \to A$). By the construction, for $\omega \in S$ let P_ω denote the integral points of $\text{Cone}(1, \omega)$ and let

$$\widetilde{\mathscr{P}}_\omega = \mathrm{Proj}_A(\mathscr{O}_A \otimes_{\mathbb{Z}[H_S]} \mathbb{Z}[P_\omega \rtimes H_S]) \subset \widetilde{\mathscr{P}}. \tag{4.3.19.6}$$

Let $X_1 \subset X$ denote the integral points of the subgroup of $X_{\mathbb{R}}$ generated by ω. Then the T_X-action on $\widetilde{\mathscr{P}}_\omega$ factors through T_{X_1}, and there exists a dense open subset $\mathscr{U}_\omega \subset \widetilde{\mathscr{P}}_\omega$ where the action is free. Furthermore the G-action on $\widetilde{\mathscr{P}}_\omega$ gives the quotient $\mathscr{U}/T_{X_1}$ the structure of an A-torsor. It follows that the two restrictions

$$f_1, f_2 : \widetilde{\mathscr{P}}_\omega \to A \tag{4.3.19.7}$$

differ by an element $a_\omega \in A(B)$ as they are determined by the two maps of A-torsors

$$f_1, f_2 : \mathscr{U}/T_{X_1} \to A. \tag{4.3.19.8}$$

In this way we obtain a mapping

$$S \to A(B), \quad \omega \mapsto a_\omega \tag{4.3.19.9}$$

with the property that if $\omega' \subset \omega$ then $a_\omega = a_{\omega'}$. Since $X_{\mathbb{R}}$ is connected it follows that all the a_ω's are equal and hence f_1 and f_2 differ globally by translation by an element of $a \in A$.

Fix one map $f : \widetilde{\mathscr{P}} \to A$ commuting with the G-action. Let $\mathscr{M}'$ denote the subsheaf of T_X-invariants in $f_*\mathscr{O}_{\widetilde{\mathscr{P}}}(1)$. If f happens to agree with the map used in the standard construction then $\mathscr{M}'$ is equal to the sheaf $\mathscr{M}$ by 4.3.6. In general $\mathscr{M}' = \mathscr{M} \otimes L_a$ for some element $a \in A$. In particular, we recover the map

$$\lambda_{\mathscr{M}} : A \to A^t. \tag{4.3.19.10}$$

We also obtain the map $c^t : X \to A$ (and hence also the map $c : X \to A^t$) by defining $c^t(x)$ to be the unique element such that the diagram

$$
\begin{array}{ccc}
\widetilde{\mathscr{P}} & \xrightarrow{\;t_x\;} & \widetilde{\mathscr{P}} \\
\downarrow & & \downarrow \\
A & \xrightarrow{\;t_{c^t(x)}\;} & A.
\end{array}
\tag{4.3.19.11}
$$

For $(d, x) \in P$ let $L^{(d,x)}$ denote the x-eigenspace of the sheaf

$$f_*\mathscr{O}_{\widetilde{\mathscr{P}}}(d) \tag{4.3.19.12}$$

We then obtain the paving S by declaring that two rational points x/d and y/s line in the same ω if and only if the map of invertible sheaves

$$L^{(d,x)} \otimes L^{(s,y)} \to L^{(d+s, x+s)} \tag{4.3.19.13}$$

is an isomorphism.

4.4 Isomorphisms over Artinian Local Rings

4.4.1. Let B be an artinian local ring with residue field k, X a group isomorphic to $\mathbb{Z}^r$, and $(A, \mathcal{M})/B$ an abelian scheme with a line bundle defining a principal polarization. Let S be an integral regular paving of $X_{\mathbb{R}}$ and let $c : X \to A^t$ be a homomorphism defining a semi–abelian scheme G/B. Let $\psi, \psi' : 1_X \to c^* \mathcal{M}^{-1}$ be two trivializations inducing trivializations τ and τ' of $(c \times c)^* \mathcal{B}$ which are compatible with the symmetric biextension structure. Also fix two maps $\rho, \rho' : H_S \to B$ and let

$$(M_B, \mathscr{P}, M_{\mathscr{P}}, L_{\mathscr{P}}, G\text{–action}) \text{ and } (M'_B, \mathscr{P}', M_{\mathscr{P}'}, L_{\mathscr{P}'}, G\text{–action}) \quad (4.4.1.1)$$

be the resulting two collections of data obtained from the standard construction.

Let Isom denote the set of pairs (ι, σ), where $\iota : M'_B \to M_B$ is an isomorphism of log structure on $\mathrm{Spec}(B)$ and $\sigma : (\mathscr{P}, M_{\mathscr{P}}, L_{\mathscr{P}}) \to (\mathscr{P}', M_{\mathscr{P}'}, L_{\mathscr{P}'})$ is an isomorphism of polarized log schemes over the morphism

$$(\mathrm{id}, \iota) : (\mathrm{Spec}(B), M_B) \to (\mathrm{Spec}(B), M'_B) \quad (4.4.1.2)$$

compatible with the G–actions. Define Isom$'$ to be the set of isomorphisms $\gamma : (\mathscr{P}, L_{\mathscr{P}}) \to (\mathscr{P}', L_{\mathscr{P}'})$ of polarized schemes over B compatible with the G–actions.

Proposition 4.4.2. *The forgetful map* Isom $\to$ Isom$'$ *is bijective.*

Proof. Let

$$(M_B, \widetilde{\mathscr{P}}, L_{\widetilde{\mathscr{P}}}, G \text{ and } X \text{ action}) \text{ and } (M'_B, \widetilde{\mathscr{P}}', M_{\widetilde{\mathscr{P}}'}, L_{\widetilde{\mathscr{P}}'}, G \text{ and } X \text{ action}) \quad (4.4.2.1)$$

be the polarized log schemes obtained from the standard construction before taking the quotient by the X–action. By [3, 4.3.1], it suffices to show that any isomorphism

$$\gamma : (\widetilde{\mathscr{P}}, L_{\widetilde{\mathscr{P}}}) \to (\widetilde{\mathscr{P}}', L_{\widetilde{\mathscr{P}}'}) \quad (4.4.2.2)$$

compatible with the G and X actions extends uniquely to an isomorphism

$$(M_B, \widetilde{\mathscr{P}}, L_{\widetilde{\mathscr{P}}}, G \text{ and } X \text{ action})$$
$$\downarrow^{(\iota, \sigma)} \quad (4.4.2.3)$$
$$(M'_B, \widetilde{\mathscr{P}}', M_{\widetilde{\mathscr{P}}'}, L_{\widetilde{\mathscr{P}}'}, G \text{ and } X \text{ action}).$$

Let $f : \widetilde{\mathscr{P}} \to A$ and $f' : \widetilde{\mathscr{P}}' \to A$ be the projections.

Lemma 4.4.3 *There exists a unique automorphism of schemes $\bar{\gamma} : A \to A$ such that the diagram*

$$\widetilde{\mathscr{P}} \xrightarrow{\ \gamma\ } \widetilde{\mathscr{P}'}$$

$$f \downarrow \qquad\qquad \downarrow f' \qquad\qquad (4.4.3.1)$$

$$A \xrightarrow{\ \bar{\gamma}\ } A$$

commutes.

Proof. The morphisms f and f' can be constructed more intrinsically as follows. Let us explain how to get the morphism f in a canonical manner. Consider first the case when B is equal to a field k (this case was essentially done in the proof of 4.2.3), and for a simplex $\omega_i \in S$ let $\mathscr{P}_i \subset \widetilde{\mathscr{P}}$ denote the closed subscheme $\mathrm{Proj}_A(\oplus_{(d,x)\in\mathrm{Cone}(1,\omega_i)}\mathscr{M}^d \otimes L_x)$. Let $\mathscr{U}_i \subset \mathscr{P}_i$ denote the maximal open subscheme where the torus T acts faithfully. Then the projection identifies A with $[\mathscr{U}_i/T]$, and the projection $\mathscr{U}_i \to [\mathscr{U}_i/T]$ extends uniquely to a morphism $\mathscr{P}_i \to [\mathscr{U}_i/T]$. It follows in particular that if γ sends $\mathscr{P}_i$ to $\mathscr{P}'_i$, then there exists a unique morphism $\bar{\gamma}_i : A \to A$ such that the diagram

$$\mathscr{P}_i \xrightarrow{\ \gamma\ } \mathscr{P}'_i$$

$$\downarrow \qquad\qquad \downarrow \qquad\qquad (4.4.3.2)$$

$$A \xrightarrow{\ \bar{\gamma}_i\ } A$$

commutes. Furthermore, if $\mathscr{P}_i$ and $\mathscr{P}_j$ have non–empty intersection, then the map $\mathscr{P}_i \cap \mathscr{P}_j \to A$ is also surjective which implies that $\bar{\gamma}_i$ and $\bar{\gamma}_j$ are equal. Since $X_{\mathbb{R}}$ is connected it follows that all the morphisms $\bar{\gamma}_i$ are equal. This proves the case when B is a field.

For the general case, note that the case of a field gives a commutative diagram of topological spaces

$$|\widetilde{\mathscr{P}}| \xrightarrow{\ \gamma\ } |\widetilde{\mathscr{P}'}|$$

$$f \downarrow \qquad\qquad \downarrow f' \qquad\qquad (4.4.3.3)$$

$$|A| \xrightarrow{\ \bar{\gamma}\ } |A|,$$

and hence it suffices to define the map $\mathscr{O}_A \to \bar{\gamma}_*\mathscr{O}_A$. For this note that by 4.3.4 we have $f_*\mathscr{O}_{\widetilde{\mathscr{P}'}} = \mathscr{O}_A$ and $f_*\mathscr{O}_{\widetilde{\mathscr{P}}} = \mathscr{O}_A$, so we can take the map

$$\mathscr{O}_A = f'_*\mathscr{O}_{\widetilde{\mathscr{P}'}} \xrightarrow{\ \gamma\ } f'_*\gamma_*\mathscr{O}_{\widetilde{\mathscr{P}}} = \bar{\gamma}_*\mathscr{O}_A. \qquad (4.4.3.4)$$

□

Lemma 4.4.4 *The morphism $\bar{\gamma} : A \to A$ is translation by an element $\alpha \in A(B)$.*

Proof. The morphism $\bar{\gamma}$ is compatible with the translation action of A. Since any morphism of A–torsors is given by translation this implies the lemma.   □

After replacing γ by the composite of γ with the action of $-\alpha$, we may therefore assume that $\bar{\gamma}$ is the identity.

Since the morphism γ is compatible with the torus action, we get that in this case γ is induced by a morphism of sheaves of algebras

$$\bar{\gamma}^* : \oplus_{(d,x)\in P}\mathscr{M}^d \otimes L_x \to \oplus_{(d,x)\in P}\mathscr{M}^d \otimes L_x. \tag{4.4.4.1}$$

obtained as the direct sum of isomorphisms $\rho(d,x) : \mathscr{M}^d \otimes L_x \to \mathscr{M}^d \otimes L_x$. These isomorphisms define a set map $\rho : P \to B^*$ such that for every $\omega_i \in S$ the restriction of ρ to the integral points N_i of $\mathrm{Cone}(1,\omega)$ is linear. In other words, the map ρ is given by a morphism (denoted by the same letter)

$$\rho : \varinjlim_{\omega_i \in S} N_i^{gp} \to B^*. \tag{4.4.4.2}$$

Furthermore, the compatibility with the X–actions implies that the restriction of ρ to $\widetilde{H}_S$ descends to a morphism $\epsilon : H_S \to B^*$ (see for example the proof of 4.2.2). It then follows that the map

$$H_S \to B^* \oplus H_S, \quad h \mapsto (\epsilon(h), h) \tag{4.4.4.3}$$

induces an isomorphism $\iota : M'_B \to M_B$. Furthermore, as discussed in 4.1.12 the maps $\rho(d,x)$ induce a natural extension of γ to an isomorphism σ of log schemes. This proves the surjectivity part of 4.4.2.

For the injectivity, let $\sigma : M_{\widetilde{\mathscr{P}}} \to M_{\widetilde{\mathscr{P}}}$ and $\iota : M_B \to M_B$ be compatible automorphisms of the log structures. By the case of a field (section 4.2), the induced automorphism $\bar{\sigma} : \overline{M}_{\widetilde{\mathscr{P}}} \to \overline{M}_{\widetilde{\mathscr{P}}}$ is the identity.

Let $\mathscr{C} \to A$ denote the cone, and let $\mathscr{C}^o$ denote the complement of the vertex so that $\widetilde{\mathscr{P}} \simeq [\mathscr{C}^o/\mathbb{G}_m]$. Let $a \in A$ be a point and U the spectrum of the completion $\widehat{\mathscr{O}}_{A,a}$, and let $\mathscr{C}_U$ be the base change of $\mathscr{C}$. Fix a global chart $\beta : P \rtimes H_S \to M_{\mathscr{C}_U^o}$ (the pullback of $M_{\widetilde{\mathscr{P}}}$). Then the restriction of σ to $\mathscr{C}_U^o$ is induced by a homomorphism

$$\tau : P \rtimes H_S \to \Gamma(\mathscr{C}_U^o, \mathscr{O}^*_{\mathscr{C}_U^o}). \tag{4.4.4.4}$$

Since σ is compatible with the T–action, the image of τ in fact lies in the invariants under $D(\mathbb{X})$ of the right side of 4.4.4.4. Looking at the reductions modulo the power of the maximal ideal of $\mathscr{O}_U$ and using 4.3.11, it follows that τ in fact takes values in $\mathscr{O}^*_U$. This implies that σ induces the identity on $M_{\mathscr{C}_U^o}$ because for any $(p,0) \in P \rtimes H_S$ with image $\zeta_{(p,0)}$ in $\mathscr{O}_{\mathscr{C}_U^o}$ the map $\mathscr{O}^*_U \to \Gamma(\mathscr{C}_U^o, \mathscr{C}_U^o)$ sending v to $v \cdot \zeta_{(p,0)}$ is injective. We conclude that σ is the identity automorphism. Since the map $M_B|_{\widetilde{\mathscr{P}}} \to M_{\widetilde{\mathscr{P}}}$ is injective this also implies that ι is the identity. This completes the proof of 4.4.2. $\square$

4.4.5. There is also a weaker version of 4.4.2 for the saturations of standard families. With notation as in 4.4.1, assume that the maps $\rho, \rho' : H_S \to B$ factor through morphisms $\rho^{\mathrm{sat}}, \rho'^{\mathrm{sat}} : H_S^{\mathrm{sat}} \to B$ (warning: these extensions are not unique), where H_S^{sat} denotes the saturation of H_S, and let

$$(M_B^{\mathrm{sat}}, \mathscr{P}, M_{\mathscr{P}}^{\mathrm{sat}}, L_{\mathscr{P}}, G\text{--action}) \quad \text{and} \quad (M_B'^{\mathrm{sat}}, \mathscr{P}', M_{\mathscr{P}'}^{\mathrm{sat}}, L_{\mathscr{P}'}, G\text{--action})$$

$$(4.4.5.1)$$

be the resulting saturations of the standard families. Let Isom'' denote the set of pairs (ι, σ) where $\iota : M_B'^{\mathrm{sat}} \to M_B^{\mathrm{sat}}$ is an isomorphism of log structures on $\mathrm{Spec}(B)$, and

$$\sigma : (\mathscr{P}, M_{\mathscr{P}}^{\mathrm{sat}}, L_{\mathscr{P}}, G\text{--action}) \to (\mathscr{P}', M_{\mathscr{P}'}^{\mathrm{sat}}, L_{\mathscr{P}'}, G\text{--action}) \qquad (4.4.5.2)$$

is an isomorphism of polarized log schemes with G–action over

$$(\mathrm{id}, \iota) : (\mathrm{Spec}(B), M_B^{\mathrm{sat}}) \to (\mathrm{Spec}(B), M_B'^{\mathrm{sat}}). \qquad (4.4.5.3)$$

Also define $\mathscr{J} \subset \mathrm{Isom}$ (where Isom is defined as in 4.4.1) to be the subset of pairs (ι, σ) such that ι extends (necessarily uniquely) to an isomorphism $\iota^{\mathrm{sat}} : M_B'^{\mathrm{sat}} \to M_B^{\mathrm{sat}}$. Since there are isomorphisms (this is easily seen from the definition of $P \rtimes H_S$)

$$M_{\mathscr{P}}^{\mathrm{sat}} = M_{\mathscr{P}} \oplus_{M_B} M_B^{\mathrm{sat}}, \quad M_{\mathscr{P}'}^{\mathrm{sat}} = M_{\mathscr{P}'} \oplus_{M_B'} M_B'^{\mathrm{sat}}, \qquad (4.4.5.4)$$

pushout defines a canonical set map $\mathscr{J} \to \mathrm{Isom}''$.

Proposition 4.4.6. *This map $F : \mathscr{J} \to \mathrm{Isom}''$ is a bijection.*

Proof. Note first that the natural maps $M_B \to M_B^{\mathrm{sat}}$, $M_B' \to M_B'^{\mathrm{sat}}$, $M_{\mathscr{P}} \to M_{\mathscr{P}}^{\mathrm{sat}}$, and $M_{\mathscr{P}'} \to M_{\mathscr{P}'}^{\mathrm{sat}}$ are inclusions. Therefore it suffices to show that for any element $(\iota, \sigma) \in \mathrm{Isom}''$ the isomorphism $\iota : M_B'^{\mathrm{sat}} \to M_B^{\mathrm{sat}}$ restricts to an isomorphism $M_B' \to M_B$, and that $\sigma^b : \sigma^* M_{\mathscr{P}'}^{\mathrm{sat}} \to M_{\mathscr{P}}^{\mathrm{sat}}$ induces an isomorphism $\sigma^* M_{\mathscr{P}'} \to M_{\mathscr{P}}$. For this in turn it suffices to show that the isomorphism $\overline{\iota} : \overline{M}_B'^{\mathrm{sat}} \to \overline{M}_B^{\mathrm{sat}}$ restricts to an isomorphism $\overline{M}_B' \to \overline{M}_B$, and that $\sigma^b : \sigma^{-1} \overline{M}_{\mathscr{P}'}^{\mathrm{sat}} \to \overline{M}_{\mathscr{P}}^{\mathrm{sat}}$ induces an isomorphism $\sigma^{-1} \overline{M}_{\mathscr{P}'} \to \overline{M}_{\mathscr{P}}$. Since these are morphisms of constructible sheaves it therefore suffices to consider the case when B is a field. In this case the result follows from 4.2.7. $\qquad \square$

4.5 Versal Families

In this section we construct what will turn out to be versal families for the stack $\overline{\mathscr{K}}_g$.

Lemma 4.5.1 *Let $f : P \to B$ be a proper flat morphism of schemes which is cohomologically flat in dimension 0 (so that the relative Picard functor $\underline{\mathrm{Pic}}_{P/B}$ exists by [5, 7.3]). Let $\lambda : B \to \underline{\mathrm{Pic}}_{P/B}$ be a morphism such that for every flat covering $B' \to B$ and every line bundle $\mathscr{L}$ on $P_{B'}$ representing $\lambda|_{B'}$ the sheaf $f_* \mathscr{L}$ is locally free of rank 1 on B' and for any point $b' \in B'$ the map $H^0(P_{B'}, \mathscr{L}) \to H^0(P_{b'}, \mathscr{L}|_{P_{b'}})$ is injective. Then there exists a pair $(\mathscr{L}, \theta \in f_* \mathscr{L})$, unique up to unique isomorphism, where*

(i) $\mathscr{L}$ is a line bundle on P representing λ.

(ii) $\theta \in f_*\mathscr{L}$ is a section nonzero in every fiber.

Proof. The uniqueness is immediate, so it suffices to prove existence fppf locally on B where the pair exists by assumption.

Remark 4.5.2. To verify the assumptions of the lemma, it suffices to base change to the completions of B at points. In particular, if $f : P \to B$ and λ has the property that for over the completion of B at any point it is isomorphic to the standard family with its polarization, then the assumptions of the lemma hold for P/B.

4.5.3. Fix $g \geq 1$ and an integer $r \leq g$, $X = \mathbb{Z}^r$, and let S be an integral regular paving of $X_{\mathbb{R}}$.

Let $\mathscr{A}_{g-r}$ denote the moduli stack of principally polarized abelian varieties of dimension $g - r$, and let $A^t \to \mathscr{A}_{g-r}$ denote the universal abelian scheme. Let

$$W_0 := \underline{\mathrm{Hom}}_{\mathscr{A}_{g-r}}(X, A^t) \qquad (4.5.3.1)$$

be the stack over $\mathscr{A}_{g-r}$ classifying homomorphisms $c : X \to A^t$. Over W_0 there is a tautological semiabelian scheme $G \to W_0$ defined by c. Define W_1 to be the stack over W_0 associating to any scheme-valued point $(A, \lambda, c) \in W_0$ the set of isomorphism classes of representative line bundles $\mathscr{M}$ for λ together with a rigidification of $\mathscr{M}$ at the origin of A (note that such representatives admit no nontrivial automorphisms). Define W_2 to be the stack over W_1 classifying data $(A, \lambda, c, \mathscr{M}) \in W_1$ together with a trivialization $\psi : 1_X \to c^*\mathscr{M}^{-1}$ such that the induced trivialization of $(c \times c)^*\mathscr{B}$ is compatible with the symmetric biextension structure. Finally define

$$W_3 := W_2 \times_{\mathrm{Spec}(\mathbb{Z})} \mathrm{Spec}(\mathbb{Z}[H_S]). \qquad (4.5.3.2)$$

Over W_3 we can then perform the standard construction to obtain

$$(\widetilde{\mathscr{P}}, M_{\widetilde{\mathscr{P}}}, L_{\widetilde{\mathscr{P}}}, G \text{ and } X \text{ actions}) \to (W_3, M_{W_3}), \qquad (4.5.3.3)$$

where M_{W_3} denote the log structure obtained by pulling back the canononical log structure on $\mathrm{Spec}(\mathbb{Z}[H_S])$.

4.5.4. Define Γ to be the pushout

$$\Gamma := H_S \oplus_{\widetilde{H}_S} (P \times \widetilde{H}_S)^{gp} \simeq H_S^{gp} \oplus_{\widetilde{H}_S^{gp}} (P \times \widetilde{H}_S)^{gp}, \qquad (4.5.4.1)$$

and let $D(\Gamma)$ denote the corresponding diagonalizable group scheme. For a ring R the group $D(\Gamma)(R)$ can be identified with the group of homomorphisms $\rho : \varinjlim_{\omega_i \in S} N_i^{gp} \to R^*$, where N_i denotes the integral points of $\mathrm{Cone}(1, \omega_i)$, such that the induced map $\widetilde{H}_S \to R^*$ descends to H_S.

There is a canonical surjection $\Gamma \to P^{gp}$ which induces an inclusion $D(X) \times \mathbb{G}_m \hookrightarrow D(\Gamma)$. Over W_0 we can then form the contracted product

$\mathcal{G} := D(\Gamma) \times^{D(X)} G$, where $G \to W_0$ is the extension of the universal abelian variety by $D(X)$ defined by the tautological homomorphism $c : X \to A^t$.

There is also a natural map $w : D(\Gamma) \to \mathbb{G}_m$ induced by the homomorphism $\mathbb{Z} \to \Gamma$ sending 1 to the image of $(1, 0) \in P$ in Γ.

Lemma 4.5.5 *The relative dimension of W_3 over $\mathbb{Z}$ is equal to $(1/2)g(g + 1) + g + 1 + \text{rk}(H_S^{gp})$.*

Proof. One computes from the definitions

$$\dim(W_0) = \dim(\mathcal{A}_{g-r}) + r(g - r)$$
$$\dim(W_1) = \dim(W_0) + g - r + 1$$
$$\dim(W_2) = \dim(W_1) + (1/2)r(r + 1) + r$$
$$\dim(W_3) = \dim(W_2) + \text{rk}(H_S^{gp}).$$

Using this and the fact that the relative dimension of $\mathcal{A}_{g-r}$ over $\mathbb{Z}$ is equal to $(1/2)(g - r)(g - r + 1)$ one obtains the result. $\square$

Lemma 4.5.6 *The map*

$$X \times X \to H_S^{gp}, \quad (x, y) \mapsto (1, x + y) * (1, 0) - (1, x) * (1, y) \qquad (4.5.6.1)$$

is bilinear and hence induces a homomorphism $S^2 X \to H_S^{gp}$, where $S^2 X$ denotes the second symmetric power of X.

Proof. The map is clearly symmetric, so it suffices to show that it is linear in the second variable. For this we compute

$$(1, x + y) * (1, 0) - (1, x) * (1, y) + (1, x + z) * (1, 0) - (1, x) * (1, z)$$

$$= (1, x + y + z) * (1, z) - (1, x + z) * (1, y + z) + (1, x + z) * (1, 0) - (1, x) * (1, z)$$

$$= (1, x + y + z) + (1, z) - (2, x + y + 2z) - (1, x + z) - (1, y + z) + (2, x + y + 2z)$$

$$\quad + (1, x + z) + (1, 0) - (2, x + z) - (1, x) - (1, z) + (2, x + z)$$

$$= (1, x + y + z) + (1, 0) - (2, x + y + z) - (1, x) - (1, y + z) + (2, x + y + z)$$

$$= (1, x + y + z) * (1, 0) - (1, x) * (1, y + z).$$
$$(4.5.6.2)$$

$\square$

4.5.7. There is an action of $D(\Gamma)$ on W_2 over W_0 defined as follows. Let R be a ring and $s := (A, \lambda, \mathcal{M}, \psi)$ an R–valued point of W_2. For an R–valued point of $D(\Gamma)$ corresponding to a homomorphism $\rho : \varinjlim N_i^{gp} \to R^*$, define a second point $s^\rho \in W_2(R)$ as follows. The point s^ρ will have the same principally polarized abelian variety and map $c : X \to A^t$ (since the action will be over W_0), and the representative line bundle is also the same but with rigidification multiplied by $w(\rho) \in \mathbb{G}_m(R^*)$ (where w is as in 4.5.4). Finally the trivialization ψ' of $c^*\mathscr{M}^{-1}$ is given by the formula

$$\psi' : 1_X \to c^*\mathscr{M}^{-1}, \quad \psi'(y) := \rho(1,0)^{-1}\rho(1,y)\psi(y) \in \mathscr{M}^{-1}(c(y)). \quad (4.5.7.1)$$

To see that this defines a second point of W_2 note that

$$\begin{aligned}
\psi'(x)\psi'(y)\psi'(x+y)^{-1} &= \rho(1,0)^{-1}\rho(1,x)\rho(1,y)\rho(1,x+y)^{-1}\tau(x,y) \\
&= \rho((1,x)*(1,y))\rho((1,0)*(1,x+y))^{-1}\tau(x,y).
\end{aligned}$$

$$(4.5.7.2)$$

Since

$$(x,y) \mapsto \rho((1,x)*(1,y))\rho((1,0)*(1,x+y))^{-1} \quad (4.5.7.3)$$

is a symmetric bilinear form on $X \times X$ by 4.5.6 this shows that the trivialization of the Poincare bundle defined by ψ' is again compatible with the symmetric biextension structure 2.2.8. We therefore obtain an action of $D(\Gamma)$ on W_2 over W_0.

The group $D(\Gamma)$ also acts on $\mathrm{Spec}(\mathbb{Z}[H_S])$ via the projection $D(\Gamma) \to D(H_S^{gp})$ induced by the inclusion $H_S^{gp} \hookrightarrow \Gamma$, so we also obtain an action of $D(\Gamma)$ on $W_3 = W_2 \times_{\mathrm{Spec}(\mathbb{Z})} \mathrm{Spec}(\mathbb{Z}[H_S])$ from the diagonal action. Note that in fact this action extends naturally to an action on the log stack (W_3, M_{W_3}) [42, paragraph preceding 5.14].

4.5.8. This action of $D(\Gamma)$ extends naturally to an action on

$$(\widetilde{\mathscr{P}}, M_{\widetilde{\mathscr{P}}}, L_{\widetilde{\mathscr{P}}}, G \text{ and } X \text{ actions}) \quad (4.5.8.1)$$

over the above constructed action on (W_3, M_{W_3}) as follows. Let $s : \mathrm{Spec}(R) \to W_3$ be a morphism defining

$$(\widetilde{\mathscr{P}}_R, M_{\widetilde{\mathscr{P}}_R}, L_{\widetilde{\mathscr{P}}_R}, G \text{ and } X \text{ actions}) \to (\mathrm{Spec}(R), M_R), \quad (4.5.8.2)$$

let $\rho : \varinjlim_i N_i^{gp} \to R^*$ be a homomorphism defining an element of $D(\Gamma)(R)$, and let

$$(\widetilde{\mathscr{P}}'_R, M_{\widetilde{\mathscr{P}}'_R}, L_{\widetilde{\mathscr{P}}'_R}, G \text{ and } X \text{ actions}) \to (\mathrm{Spec}(R), M'_R) \quad (4.5.8.3)$$

be the data obtained from $s^\rho : \mathrm{Spec}(R) \to W_3$. Let $\iota : H_S \to R^*$ denote the homomorphism defined by ρ, and let $\beta : H_S \to M_R$ and $\beta' : H_S \to M'_R$ be the natural charts. The map ι defines an isomorphism of log structures $\delta : M_R \to M'_R$ which sends $\beta(h)$ to $\lambda(\iota(h)) + \beta'(h)$. Furthermore, for every $(d,x) \in P$ we obtain an isomorphism

$$\rho(d,x) : \mathscr{M}^d \otimes L_x \to \mathscr{M}^d \otimes L_x. \quad (4.5.8.4)$$

These isomorphisms induce an isomorphism of graded algebras which in turn gives an isomorphism

$$\sigma : (\widetilde{\mathscr{P}}'_R, M_{\widetilde{\mathscr{P}}'_R}, L_{\widetilde{\mathscr{P}}'_R}, G \text{ and } X \text{ actions}) \to (\widetilde{\mathscr{P}}_R, M_{\widetilde{\mathscr{P}}_R}, L_{\widetilde{\mathscr{P}}_R}) \qquad (4.5.8.5)$$

over the isomorphism

$$(\mathrm{Spec}(R), M'_R) \to (\mathrm{Spec}(R), M_R) \qquad (4.5.8.6)$$

defined by the identity map on $\mathrm{Spec}(R)$ and δ.

It follows immediately from the construction that σ is compatible with the G-action. The map σ is also compatible with the X-action. This follows from the commutativity of the diagrams

$$
\begin{array}{ccc}
t_y^*(\mathscr{M}^d \otimes L_x) & \xrightarrow{\psi'(y)^d \tau'(y,x)} & \mathscr{M}^d \otimes L_{x+dy} \\
\downarrow{\scriptstyle \rho(d,x)} & & \downarrow{\scriptstyle \rho(d,x+dy)} \\
t_y^*(\mathscr{M}^d \otimes L_x) & \xrightarrow{\psi(y)^d \tau(y,x)} & \mathscr{M}^d \otimes L_{x+dy},
\end{array}
\qquad (4.5.8.7)
$$

where ψ' is defined as in 4.5.7 and τ' is the associated trivialization of the Poincare biextension. The commutativity of 4.5.8.7 in turns follows from the following lemma:

Lemma 4.5.9 Let $\chi : S^2 X \to R^*$ denote the homomorphism obtained from ρ (and 4.5.6). Then

$$\rho(1,0)^d \rho(1,y)^{-d} \chi(y \otimes x) \rho(d, x+dy) \rho(d, x)^{-1} = 1. \qquad (4.5.9.1)$$

Proof. We prove this by induction on $d \geq 1$.

The case $d = 1$ is immediate from the definition of χ.

For the inductive step, assume the result holds for $d - 1$. This gives the formula

$$\rho(1,0)^{d-1} \rho(1,y)^{-(d-1)} \chi(y, x) = \rho(d-1, x+(d-1)y)^{-1} \rho(d-1, x) \qquad (4.5.9.2)$$

which enables us to rewrite the left side of 4.5.9.1 as

$$\rho(1,0)\rho(1,y)^{-1}\rho(d-1, x+(d-1)y)^{-1}\rho(d-1,x)\rho(d, x+dy)\rho(d, x)^{-1} \qquad (4.5.9.3)$$

which in turn is equal to

$$\rho((1,y) * (d-1, x+(d-1)y))^{-1}\rho((1,0) * (d-1,x)). \qquad (4.5.9.4)$$

Since the map $\widetilde{H}_S^{gp} \to R^*$ defined by ρ descends to H_S^{gp} by assumption, this expression is equal to 1 as desired. $\square$

4.5.10. There is also an action of the tautological semiabelian group scheme $G \to W_0$ on W_2 over W_0 (and hence also an action on W_3).

Let T be a scheme, and consider a point of $W_2(T)$ defined by $(A, \lambda, \mathscr{M}, \psi)$, and let $g \in G(T)$ be a point corresponding to data $(a \in A, \iota_x : t_a^* L_x \to L_x)$. We define a new point of $W_2(T)$ as follows. Note first that giving the isomorphisms ι_x is equivalent to giving sections of

$$L_x(a) = \mathcal{M}(a+x) \otimes \mathcal{M}(a)^{-1} \otimes \mathcal{M}(x)^{-1} = L_a(x), \qquad (4.5.10.1)$$

where L_a denotes the rigidified line bundle $L_{\lambda(a)}$. Let $\mathcal{M}'$ denote $\mathcal{M} \otimes L_a$, and define

$$\psi'(x) \in \mathcal{M}'^{-1}(x) = \mathcal{M}^{-1}(x) \otimes L_a^{-1}(x) \qquad (4.5.10.2)$$

to be the element $\psi(x) \otimes (\iota_x)^{-1}$. The following lemma implies that $(A, \lambda, \mathcal{M}', \psi')$ defines a point of W_2.

Lemma 4.5.11 *The trivialization $\Lambda(\psi') : 1_{X \times X} \to (c \times c)^* \mathcal{B}$ is equal to the trivialization $\Lambda(\psi)$, and in particular is compatible with the symmetric biextension structure.*

Proof. This amounts to the statement that the section

$$\psi(x+y)^{-1}\psi(x)\psi(y)\iota_{x+y}^{-1}\iota_x\iota_y \qquad (4.5.11.1)$$

of

$$\mathcal{M}(x+y) \otimes \mathcal{M}(x)^{-1} \otimes \mathcal{M}(y)^{-1} \otimes L_a^{-1}(x+y) \otimes L_a(x) \otimes L_a(y) \quad (4.5.11.2)$$

maps to

$$\psi(x+y)^{-1}\psi(x)\psi(y) \in \mathcal{M}(x+y) \otimes \mathcal{M}(x)^{-1} \otimes \mathcal{M}(y)^{-1} \qquad (4.5.11.3)$$

under the canonical isomorphism

$$\mathcal{M}(x+y) \otimes \mathcal{M}(x)^{-1} \otimes \mathcal{M}(y)^{-1} \otimes L_a^{-1}(x+y) \otimes L_a(x) \otimes L_a(y) \quad (4.5.11.4)$$

$$\downarrow$$

$$\mathcal{M}(x+y) \otimes \mathcal{M}(x)^{-1} \otimes \mathcal{M}(y)^{-1}.$$

This follows from the commutativity of the diagram 2.1.8.6. □

4.5.12. The action of G on W_2 also defines an action on $W_3 = W_2 \times \mathrm{Spec}(\mathbb{Z}[H_S])$ by having G act on the first factor of this product, and even on the log scheme (W_3, M_{W_3}). This action of G lifts to an action on the log scheme $(\widetilde{\mathscr{P}}, M_{\widetilde{\mathscr{P}}})$ over (W_3, M_{W_3}). Namely, given $g = (a, \{\iota_x\}) \in G(T)$ for some scheme T, and $(A, \lambda, \mathcal{M}, \psi) \in W_2(T)$ and a lifting $T \to \mathrm{Spec}(\mathbb{Z}[H_S])$ defining a log structure M_T on T, we get for any choice of isomorphism $\sigma : t_a^* \mathcal{M} \simeq \mathcal{M} \otimes L_a$ an isomorphism of sheaves of algebras on A

$$t_a^*(\oplus_{(d,x)} \mathcal{M}^d \otimes L_x) \to \oplus_{(d,x)} \mathcal{M}'^d \otimes L_x \qquad (4.5.12.1)$$

from the isomorphisms

$$\sigma^{\otimes d} \otimes \iota_x : t_a^*(\mathcal{M}^d \otimes L_x) \to \mathcal{M}'^d \otimes L_x. \tag{4.5.12.2}$$

This isomorphism depends on the choice of σ. However, the induced morphism of log schemes

$$(\widetilde{\mathscr{P}}, M_{\widetilde{\mathscr{P}}}) \to (\widetilde{\mathscr{P}}, M_{\widetilde{\mathscr{P}}}) \times_{A, t_a} A \tag{4.5.12.3}$$

is independent of this choice. Since an isomorphism σ exists fppf locally on any base scheme T it follows that the isomorphism can be defined globally. We leave to the reader the verification that this action is compatible with the X-actions.

4.5.13. It follows from the construction that the restriction of the G–action on W_2 to $D(X)$ is equal to the restriction of the $D(\Gamma)$–action to $D(X)$. We therefore obtain an action of $\mathscr{G} := D(\Gamma) \times^T G$ on the log scheme (W_3, M_{W_3}) over W_0. This action even lifts to an action on $(\widetilde{\mathscr{P}}, M_{\widetilde{\mathscr{P}}})$.

4.5.14. It follows that over the quotient stack $\mathscr{W}_3 := [W_3/\mathscr{G}]$ with its log structure $M_{\mathscr{W}_3}$ defined by M_{W_3}, there is a family

$$(\widetilde{\mathscr{P}}, M_{\widetilde{\mathscr{P}}}, G \text{ and } X \text{ actions}) \to (\mathscr{W}_3, M_{\mathscr{W}_3}). \tag{4.5.14.1}$$

Let $I \subset \mathcal{O}_{W_3}$ denote the ideal defined by the images of the non-zero elements of H_S, and let $W_{3,n} \subset W_3$ denote the closed substack defined by I^n. The action of $\mathscr{G}$ on W_3 preserves the substacks $W_{3,n}$, and hence by taking the quotient we obtain closed substacks $\mathscr{W}_{3,n} \subset \mathscr{W}_3$. Over each $\mathscr{W}_{3,n}$ the action of X on $(\widetilde{\mathscr{P}}, M_{\widetilde{\mathscr{P}}}, L_{\widetilde{\mathscr{P}}})$ is properly discontinuous, and hence as in 4.1.22 we can take the quotient by the X–action to obtain morphisms

$$f_n : (\mathscr{P}_n, M_{\mathscr{P}_n}, G\text{–action}) \to (\mathscr{W}_{3,n}, M_{\mathscr{W}_{3,n}}). \tag{4.5.14.2}$$

Because the action of $\mathscr{G}$ does not lift globally to an action on the line bundle over $\widetilde{\mathscr{P}}$, but does lift locally to the line bundle, we obtain furthermore a section $\mathscr{W}_{3,n} \to \underline{\mathrm{Pic}}_{\mathscr{P}_n/\mathscr{W}_{3,n}}$ which is locally defined by the line bundle arising in the standard construction. By 4.5.1 it follows that there exists a unique pair $(\mathscr{L}_n, \theta_n \in f_{n*}\mathscr{L}_n)$ consisting of a line bundle on $\mathscr{P}_n$ and a section $\theta_n \in f_{n*}\mathscr{L}_n$ which is nonvanishing in each fiber such that $\mathscr{L}_n$ defines the map $\mathscr{W}_{3,n} \to \underline{\mathrm{Pic}}_{\mathscr{P}_n/\mathscr{W}_{3,n}}$.

Remark 4.5.15. An examination of section 4.4 shows that if we consider two points of $W_{3,n}(T)$ as in 4.4.1, where T is the spectrum of an artinian local ring, then any isomorphism between the two collections of data is obtained from a point of the group scheme $\mathscr{G}$.

4.5.16. Let $\mathscr{A}_g^{\mathrm{Alex}}$ denote Alexeev's stack (with $d = 1$) defined in 2.4.9

Let $\mathcal{Q} \subset \mathscr{A}_g^{\mathrm{Alex}}$ denote the closure (with the reduced substack) of the open substack classifying objects of $\mathscr{A}_g^{\mathrm{Alex}}$ with G an abelian variety, and let $\widetilde{\mathcal{Q}}$ denote its normalization. Let $\mathcal{Q}_n$ denote the n–th infinitesimal neighborhood of the reduced closed substack of $\mathcal{Q}$ classifying objects where G has non–trivial toric part.

Lemma 4.5.17 *For every n, the forgetful map $\mathcal{W}_{3,n} \to \mathcal{A}_g^{\mathrm{Alex}}$ factors through a morphism $j_n : \mathcal{W}_{3,n} \to \mathcal{Q}_n$.*

Proof. Let V_n denote $W_{3,n}$. Since $V_n \to \mathcal{A}_{g-r}$ is representable the stack V_n is a Deligne–Mumford stack. Let $g_0 : \mathrm{Spec}(A_0) \to V_0$ be an étale surjection, and for every n let $g_n : \mathrm{Spec}(A_n) \to V_n$ be the unique étale morphism lifting g_0. Define $A := \varprojlim A_n$. The maps $\mathscr{V}_n \to \mathcal{A}_g^{\mathrm{Alex}}$ then induce a morphism $\mathrm{Spec}(A) \to \mathcal{A}_g^{\mathrm{Alex}}$. It suffices to show that this morphism factors through $\mathcal{Q}$. This is clear because $\mathrm{Spec}(A)$ is reduced, and every generic point of $\mathrm{Spec}(A)$ maps to $\mathcal{Q}$. $\square$

Lemma 4.5.18 *The map $j_n : \mathcal{W}_{3,n} \to \mathcal{Q}_n$ is fully faithful. In particular, $\mathcal{W}_{3,n}$ has finite diagonal.*

Proof. Consider two morphisms $f, g : T \to \mathcal{W}_{3,n}$ from a scheme T, and the induced morphism

$$\gamma : T \times_{f, \mathcal{W}_{3,n}, g} T \to T \times_{j_n \circ f, \mathcal{Q}_n, j_n \circ g} T. \qquad (4.5.18.1)$$

This is a morphism of algebraic spaces of finite presentation over T, and hence to prove that it is an isomorphism it suffices to show that the map induces a bijection on scheme–valued points with values in spectra of artinian local algebras. This follows from 4.5.15. $\square$

Lemma 4.5.19 *The map $j_n : \mathcal{W}_{3,n} \to \mathcal{Q}_n$ is an immersion.*

Proof. The follows from the fact that it is a finite type fully faithful morphism of algebraic stacks. $\square$

Lemma 4.5.20 *The morphism on normalizations $j_n : (\widetilde{\mathcal{W}_3})_n \to (\widetilde{\mathcal{Q}})_n$ is an open immersion.*

Proof. Let $V \to \mathcal{Q}$ be a smooth surjection with V a scheme, and let $v \in V$ be a point mapping to the image of $(\mathcal{W}_3)_0$ in $\mathcal{Q}_0$. Let $\widehat{V}$ denote the spectrum of the completion of the normalization of $\widehat{\mathcal{O}}_{V,v}$. For each n, let $Z_n \subset \widehat{V}$ denote the inverse image of $(\mathcal{W}_3)_n$. The system $\{Z_n\}$ defines a closed subscheme $\widehat{Z} \subset \widehat{V}$. Furthermore, since each Z_n is smooth over $(\mathcal{W}_3)_n$, the subscheme $\widehat{Z}$ is integral. Thus $\widehat{Z} \subset \widehat{V}$ is a closed immersion of normal schemes. Hence to prove that it is an isomorphism it suffices to show that the two schemes have the same dimension. We already know that $\widehat{V}$ has dimension $(1/2)g(g+1) + h + 1$, where h is the relative dimension of V over $\mathcal{Q}$. This is because $\mathcal{A}_g$ has relative dimension $(1/2)g(g+1)$ over $\mathbb{Z}$.

On the other hand, by 4.5.5 and the fact that

$$\dim(\mathcal{G}) = \mathrm{rk}(H_S^{gp}) + g + 1 \qquad (4.5.20.1)$$

we find that the stack $\mathcal{W}_3$ has relative dimension $(1/2)g(g+1)$ over $\mathbb{Z}$. This implies that $\widehat{Z}$ has dimension $(1/2)g(g+1) + h + 1$ also. $\square$

4.6 Definition of the Moduli Problem

4.6.1. Define $\mathscr{K}_g$ to be the fibered category over $\mathbb{Z}$ which to any scheme B associates the groupoid of data as follows

$$(G, M_B, f : (X, M_X) \to (B, M_B), L, \theta, \rho) \qquad (4.6.1.1)$$

where

(i) M_B is a fine log structure on B.
(ii) $f : (X, M_X) \to (B, M_B)$ is a log smooth morphism whose underlying morphism of schemes $X \to B$ is proper,
(iii) L is a relatively ample invertible sheaf on X/B,
(iv) G is a semiabelian scheme over B of relative dimension g, and ρ is an action of G on (X, M_X) over (B, M_B),
(v) $\theta \in f_*L$ is a global section,

such that for every geometric point $\bar{s} \to B$ the following hold:

(vi) the zero–local of the section $\theta_s \in \Gamma(X_{\bar{s}}, L_{\bar{s}})$ does not contain any G–orbits.
(vii) the data $(G_{\bar{s}}, M_B|_{\bar{s}}, X_{\bar{s}}, M_{X_{\bar{s}}}, L_{\bar{s}}, G_{\bar{s}}\text{–action})$ is isomorphic to the saturation of a standard family over $\operatorname{Spec}(k(\bar{s}))$.

There is a natural log structure $M_{\mathscr{K}_g}$ on $\mathscr{K}_g$ (see [46, Appendix B] for the notion of a log structure on a stack).

The main theorem of this chapter, which will be proven in the following sections, is the following:

Theorem 4.6.2 (i) *The fibered category $\mathscr{K}_g$ is a proper algebraic stack over $\mathbb{Z}$ with finite diagonal and containing $\mathscr{A}_g$ as a dense open substack (in fact $\mathscr{A}_g$ is isomorphic to the open substack of $\mathscr{K}_g$ where the log structure $M_{\mathscr{K}_g}$ is trivial).*
(ii) *The log stack $(\mathscr{K}_g, M_{\mathscr{K}_g})$ is log smooth over $\operatorname{Spec}(\mathbb{Z})$. In particular, the stack $\mathscr{K}_g$ has toroidal singularities.*
(iii) *The stack $\mathscr{K}_g$ is isomorphic to the normalization of the main component (the closure of $\mathscr{A}_g$) of Alexeev's stack $\mathscr{A}_g^{\mathrm{Alex}}$.*

4.7 The Valuative Criterion for Properness

4.7.1. Let R be a complete discrete valuation ring with field of fractions K. Fix a uniformizer $\pi \in R$, let $\mathfrak{m}$ denote the maximal ideal of R, set $S := \operatorname{Spec}(R)$, and let η (resp. s) denote the generic (resp. closed) point of S. Let A_K be an abelian scheme over K and (P_K, L_K, Θ_K) a principally polarized A_K–torsor with a non–zero section $\Theta_K \in \Gamma(P_K, L_K)$.

Proposition 4.7.2. *After possibly replacing R by a finite extension, there exists a compatible family of objects $\{F_n\} \in \mathscr{K}_g(R_n)$ such that the morphism $\operatorname{Spec}(R) \to Q$ defined by forgetting the log structures restricts on the generic fiber to the map $\operatorname{Spec}(K) \to Q$ defined by (P_K, L_K, θ_K)*

Remark 4.7.3. It follows from 4.8.1 in the next section that the compatible family $\{F_n\}$ is induced by a unique object $F \in \mathcal{K}_g(R)$ restricting to $(A_K, P_K, L_K, \theta_K)$ on the generic fiber.

Proof. This essentially follows from the proof of [3, 5.7] (see also [4, §3]).

First of all, after performing an extension of R we may assume that P_K is trivial and hence (P_K, L_K, Θ_K) is given simply by a principally polarized abelian variety (A_K, L_K) with a global nonvanishing section $\Theta \in \Gamma(A_K, L_K)$. Note that the dimension of $\Gamma(A_K, L_K)$ is 1 by [35, combine Theorem on p. 150 with Corollary on p. 159], so Θ is unique up to multiplication by K^*. Thus any other nonvanishing section Θ' can be obtained from Θ by applying an automorphism of (P_K, L_K). We can therefore ignore the section.

By the semi–stable reduction [19, IX.3.6], there exists after possibly making another finite base change a semiabelian group scheme with an invertible sheaf $(G, L_G)/S$ restricting to (A_K, L_K) over η. After making another finite extension we may assume that the closed fiber G_s is split. In this case, by [13, II.6.2] the semiabelian group scheme $(G, L_G)/S$ is obtained from data as follows:

(i) An abelian scheme A/S with a rigidified line bundle $\mathcal{M}$ defining a principal polarization, and a split torus $T = D(X)$ (where $X \simeq \mathbb{Z}^r$ for some r).

(ii) A homomorphism $c : X \to A^t$ defining a semi–abelian group scheme over S

$$1 \to T \to \widetilde{G} \to A \to 0. \qquad (4.7.3.1)$$

(iii) A trivialization $a : 1_X \to c^* \mathcal{M}_\eta^{-1}$ inducing a trivialization $\tilde{\tau} : 1_{X \times X} \to (c \times c)^* \mathcal{B}_\eta^{-1}$ of the Poincare sheaf which is compatible with the symmetric biextension structure. The trivialization $\tilde{\tau}$ is required to satisfy the following positivity condition: for every nonzero $x \in X$, the section $\tilde{\tau}(x, x) \in \mathcal{B}_{(x,x)}^{-1} \otimes_R K$ extends to a section of $\mathcal{B}_{(x,x)}^{-1}$ which is congruent to zero modulo $\mathfrak{m}$.

For any $x \in X$, the module $\mathcal{M}^{-1}(c(x))$ is free of rank 1 over R. We can therefore speak of the valuation of any element $s \in \mathcal{M}^{-1}(c(x)) \otimes_R K$. In particular, we obtain a function

$$A : X \to \mathbb{Z}, \quad x \mapsto \text{valuation of } a(x) \in \mathcal{M}^{-1}(c(x)) \otimes_R K. \qquad (4.7.3.2)$$

By the same method we can define a function

$$B : S^2 X \to \mathbb{Z}, \quad x \otimes y \mapsto \text{valuation of } \tilde{\tau}(x, y) \in \mathcal{B}(x, y) \otimes_R K. \qquad (4.7.3.3)$$

The compatibility with the symmetric biextension property implies that B is a quadratic form, and the positivity assumption implies that B is positive definite. Finally it follows from the definitions that the function

$$x \mapsto A(x) - (1/2)B(x, x) \qquad (4.7.3.4)$$

is a linear function.

Define $\psi : 1_X \to \mathscr{M}^{-1}$ to be the trivialization characterized by the formula

$$a(x) = \psi(x)\pi^{A(x)}. \qquad (4.7.3.5)$$

Then the associated trivialization $\tau : 1_{X \times X} \to (c \times c)^* \mathscr{B}^{-1}$ satisfies

$$\tilde{\tau}(x, y) = \tau(x, y)\pi^{B(x,y)}. \qquad (4.7.3.6)$$

In particular, τ is compatible with the symmetric biextension structure.

Let $g : X_{\mathbb{R}} \to \mathbb{R}$ be the piecewise linear function whose graph is the lower envelope of the convex hull of the set

$$\{(x, A(x)), |x \in X\}. \qquad (4.7.3.7)$$

The function g may not be integer valued on X. However, if $R \to R'$ is a finite extension of discrete valuation rings with ramification e, and if g' denotes the function obtained by doing the preceding construction after first base changing to R', then $g' = eg$. It follows that after making a suitable finite extension of R we may assume that g is integer valued on X.

Let S be the paving of $X_{\mathbb{R}}$ defined by g (the domains of linearity), so that we have P and H_S as in the standard construction. By the definition of S, the function g induces a morphism

$$\varinjlim_{\omega_i \in S} N_i^{gp} \to \mathbb{Z}, \qquad (4.7.3.8)$$

which restricts to a map $\widetilde{H}_S \to \mathbb{N}$. Furthermore, for any $y \in X$ the function $x \mapsto A(x + y)$ differs from $A(x)$ by a linear function. From this it follows that the map $\widetilde{H}_S \to \mathbb{N}$ descends to a morphism $\beta : H_S \to \mathbb{N}$. Define $\rho : H_S \to R$ to be the map $h \mapsto \pi^{\beta(h)}$. Note that since $\mathbb{N}$ is saturated this extends uniquely to a morphism $H_S^{\text{sat}} \to R$. Applying the standard construction we then obtain a compatible family of objects $\{F_n\} \in \varprojlim \mathscr{K}_g(R_n)$. We claim that this collection is the family sought for in the proposition.

For this note first that after forgetting the log structures, the F_n is obtained by reduction from the graded algebra with X and G action (defined as in section 4.1)

$$\mathcal{R} := \oplus_{(d,x) \in P} \mathscr{M}^d \otimes L_x. \qquad (4.7.3.9)$$

We claim that this algebra is canonically isomorphic to the algebra also denoted $\mathcal{R}$ in [4, 3.2]. By [loc. cit., 3.24] this will complete the proof of the proposition.

To establish this identification, let $\mathcal{S}_2$ denote the graded algebra over A_η

$$\oplus_{(d,x)} \mathscr{M}_\eta^d \otimes L_{x,\eta} \qquad (4.7.3.10)$$

with algebra structure given by the canonical isomorphisms

$$\text{can} : (\mathscr{M}_\eta^d \otimes L_{x,\eta}) \otimes (\mathscr{M}_\eta^e \otimes L_{y,\eta}) \to \mathscr{M}_\eta^{d+e} \otimes L_{x+y,\eta}. \qquad (4.7.3.11)$$

By the same reasoning as in section 4.1 there is a canonical action of G_η on S_2 over the translation action on A_η, and the trivialization $a : 1_X \to \mathscr{M}_\eta^{-1}$ also defines an action of X on S_2.

There is a natural map from the algebra $\mathcal{R} \otimes K$ 4.7.3.9 to S_2 defined as follows. This map is induced by the morphisms

$$\pi^{g(x)} \cdot \mathrm{id} : \mathscr{M}^d \otimes L_x \to \mathscr{M}_\eta^d \otimes L_{x,\eta}. \tag{4.7.3.12}$$

Then it follows from the construction that the induced map $\mathcal{R} \otimes K \to S_2$ is compatible with the algebra structures as well as the G and X actions. Finally, from [3, 3.7] it follows that our $\mathcal{R}$ agrees with the one constructed there. This completes the proof of 4.7.2. $\qquad\square$

4.7.4. Let $\tilde{Q}$ denote the normalization of the main component of Alexeev's compactification, and let $\mathscr{P}_{\tilde{Q}} \to \tilde{Q}$ be the base change of the universal scheme with semi–abelian group action over Q. Let $G_{\tilde{Q}} \to \tilde{Q}$ denote the semiabelian group scheme which acts on $\mathscr{P}_{\tilde{Q}}$. Let $j : \tilde{Q}^o \subset \tilde{Q}$ denote the maximal open dense substack where $\mathscr{P}$ is smooth, and let $\tilde{j} : \mathscr{P}^o \hookrightarrow \mathscr{P}$ denote the inverse image in $\mathscr{P}$. Define $N_{\mathscr{P}}$ (resp. $N_{\tilde{Q}}$) to be the log structure $\tilde{j}_*^{log} \mathscr{O}_{\mathscr{P}^o}^*$ (resp. $j_*^{log} \mathscr{O}_{\tilde{Q}^o}^*$) so that there is a natural morphism of log stacks

$$(\mathscr{P}, N_{\mathscr{P}}) \to (\tilde{Q}, N_{\tilde{Q}}). \tag{4.7.4.1}$$

Proposition 4.7.5. (i) *The log structures $N_{\mathscr{P}}$ and $N_{\tilde{Q}}$ are fs and the morphism 4.7.4.1 is log smooth.*
(ii) *For every geometric point $\bar{s} \to \tilde{Q}$, the data*

$$(G_{\bar{s}}, N_{\tilde{Q}}|_{\bar{s}}, \mathscr{P}_{\bar{s}}, M_{\mathscr{P}}|_{\mathscr{P}_{\bar{s}}}, L_{\mathscr{P}_{\bar{s}}}, \theta) \tag{4.7.5.1}$$

defines an object of $\mathscr{K}_g$, and hence we obtain a morphism of stacks $\tilde{Q} \to \mathscr{K}_g$.
(iii) *For any artinian local ring B, the functor $\tilde{Q}(B) \to \mathscr{K}_g(B)$ is an equivalence of categories.*

Proof. By 4.7.2, for any field valued point $s : \mathrm{Spec}(k) \to \tilde{Q}$ the pullback $\mathscr{P}_k \to \mathrm{Spec}(k)$ is isomorphic to the scheme obtained by applying the standard construction for some choice of data S, ψ etc. By the deformation theory 4.3.3 it then follows that locally in the smooth topology $\tilde{Q}$ is smooth over $\mathrm{Spec}(\mathbb{Z}[H_S^{\mathrm{sat}}])$ and that $\tilde{Q}^o$ is equal to the inverse image of $\mathrm{Spec}(\mathbb{Z}[H_S^{\mathrm{sat, gp}}]) \subset \mathrm{Spec}(\mathbb{Z}[H_S^{\mathrm{sat}}])$. From this and [25, 11.6] it follows that $N_{\tilde{Q}}$ is fine. Furthermore, the deformation theory also shows that étale locally on $\tilde{Q}$ and $\mathscr{P}$ there exists a commutative diagram

$$\begin{array}{ccc}
\mathscr{P} & \xrightarrow{g} & \mathrm{Spec}(\mathbb{Z}[Q]) \\
\downarrow & & \downarrow{d^*} \\
\tilde{Q} & \xrightarrow{f} & \mathrm{Spec}(\mathbb{Z}[H_S^{\mathrm{sat}}]),
\end{array} \tag{4.7.5.2}$$

where f is smooth as above, g induces a smooth map

$$\mathscr{P} \to \widetilde{Q} \times_{\mathrm{Spec}(\mathbb{Z}[H_S^{\mathrm{sat}}])} \mathrm{Spec}(\mathbb{Z}[Q]), \tag{4.7.5.3}$$

and d^* is induced by an integral morphism of monoids $d : H_S^{\mathrm{sat}} \to Q$ with $\mathrm{Coker}(H_S^{gp} \to Q^{gp})$ torsion free, such that the map

$$H_S^{gp} \oplus_{H_S^{\mathrm{sat}}} Q \to Q^{gp} \tag{4.7.5.4}$$

is an isomorphism [24, 3.5]. From this and [25, 11.6] it follows that $N_{\mathscr{P}}$ is also fine and that 4.7.4.1 is log smooth. This also proves (ii).

Statement (iii) follows from 4.5.20. □

4.8 Algebraization

Let B be a complete noetherian local ring, $\mathfrak{m} \subset B$ the maximal ideal, and set $B_n := B/\mathfrak{m}^n$.

Proposition 4.8.1. *The natural functor*

$$\mathscr{K}_g(B) \to \varprojlim \mathscr{K}_g(B_n) \tag{4.8.1.1}$$

is an equivalence of categories.

The proof of 4.8.1 will be in several steps 4.8.2–4.8.11.

4.8.2. Let $I \subset B$ be a square-zero ideal. Set $\overline{B} := B/I$ and define

$$\overline{B}_n := B_n \otimes_B \overline{B}, \quad I_n := \mathrm{Ker}(B_n \to \overline{B}_n). \tag{4.8.2.1}$$

Fix a free abelian group X, an integral regular paving S of $X_{\mathbb{R}}$, and a map $H_S \to B$ sending all non–zero elements to $\mathfrak{m}$. This map defines log structures M_B, $M_{\overline{B}}$, etc. Fix also a principally polarized semiabelian scheme G/B whose reduction $G \otimes_B B_0$ is split. Choose a splitting, and let

$$0 \to T \to \widetilde{G} \to A \to 0 \tag{4.8.2.2}$$

be the extension over B obtained as in [13, II §1]. Here T is the torus corresponding to X, and the extension is given by a map $c : X \to A^t$. Fix descent data for the polarization on G to A giving a line bundle $\mathscr{M}$ on A defining a principal polarization. We also choose a rigidification of $\mathscr{M}$ on A, and a trivialization $\psi : 1_X \to c^* \mathscr{M}^{-1}$ such that the induced trivialization of $(c \times c)^* \mathscr{B}^{-1}$ is compatible with the symmetric biextension structure. We can then apply the standard construction and saturation to obtain compatible systems

$$\{(\mathscr{P}_n, M_{\mathscr{P}_n}, L_{\mathscr{P}_n})/(B, M_B)\} \quad \text{and} \quad \{(\overline{\mathscr{P}}_n, M_{\overline{\mathscr{P}}_n}, L_{\overline{\mathscr{P}}_n})/(\overline{B}, M_{\overline{B}})\}$$

$$\tag{4.8.2.3}$$

of saturated standard families.

Lemma 4.8.3 *Let F be a coherent sheaf on $\overline{\mathscr{P}}$ flat over $\overline{B}$, and let F_n denote the pullback to $\overline{\mathscr{P}} \otimes B_n$. Then for every integer i the natural map*

$$\varprojlim H^i(\overline{\mathscr{P}}, I \otimes F_n) \to \varprojlim H^i(\overline{\mathscr{P}}, I_n \otimes_B F_n) \qquad (4.8.3.1)$$

is a topological isomorphism.

Proof. Consider the exact sequences

$$0 \to I \cap \mathfrak{m}^n / \mathfrak{m}^n I \to I \otimes_{\overline{B}} \overline{B}_n \to I_n \to 0. \qquad (4.8.3.2)$$

By the Artin-Rees lemma there exists an integer k such that

$$I \cap \mathfrak{m}^n = \mathfrak{m}^{n-k}(\mathfrak{m}^k \cap I) \qquad (4.8.3.3)$$

and hence $I \cap \mathfrak{m}^n / \mathfrak{m}^n I$ is annihilated by $\mathfrak{m}^k$. Since F is flat over $\overline{B}$, 4.8.3.2 induces short exact sequences

$$0 \to (I \cap \mathfrak{m}^n / \mathfrak{m}^n I) \otimes F \to I \otimes F_n \to I_n \otimes F_n \to 0. \qquad (4.8.3.4)$$

Looking at the associated long exact sequences it follows that the kernel and cokernel of

$$H^i(\overline{\mathscr{P}}, I \otimes F_n) \to H^i(\overline{\mathscr{P}}, I_n \otimes_B F_n) \qquad (4.8.3.5)$$

is annihilated by $\mathfrak{m}^k$, and that 4.8.3.1 is an isomorphism. $\square$

Lemma 4.8.4 *There exists a sequence of integers $\nu(n)$ with $\varinjlim_n \nu(n) = \infty$ such that for every n the image of the map*

$$H^i(\overline{\mathscr{P}}, I \otimes F) \to H^i(\overline{\mathscr{P}}, I_{\nu(n)} \otimes F_{\nu(n)}) \qquad (4.8.4.1)$$

is equal to the image of the map

$$H^i(\overline{\mathscr{P}}, I_n \otimes F_n) \to H^i(\overline{\mathscr{P}}, I_{\nu(n)} \otimes F_{\nu(n)}). \qquad (4.8.4.2)$$

Proof. This follows from the preceding lemma and [17, III.4.1.7] which implies that the natural map

$$H^i(\overline{\mathscr{P}}, I \otimes F) \to \varprojlim H^i(\overline{\mathscr{P}}, I \otimes F_n) \qquad (4.8.4.3)$$

is a topological isomorphism. $\square$

Lemma 4.8.5 *Let $(\overline{\mathscr{P}}, M_{\overline{\mathscr{P}}}) \to (\mathrm{Spec}(\overline{B}), M_{\overline{B}})$ be a log smooth proper morphism and $L_{\overline{\mathscr{P}}}$ a polarization on $\overline{\mathscr{P}}$. Assume given an isomorphism of projective systems*

$$\bar{\gamma} : \{(\overline{\mathscr{P}}, M_{\overline{\mathscr{P}}}, L_{\overline{\mathscr{P}}}) \otimes \overline{B}_n\} \to \{(\overline{\mathscr{P}}_n, M_{\overline{\mathscr{P}}_n}, L_{\overline{\mathscr{P}}_n})\}. \qquad (4.8.5.1)$$

Then there exists a unique log smooth lifting $(\mathscr{P}, M_{\mathscr{P}}) \to (\mathrm{Spec}(B), M_B)$ of $(\overline{\mathscr{P}}, M_{\overline{\mathscr{P}}})$ together with a lifting $L_{\mathscr{P}}$ of the polarization and an isomorphism of projective systems

$$\gamma : \{(\mathscr{P}, M_{\mathscr{P}}, L_{\mathscr{P}}) \otimes B_n\} \to \{(\mathscr{P}_n, M_{\mathscr{P}_n}, L_{\mathscr{P}_n})\} \qquad (4.8.5.2)$$

lifting $\bar{\gamma}$.

Proof. Applying 4.8.4 with $i = 0$ to $F = T_{(\overline{\mathscr{P}}, M_{\overline{\mathscr{P}}})/(\overline{B}, M_{\overline{B}})}$ and using [24, 3.14], it follows that to construct $(\mathscr{P}, M_{\mathscr{P}})$ it suffices to find a log smooth deformation $(\mathscr{P}, M_{\mathscr{P}})$ such that for all n the reduction $(\mathscr{P}, M_{\mathscr{P}}) \otimes B_n$ is isomorphic to $(\mathscr{P}_n, M_{\mathscr{P}_n})$. For then in fact we can choose the isomorphisms to be compatible. To see this last statement, assume we have chosen a collection of compatible isomorphisms up to $n - 1$. Then the lemma implies that we can choose an isomorphism $(\mathscr{P}, M_{\mathscr{P}}) \otimes B_n \to (\mathscr{P}_n, M_{\mathscr{P}_n})$ which agrees with the previously chosen isomorphisms for $m \leq \nu(n - 1)$. Since $\nu(n) \to \infty$ we can inductively choose the isomorphisms to be compatible.

The obstruction to finding a log smooth deformation $(\mathscr{P}, M_{\mathscr{P}})$ of $(\overline{\mathscr{P}}, M_{\overline{\mathscr{P}}})$ is by [24, 3.14] a class in $H^2(\overline{\mathscr{P}}, I \otimes T_{(\overline{\mathscr{P}}, M_{\overline{\mathscr{P}}})/(\overline{B}, M_{\overline{B}})})$. Furthermore this class is functorial. Since

$$H^2(\overline{\mathscr{P}}, I \otimes T_{(\overline{\mathscr{P}}, M_{\overline{\mathscr{P}}})/(\overline{B}, M_{\overline{B}})}) \to \varprojlim H^2(\overline{\mathscr{P}}_n, I_n \otimes T_{(\overline{\mathscr{P}}_n, M_{\overline{\mathscr{P}}_n})/(\overline{B}_n, M_{\overline{B}_n})})$$

(4.8.5.3)

is an isomorphism and the obstruction maps to zero in each $H^2(\overline{\mathscr{P}}_n, I_n \otimes T_{(\overline{\mathscr{P}}_n, M_{\overline{\mathscr{P}}_n})/(\overline{B}_n, M_{\overline{B}_n})})$ this implies that there exists a log smooth deformation $(\mathscr{P}', M_{\mathscr{P}'})$.

The isomorphism classes of log smooth deformations are by [24, 3.14] a torsor under the group $H^1(\overline{\mathscr{P}}, I \otimes T_{(\overline{\mathscr{P}}, M_{\overline{\mathscr{P}}})/(\overline{B}, M_{\overline{B}})})$. Since

$$H^1(\overline{\mathscr{P}}, I \otimes T_{(\overline{\mathscr{P}}, M_{\overline{\mathscr{P}}})/(\overline{B}, M_{\overline{B}})}) \to \varprojlim H^1(\overline{\mathscr{P}}_n, I_n \otimes T_{(\overline{\mathscr{P}}_n, M_{\overline{\mathscr{P}}_n})/(\overline{B}_n, M_{\overline{B}_n})})$$

(4.8.5.4)

is an isomorphism we obtain a unique class $\tau \in H^1(\overline{\mathscr{P}}, I \otimes T_{(\overline{\mathscr{P}}, M_{\overline{\mathscr{P}}})/(\overline{B}, M_{\overline{B}})})$ from the collection

$$([(\mathscr{P}', M_{\mathscr{P}'}) \otimes B_n] - [(\mathscr{P}_n, M_{\mathscr{P}_n})]) \in \varprojlim H^1(\overline{\mathscr{P}}_n, I_n \otimes T_{(\overline{\mathscr{P}}_n, M_{\overline{\mathscr{P}}_n})/(\overline{B}_n, M_{\overline{B}_n})}).$$

(4.8.5.5)

Changing our choice of $(\mathscr{P}', M_{\mathscr{P}'})$ we then obtain the desired lifting $(\mathscr{P}, M_{\mathscr{P}})$. Finally the lifting of the polarization is obtained from the Grothendieck existence theorem [17, III.5.1.4] and the sheaves $L_{\mathscr{P}_n}$. This completes the proof of 4.8.5. □

4.8.6. Next we make some observations about formal log schemes. Let R be a regular ring, and Q a fine saturated monoid with $Q^* = \{0\}$. Let $\theta : Q \to P$ be an integral morphism of fine monoids such that $\mathrm{Coker}(Q^{gp} \to P^{gp})$ is torsion free. Note that this implies that P^{gp} is torsion free as there is a short exact sequence

$$0 \longrightarrow Q^{gp} \overset{s}{\longrightarrow} P^{gp} \overset{t}{\longrightarrow} \mathrm{Coker}(Q^{gp} \to P^{gp}) \longrightarrow 0, \qquad (4.8.6.1)$$

where s is injective since $Q \to P$ is integral, and Q^{gp} is torsion free since Q is saturated with $Q^* = \{0\}$.

We then obtain a log smooth morphism of fine log schemes

$$\text{Spec}(P \to R[P]) \to \text{Spec}(Q \to R[Q]). \qquad (4.8.6.2)$$

Let $J \subset R[Q]$ be an ideal, and let

$$f : (\mathscr{X}, M_{\mathscr{X}}) \to (\mathscr{Y}, M_{\mathscr{Y}}) \qquad (4.8.6.3)$$

be the morphism of fine formal log schemes obtained by completing 4.8.6.2 along J.

Let $\bar{x} \to \mathscr{X}$ be a geometric point, $\mathscr{O}_{\mathscr{X},\bar{x}}$ the local ring at $\bar{x}$, and $N_{\bar{x}}$ the log structure on $\text{Spec}(\mathscr{O}_{\mathscr{X},\bar{x}})$ associated to the map $M_{\mathscr{X},\bar{x}} \to \mathscr{O}_{\mathscr{X},\bar{x}}$. Let $\widehat{\mathscr{O}}_{\mathscr{X},\bar{x}}$ denote the completion along the maximal ideal of the local ring $\mathscr{O}_{\mathscr{X},\bar{x}}$.

Lemma 4.8.7 *The log scheme* $(\text{Spec}(\mathscr{O}_{\mathscr{X},\bar{x}}), N_{\bar{x}})$ *is log regular in the sense of* [25, 2.1].

Proof. By [25, 3.1] and the openness of the log regular locus [25, 7.1], it suffices to show that the log scheme

$$(\text{Spec}(\widehat{\mathscr{O}}_{\mathscr{X},\bar{x}}), N_{\bar{x}}|_{\text{Spec}(\widehat{\mathscr{O}}_{\mathscr{X},\bar{x}})}) \qquad (4.8.7.1)$$

is log regular. But this log scheme is isomorphic to the scheme

$$\text{Spec}(\widehat{\mathscr{O}}_{\text{Spec}(R[P]),\bar{x}}) \qquad (4.8.7.2)$$

with log structure induced by the given map $P \to \widehat{\mathscr{O}}_{\text{Spec}(R[P]),\bar{x}}$. The lemma therefore follows from the fact that $\text{Spec}(P \to R[P])$ is log regular being log smooth over a regular ring [25, 8.2]. $\square$

Lemma 4.8.8 *Let* $U_{\bar{x}} \subset \text{Spec}(\mathscr{O}_{\mathscr{X},\bar{x}})$ *be the open set where* $N_{\bar{x}}$ *is trivial. Then*

$$M_{\mathscr{X},\bar{x}} = \{f \in \mathscr{O}_{\mathscr{X},\bar{x}}|\ f \text{ maps to a unit in } \Gamma(U_{\bar{x}}, \mathscr{O}_{U_{\bar{x}}})\}. \qquad (4.8.8.1)$$

Proof. This follows from the preceding lemma and [25, 11.6]. $\square$

Corollary 4.8.9 *Assume further that the natural map* $Q^{gp} \oplus_Q P \to P^{gp}$ *is an isomorphism. Then*

$$M_{\mathscr{X},\bar{x}} = \{f \in \mathscr{O}_{\mathscr{X},\bar{x}}|\ f \text{ maps to a unit in } \mathscr{O}_{\mathscr{X},\bar{x}} \otimes_{\mathbb{Z}[Q]} \mathbb{Z}[Q^{gp}]\}. \qquad (4.8.9.1)$$

Lemma 4.8.10 *With notation as in 4.8.2, let* $\omega \in S$ *be a simplex,* $(P \rtimes H_S)_\omega$ *the monoid obtained by inverting elements which lie in* $\text{Cone}(1, \omega)$, *and let* $(P \rtimes H_S)_{\omega,0}$ *denote the degree* 0 *elements of* $(P \rtimes H_S)_\omega$. *Then the natural map*

$$H_S^{gp} \oplus_{H_S} (P \rtimes H_S)_{\omega,0} \to (P \rtimes H_S)_{\omega,0}^{gp} \qquad (4.8.10.1)$$

is an isomorphism.

Proof. Since everything is invariant under the translation by X–action assume that $(1, 0) \in \text{Cone}(1, \omega) \subset \mathbb{X}$, in which case for any $(d, x) \in P$ we have

$$((d, x), 0) + ((d, -x), 0) = (2d(1, 0), 0) + h \qquad (4.8.10.2)$$

for some $h \in H_S$. $\square$

4.8.11. We can now complete the proof of 4.8.1.

That every object of the right side of 4.8.1.1 is in the essential image follows immediately from 4.5.20 and the fact that the map

$$\widetilde{\mathcal{Q}}(B) \to \varprojlim \widetilde{\mathcal{Q}}(B_n) \qquad (4.8.11.1)$$

is an equivalence of categories since $\widetilde{\mathcal{Q}}$ is algebraic (see for example [7, remark on p. 182]). Also, since the stack $\mathscr{A}_{g,1}^{\mathrm{Alex}}$ is algebraic, for any compatible family of objects

$$(G_n, M_{B_n}, \mathscr{P}_n, M_{\mathscr{P}_n}, L_{\mathscr{P}_n}, G_n\text{–action}) \in \varprojlim \mathscr{K}_g(B_n) \qquad (4.8.11.2)$$

the underlying compatible family $(G_n, \mathscr{P}_n, L_{\mathscr{P}_n}, G_n\text{–action})$ of polarized schemes with semi–abelian group action is obtained from a unique polarized scheme with group action

$$(G, \mathscr{P}, L_{\mathscr{P}}, G\text{–action}). \qquad (4.8.11.3)$$

Furthermore, since the stack $\mathcal{L}og_{(\mathrm{Spec}(\mathbb{Z}), \mathscr{O}_{\mathrm{Spec}(\mathbb{Z})}^*)}$ defined in [42, 1.1] is algebraic, the functor

$$(\text{fine log structures on } \mathrm{Spec}(B)) \to \varprojlim(\text{fine log structures on } \mathrm{Spec}(B_n)) \qquad (4.8.11.4)$$

is an equivalence of categories. Therefore the log structures M_{B_n} are obtained from a unique log structure M_B on $\mathrm{Spec}(B)$. Let $H_S \to B$ be a chart. Then to prove 4.8.1 it suffices to show that if $M_{\mathscr{P}}$ and $M'_{\mathscr{P}}$ are log structures on $\mathscr{P}$ under $M_B|_{\mathscr{P}}$ with an extension of the action of G defining an object of $\mathscr{K}_g(B)$, then any compatible family of isomorphisms $\epsilon_n : M_{\mathscr{P}}|_{\mathscr{P}_n} \to M'_{\mathscr{P}}|_{\mathscr{P}_n}$ under M_{B_n} and respecting the G action is induced by a unique isomorphism $\epsilon : M_{\mathscr{P}} \to M'_{\mathscr{P}}$. The uniqueness is immediate since $\mathscr{P}$ is proper so that ϵ is determined by the maps on stalks at geometric points in the closed fiber.

For the existence of ϵ, note that by the deformation theory 4.3.3, there exists a regular local ring R and a surjection $\delta : R[[H_S]] \to B$ such that $(G, \mathscr{P}, L_{\mathscr{P}}, G\text{–action})$ is obtained by base change from a family

$$(G^\dagger, \mathscr{P}^\dagger, L_{\mathscr{P}^\dagger}, G^\dagger\text{–action}) \qquad (4.8.11.5)$$

over $R[[H_S]]$. Let $J \subset R[[H_S]]$ denote the kernel of δ, let $\mathfrak{a} \subset R[[H_S]]$ denote the inverse image of $\mathfrak{m} \subset B$, and let $\mathscr{P}_n^\dagger$ denote the reduction of $\mathscr{P}^\dagger$ modulo $\mathfrak{a}^n$. Denote by $\widehat{\mathscr{P}^\dagger}$ the formal scheme obtained by taking the J–adic completion of $\mathscr{P}^\dagger$. Then by 4.8.5 there exist log structures $M_{\widehat{\mathscr{P}^\dagger}}$ and $M'_{\widehat{\mathscr{P}^\dagger}}$ on $\widehat{\mathscr{P}^\dagger}$ lifting $M_{\mathscr{P}}$ and $M'_{\mathscr{P}}$ and isomorphisms

$$\epsilon_n^\dagger : M_{\mathscr{P}_n^\dagger} \to M'_{\mathscr{P}_n^\dagger} \qquad (4.8.11.6)$$

reducing to the isomorphisms ϵ_n. Combining 4.8.9 and 4.8.10 we then obtain an isomorphism $\epsilon^\dagger : M_{\widehat{\mathscr{P}^\dagger}} \to M'_{\widehat{\mathscr{P}^\dagger}}$ inducing the isomorphisms $\epsilon_n^\dagger$. The reduction modulo J of $\epsilon^\dagger$ then gives the desired isomorphism $M_{\mathscr{P}} \to M'_{\mathscr{P}^\dagger}$.

Remark 4.8.12. In section 5.7 we will prove more general results about algebraization of log structures.

4.9 Completion of Proof of 4.6.2

Since both $\mathcal{K}_g$ and $\tilde{\mathcal{Q}}$ are stacks with respect to the étale topology, the following proposition completes the proof of 4.6.2.

Proposition 4.9.1. (i) *For any scheme T and object $F \in \mathcal{K}_g(T)$, there exists an étale cover $U \to T$ such that $F|_U \in \mathcal{K}_g(U)$ is in the essential image of $\tilde{\mathcal{Q}}(U) \to \mathcal{K}_g(U)$.*
(ii) *The functor $\tilde{\mathcal{Q}} \to \mathcal{K}_g$ is fully faithful.*

Proof. By a standard reduction we may assume that T is noetherian.
First we prove (i). Let

$$(G, M_T, \mathscr{P}, M_{\mathscr{P}}, L_{\mathscr{P}}, G\text{-action}) \tag{4.9.1.1}$$

denote the data corresponding to F. Forgetting the log structures we obtain a morphism $T \to \mathscr{A}_{g,1}^{\text{Alex}}$. By the description of the deformation rings of $\mathcal{K}_g$ provided by 4.3.3 (see also the proof of 4.5.17) this morphism in fact factors through a morphism $s : T \to \tilde{\mathcal{Q}}$. Let M_T' and $M_{\mathscr{P}}'$ denote the log structures on T and $\mathscr{P}$ respectively obtained by pulling back the log structures on $\tilde{\mathcal{Q}}$ and the universal scheme over this stack. We then obtain a second object of $\mathcal{K}_g(T)$

$$(G, M_T', \mathscr{P}, M_{\mathscr{P}}', L_{\mathscr{P}}, G\text{-action}) \tag{4.9.1.2}$$

with the same underlying polarized scheme with semi–abelian action as F. We claim that there exists a unique pair of isomorphisms of log structures

$$\iota : M_T \to M_T', \quad \sigma : M_{\mathscr{P}} \to M_{\mathscr{P}}', \tag{4.9.1.3}$$

where σ is compatible with the G–action. Note that by 4.4.2 and 4.4.6 we have already shown this when T is the spectrum of an Artinian local ring. This implies in particular that the pair (ι, σ) is unique if it exists.

To construct (ι, σ) in general we proceed by noetherian induction. So we assume that for any proper closed subscheme $Z \subset T$ there exists a pair (ι_Z, σ_Z) over Z. Then we must show that the pair (ι, σ) also exists over T.

To construct (ι, σ), we first construct the isomorphisms $\bar{\iota} : \overline{M}_T \to \overline{M}_T'$ and $\bar{\sigma} : \overline{M}_{\mathscr{P}} \to \overline{M}_{\mathscr{P}}'$.

For this recall [32, II.3.10] that if $j : U \hookrightarrow T$ is an open set with complement $i : Z \hookrightarrow T$, then the category of sheaves in T_{et} is equivalent to the category of triples (F_1, F_2, ϕ), where F_1 is a sheaf on Z_{et}, F_2 is a sheaf on U_{et}, and $\phi : F_1 \to i^* j_* F_2$ is a morphism of sheaves on Z. This equivalence is induced by the functor sending a sheaf F on T to the triple with $F_1 = i^* F$, $F_2 = j^* F$, and ϕ the map $i^* F \to i^* j_* j^* F$ induced by adjunction.

By the case of an Artinian local ring, we can find (ι, σ) over the generic points of T, and hence by a standard limit argument there exists a dense open set $j : U \subset T$ and a pair (ι_U, σ_U) over U. Let $i : Z \subset T$ be the complement of

U. By the noetherian induction hypothesis we also have isomorphisms (ι_Z, σ_Z) over Z (with the reduced structure). To construct $\bar{\iota}$ and $\bar{\sigma}$ it therefore suffices to show that the following diagrams commute

$$
\begin{array}{ccc}
i^*\overline{M}_T & \longrightarrow & i^*j_*j^*\overline{M}_T \\
{\scriptstyle \iota_Z}\downarrow & & \downarrow{\scriptstyle \iota_U} \\
i^*\overline{M}'_T & \longrightarrow & i^*j_*j^*\overline{M}'_T,
\end{array}
\qquad (4.9.1.4)
$$

$$
\begin{array}{ccc}
i^*_{\mathscr{P}}\overline{M}_{\mathscr{P}} & \longrightarrow & i^*_{\mathscr{P}}j_{\mathscr{P}*}j^*_{\mathscr{P}}\overline{M}_{\mathscr{P}} \\
{\scriptstyle \sigma_Z}\downarrow & & \downarrow{\scriptstyle \sigma_U} \\
i^*_{\mathscr{P}}\overline{M}'_{\mathscr{P}} & \longrightarrow & i^*_{\mathscr{P}}j_{\mathscr{P}*}j^*_{\mathscr{P}}\overline{M}'_{\mathscr{P}}.
\end{array}
\qquad (4.9.1.5)
$$

Here the horizontal arrows are the adjunction maps, and $j_{\mathscr{P}} : \mathscr{P}_U \subset \mathscr{P}$ and $i_{\mathscr{P}} : \mathscr{P}_Z \subset \mathscr{P}$ are the inclusions. To verify this we may work étale locally on T, and hence may assume that T is the spectrum of a strictly henselian local ring. Let $\widehat{T}$ denote the spectrum of the completion of $\Gamma(T, \mathscr{O}_T)$ at the maximal ideal. We then have a commutative diagram with cartesian squares

$$
\begin{array}{ccccc}
\widehat{U} & \xrightarrow{\hat{j}} & \widehat{T} & \xleftarrow{\hat{i}} & \widehat{Z} \\
{\scriptstyle \alpha_U}\downarrow & & \downarrow{\scriptstyle \alpha_T} & & \downarrow{\scriptstyle \alpha_Z} \\
U & \xrightarrow{j} & T & \xleftarrow{i} & Z.
\end{array}
\qquad (4.9.1.6)
$$

Let $M_{\widehat{T}}$ and $M'_{\widehat{T}}$ denote the pullbacks of M_T and M'_T respectively to $\widehat{T}$. We then obtain a commutative diagram

$$
\begin{array}{ccccc}
\hat{i}^*\overline{M}_{\widehat{T}} = \alpha_Z^* i^*\overline{M}_T & \longrightarrow & \alpha_Z^* i^*j_*j^*\overline{M}_T = \hat{i}^*\alpha_T^* j_*j^*\overline{M}_T & \xrightarrow{\gamma} & \hat{i}^*\hat{j}_*\hat{j}^*\overline{M}_{\widehat{T}} \\
{\scriptstyle \iota_Z}\downarrow & & \downarrow{\scriptstyle \iota_U} & & \downarrow{\scriptstyle \iota_{\widehat{U}}} \\
\hat{i}^*\overline{M}'_{\widehat{T}} = \alpha_Z^* i^*\overline{M}'_T & \longrightarrow & \alpha_Z^* i^*j_*j^*\overline{M}'_T = \hat{i}^*\alpha_T^* j_*j^*\overline{M}'_T & \xrightarrow{\gamma'} & \hat{i}^*\hat{j}_*\hat{j}^*\overline{M}'_{\widehat{T}},
\end{array}
$$
$$(4.9.1.7)$$

where $\iota_{\widehat{U}}$ is the pullback of ι_U. Since the morphism α_T is surjective, the maps γ and γ' are injective. It follows that to prove that 4.9.1.4 commutes it suffices to consider $\widehat{T}$. A similar argument shows that to prove that 4.9.1.5 commutes we may replace T by $\widehat{T}$. This reduces the construction of $\bar{\iota}$ and $\bar{\sigma}$ to the case when T is the spectrum of a complete noetherian local ring. In this case the result follows from 4.8.1.

So now we fix $(\bar{\iota}, \bar{\sigma})$ and construct (ι, σ). Let Q be the functor over $\mathscr{P}$–schemes which to any $f : S \to \mathscr{P}$ associates the set of isomorphisms of log structures $\lambda : f^*M_{\mathscr{P}} \to f^*M'_{\mathscr{P}}$ inducing $f^*\bar{\sigma} : f^*\overline{M}_{\mathscr{P}} \to f^*\overline{M}'_{\mathscr{P}}$.

Lemma 4.9.2 *The functor Q is an algebraic space separated over $\mathscr{P}$.*

Proof. The assertion is étale local on $\mathscr{P}$. We may therefore assume that there exists a fine monoid P and charts $\beta_1 : P \to M_{\mathscr{P}}$ and $\beta_2 : P \to M'_{\mathscr{P}}$ such that the diagram

$$
\begin{array}{ccc}
P & \overline{} & P \\
\bar{\beta}_1 \downarrow & & \downarrow \bar{\beta}_2 \\
\overline{M}_{\mathscr{P}} & \xrightarrow{\bar{\sigma}} & \overline{M}'_{\mathscr{P}}
\end{array}
\tag{4.9.2.1}
$$

commutes. Let $\alpha_1 : P \to \mathscr{O}_{\mathscr{P}}$ and $\alpha_2 : P \to \mathscr{O}_{\mathscr{P}}$ be the morphisms induced by β_1 and β_2. Then as in 3.4.4 Q is represented by the scheme

$$
\mathrm{Spec}_{\mathscr{P}}(\mathscr{O}_{\mathscr{P}}[P^{\mathrm{gp}}]/(\alpha_1(p)u_p = \alpha_2(p))_{p \in P},
\tag{4.9.2.2}
$$

where for $p \in P$ we write $u_p \in \mathscr{O}_{\mathscr{P}}[P^{\mathrm{gp}}]$ for the corresponding element of the monoid algebra. $\quad\square$

Let $\underline{\mathrm{Sec}}(Q/\mathscr{P})$ denote the functor over T which to any $S \to T$ associates the set of sections of the map $Q_S \to \mathscr{P}_S$ obtained by base change. As explained in [45, 1.5], the functor $\underline{\mathrm{Sec}}(Q/\mathscr{P})$ is an algebraic space. Also, the group scheme G acts naturally on $\underline{\mathrm{Sec}}(Q/\mathscr{P})$. This is because by the construction the isomorphism $\bar{\sigma}$ is compatible with the G–action, and hence there is a natural action of G on Q over $\mathscr{P}$. Precisely, if $f : S \to \mathscr{P}$ is a morphism, $\lambda \in Q(S)$ an isomorphism lifting σ, and $g \in G(T)$, then $\lambda^g \in Q(S)$ is the isomorphism $(t_g \circ f)^* M_{\mathscr{P}} \to (t_g \circ f)^* M'_{\mathscr{P}}$ making the diagram

$$
\begin{array}{ccc}
(t_g \circ f)^* M_{\mathscr{P}} & \xrightarrow{\simeq} & f^* M_{\mathscr{P}} \\
\lambda^g \downarrow & & \downarrow \lambda \\
(t_g \circ f)^* M'_{\mathscr{P}} & \xrightarrow{\simeq} & f^* M'_{\mathscr{P}}
\end{array}
\tag{4.9.2.3}
$$

commute. It follows that the group scheme G also acts on $\underline{\mathrm{Sec}}(Q/\mathscr{P})$. Define I to be the fiber product of the diagram

$$
\begin{array}{c}
\underline{\mathrm{Sec}}(Q/\mathscr{P}) \\
\Delta \downarrow \\
G \times \underline{\mathrm{Sec}}(Q/\mathscr{P}) \xrightarrow{\mathrm{pr}_2 \times \rho} \underline{\mathrm{Sec}}(Q/\mathscr{P}) \times_T \underline{\mathrm{Sec}}(Q/\mathscr{P}),
\end{array}
\tag{4.9.2.4}
$$

where ρ denotes the action. Then we wish to show that the map $I \to T$ is an isomorphism. Since I is an algebraic space, this follows from the fact that for any artinian local ring B the map $I(B) \to S(B)$ is an isomorphism. We thus obtain the morphism σ.

To obtain the morphism ι, choose (after replacing T by an étale cover) charts $\beta : H \to M_T$ and $\beta' : H \to M'_T$ such that the diagram

$$
\begin{array}{ccc}
H & \overline{} & H \\
\bar{\beta} \downarrow & & \downarrow \bar{\beta}' \\
\overline{M}_T & \xrightarrow{\iota} & \overline{M}'_T
\end{array}
\tag{4.9.2.5}
$$

commutes. Then for any $h \in H$ there exists a unique unit $u(h) \in \Gamma(\mathscr{P}, \mathcal{O}^*_{\mathscr{P}})$ such that the image of h in $M_{\mathscr{P}}$ maps under σ to $\lambda(u(h))$ plus the image of h in $M'_{\mathscr{P}}$ under the composite $H \to M'_T \to M'_{\mathscr{P}}$. Since $\Gamma(\mathscr{P}, \mathcal{O}^*_{\mathscr{P}}) = \Gamma(T, \mathcal{O}^*_T)$ this defines the isomorphism ι. This completes the proof of (i).

To prove (ii), let $o_1, o_2 \in \tilde{\mathcal{Q}}(T)$ be two objects over some scheme T. We then obtain inclusions of sheaves on the big étale site of T

$$\underline{\mathrm{Isom}}_{\tilde{\mathcal{Q}}}(o_1, o_2) \subset \underline{\mathrm{Isom}}_{\mathscr{K}_g}(o_1, o_2) \subset \underline{\mathrm{Isom}}_{\mathscr{A}^{\mathrm{Alex}}_{g,1}}(o_1, o_2), \qquad (4.9.2.6)$$

where we abusively also write o_i for the image of o_i in $\mathscr{K}_g$ and $\mathscr{A}^{\mathrm{Alex}}_{g,1}$ (to verify that the second map is inclusion note that it suffices to do so for scheme-valued points over artinian local rings where it follows from the results of section 4.4). We then wish to show that the first inclusion is an isomorphism. For this it suffices to show that if $\iota \in \underline{\mathrm{Isom}}_{\mathscr{A}^{\mathrm{Alex}}_{g,1}}(o_1, o_2)(T)$ is a section in the image of $\underline{\mathrm{Isom}}_{\mathscr{K}_g}(o_1, o_2)(T)$ then in fact ι lies in $\underline{\mathrm{Isom}}_{\tilde{\mathcal{Q}}}(o_1, o_2)(T)$. Since both $\underline{\mathrm{Isom}}_{\tilde{\mathcal{Q}}}(o_1, o_2)$ and $\underline{\mathrm{Isom}}_{\mathscr{A}^{\mathrm{Alex}}_{g,1}}(o_1, o_2)$ are representable by schemes over T, to verify this it suffices to consider the case when T is an artinian local ring.

Let $\mathscr{W}_{3,n}$ be as in 4.5.14, and let $(\mathscr{W}^{\mathrm{sat}}_{3,n}, M_{\mathscr{W}^{\mathrm{sat}}_{3,n}})$ denote the saturation of the log scheme $(\mathscr{W}_{3,n}, M_{\mathscr{W}_{3,n}})$.$=$ (so by 4.5.20 we have a locally closed immersion $\mathscr{W}_{3,n} \hookrightarrow \tilde{\mathcal{Q}}$).

We then obtain a commutative diagram

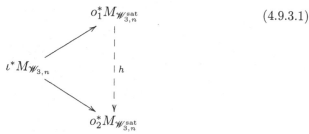

$$(4.9.2.7)$$

The following lemma therefore completes the proof of 4.9.1 as it shows that to give an isomorphism between o_1 and o_2 in $\mathscr{K}_g$ is precisely equivalent to giving an isomorphism in $\tilde{\mathcal{Q}}$. $\square$

Lemma 4.9.3 *There exists an isomorphism $h : o_1^* M_{\mathscr{W}^{\mathrm{sat}}_{3,n}} \to o_2^* M_{\mathscr{W}^{\mathrm{sat}}_{3,n}}$ filling in the diagram*

$$o_1^* M_{\mathscr{W}^{\mathrm{sat}}_{3,n}} \qquad (4.9.3.1)$$

$$\iota^* M_{\mathscr{W}_{3,n}} \qquad \downarrow h$$

$$o_2^* M_{\mathscr{W}^{\mathrm{sat}}_{3,n}}$$

if and only if ι is induced by an isomorphism in $\mathscr{W}^{\mathrm{sat}}_{3,n}$, in which case h is unique.

Proof. The uniqueness of h is clear as the maps $\iota^* M_{\mathscr{W}_{3,n}} \to o_i^* M_{\mathscr{W}_{3,n}^{\mathrm{sat}}}$ identify the sheaves $o_i^* M_{\mathscr{W}_{3,n}^{\mathrm{sat}}}$ with the saturation of $\iota^* M_{\mathscr{W}_{3,n}}$. It is also clear that if ι is induced by an isomorphism in $\mathscr{W}_{3,n}$ then there is a map h filling in the diagram.

Conversely, assume that h exists and let $(T^{\mathrm{sat}}, M_{\mathrm{sat}})$ denote the saturation of the log scheme $(T, \iota^* M_{\mathscr{W}_{3,n}})$. Then the diagram

$$
\begin{array}{ccc}
T^{\mathrm{sat}} & \longrightarrow & \mathscr{W}_{3,n}^{\mathrm{sat}} \\
\downarrow & & \downarrow \\
T & \longrightarrow & \mathscr{W}_{3,n}
\end{array}
\qquad (4.9.3.2)
$$

is cartesian, so the maps o_i are obtained from a unique set of sections $s_i :$ $T \to T^{\mathrm{sat}}$. The existence of h then implies that we can extend these sections to obtain a commutative diagram of log schemes

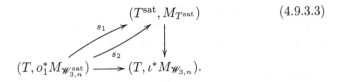

$$(4.9.3.3)$$

By the universal property of the saturation of a log scheme it follows that $s_1 = s_2$. $\square$

5

Moduli of Abelian Varieties with Higher Degree Polarizations

Fix integers $g, d \geq 1$ and let $\mathscr{A}_{g,d}$ denote the moduli stack associating to a scheme T the groupoid of pairs (A, λ), where A is an abelian scheme over T and $\lambda : A \to A^t$ is a polarization of degree d. Recall that this means that the kernel of λ is a finite flat group scheme over T of rank d^2, and that fppf locally on T there exists an ample line bundle L on A such that the map

$$\lambda_L : A \to A^t, \quad a \mapsto [t_a^* L \otimes L^{-1}]$$

is equal to λ. In this case if $f : A \to T$ is the structural morphism then $f_* L$ is locally free of rank d on T and its formation commutes with arbitrary base change on T (see for example [36, I, §1] for a summary of basic properties of ample line bundles on abelian varieties).

The stack $\mathscr{A}_{g,d}$ is a Deligne–Mumford stack over $\mathbb{Z}$. Our aim in this section is to modify the techniques of the previous section to give a compactification $\mathscr{A}_{g,d} \hookrightarrow \overline{\mathscr{A}}_{g,d}$. For $d = 1$ this compactification agrees with the one constructed in the previous chapter (though as mentioned in the introduction the moduli interpretation is different, as reflected in the notation).

5.1 Rethinking $\mathscr{A}_{g,d}$

5.1.1. Let $\mathscr{T}_{g,d}$ denote the stack over $\mathbb{Z}$ associating to any scheme T the groupoid of triples (A, P, L), where A is an abelian scheme of dimension g over T, $f : P \to T$ is an A-torsor, and L is an ample invertible sheaf on P such that $f_* L$ is locally free of rank d on T.

Note that for such a triple $(A, P, L)/T$ we obtain by 3.1.1 a morphism

$$\lambda_L : A \to A^t. \tag{5.1.1.1}$$

The kernel of this morphism is a finite flat group scheme over T of rank d^2. Indeed this can be verified fppf locally on T when P is trivial where it follows

M.C. Olsson, *Compactifying Moduli Spaces for Abelian Varieties*. Lecture Notes in Mathematics 1958.
© Springer-Verlag Berlin Heidelberg 2008

from the theory of abelian varieties. We will denote this group scheme by $H(P, L)$, or just $H(L)$ if no confusion seems likely to arise.

Let $\mathscr{G}_{(A,P,L)}$ denote the functor which to any T–scheme $T' \to T$ associates the group of automorphisms of the base change $(P_{T'}, L_{T'})$ which commute with the action of $A_{T'}$. We refer to $\mathscr{G}_{(A,P,L)}$ as the *theta group* of (A, P, L).

Lemma 5.1.2 *For any section* $\alpha \in \mathscr{G}_{(A,P,L)}(T)$ *there exists a unique section* $x \in H(L)(T)$ *such that the automorphism* $\alpha_P : P \to P$ *is given by translation by* x.

Proof. By the uniqueness part of the lemma and descent theory, we may work fppf–locally on T and may therefore assume that P is trivial. In this case α_P is given by an automorphism of A (not necessarily respecting the group structure) which commutes with translation. By [34, 6.4] there exists a section $x \in A$ such that $\alpha_P = t_x \circ h$, where h is a homomorphism and t_x denotes translation by $x \in A(B)$ (note that x is uniquely defined by α_P as $x = \alpha_P(e)$). Since α_P commutes with the A-action, we have for any sections $y, z \in A(T)$ an equality

$$z + x + h(y) = z + \alpha_P(y) = \alpha_P(y + z) = x + h(y + z) = x + h(y) + h(z). \quad (5.1.2.1)$$

This implies that $h(z) = z$ and hence h is the identity. It follows that $\alpha_P = t_x$ for a unique section $x \in A(B)$. And since $t_x^* L$ is isomorphic to L this section x must be an element of $H(L)$. □

5.1.3. By the lemma there is a canonical homomorphism $\mathscr{G}_{(A,P,L)} \to H(L)$. This morphism is surjective. Indeed by the definition of $H(L)$, for any $x \in H(L)$ there exists fppf locally on T an isomorphism $t_x^* L \simeq L$. The choice of such an isomorphism defines a lifting of x to $\mathscr{G}_{(A,P,L)}$. It follows that the group scheme $\mathscr{G}_{(A,P,L)}$ sits in a short exact sequence

$$1 \to \mathbb{G}_m \to \mathscr{G}_{(A,P,L)} \to H(L) \to 1. \quad (5.1.3.1)$$

Let $\underline{\mathrm{Aut}}_{\mathscr{I}_{g,d}}(A, P, L)$ denote the functor of automorphisms of the triple (A, P, L), and let $\underline{\mathrm{Aut}}(A, \lambda)$ denote the group scheme of automorphisms $\sigma : A \to A$ of the abelian scheme A such that the diagram

$$
\begin{array}{ccc}
A & \xrightarrow{\sigma} & A \\
\lambda_L \downarrow & & \downarrow \lambda_L \\
A^t & \xleftarrow{\sigma^*} & A^t
\end{array}
\qquad (5.1.3.2)
$$

commutes. There is a natural short exact sequence

$$1 \to \mathscr{G}_{(A,P,L)} \to \underline{\mathrm{Aut}}_{\mathscr{I}_{g,d}}(A, P, L) \to \underline{\mathrm{Aut}}(A, \lambda). \quad (5.1.3.3)$$

Proposition 5.1.4. *The natural functor*

$$F : \mathscr{T}_{g,d} \to \mathscr{A}_{g,d}, \quad (A, P, L) \mapsto (A, \lambda_L) \tag{5.1.4.1}$$

induces an isomorphism between $\mathscr{A}_{g,d}$ and the rigidification of $\mathscr{T}_{g,d}$ with respect to the normal subgroups $\mathscr{G}_{(A,P,L)} \subset \underline{\mathrm{Aut}}_{\mathscr{T}_{g,d}}(A, P, L)$.

Proof. For any scheme T and point $x \in \mathscr{A}_{g,d}(T)$, there exists evidently an fppf cover of T such that x is in the essential image of F. It therefore suffices to show that F induces a fully faithful functor from the rigidication of $\mathscr{T}_{g,d}$. For this in turn it suffices to show that for any two objects $(A_i, P_i, L_i) \in \mathscr{T}_{g,d}(T)$ $(i = 1, 2)$ the natural map

$$\mathscr{G}_{(A_1,P_1,L_1)} \backslash \underline{\mathrm{Isom}}_{\mathscr{T}_{g,d}}((A_1, P_1, L_1), (A_2, P_2, L_2))$$

$$\downarrow \tag{5.1.4.2}$$

$$\underline{\mathrm{Isom}}_{\mathscr{A}_{g,d}}((A_1, \lambda_{L_1}), (A_2, \lambda_{L_2}))$$

is an isomorphism of sheaves for the fppf topology. This map is injective by the definition of $\mathscr{G}_{(A_1,P_1,L_1)}$. Thus what remains to show is that any isomorphism $h : A_1 \to A_2$ such that the diagram

$$
\begin{array}{ccc}
A_1 & \xrightarrow{\ h\ } & A_2 \\
{\scriptstyle \lambda_{L_1}}\downarrow & & \downarrow{\scriptstyle \lambda_{L_2}} \\
A_1^t & \xleftarrow{\ h^*\ } & A_2^t
\end{array}
\tag{5.1.4.3}
$$

commutes extends fppf locally on T to an isomorphism $(A_1, P_1, L_1) \to (A_2, P_2, L_2)$. To prove this we may as well assume that P_1 and P_2 are trivial and can assume that T is the spectrum of a strictly henselian local ring. The commutativity of 5.1.4.3 then implies that for every scheme–valued point $x \in A_1$

$$t_x^* h^* L_2 \otimes h^* L_2^{-1} \simeq h^*(t_{h(x)}^* L_2 \otimes L_2^{-1}) \simeq t_x^* L_1 \otimes L_1^{-1}. \tag{5.1.4.4}$$

Rearranging the equation we find that

$$t_x^*(h^* L_2 \otimes L_1^{-1}) \simeq h^* L_2 \otimes L_1^{-1}. \tag{5.1.4.5}$$

This implies that $h^* L_2 \otimes L_1^{-1}$ is isomorphic to $t_a^* L_1 \otimes L_1^{-1}$ for some section $a \in A_1$ (see for example [35, Chapter II, §8]). Therefore $h^* L_2 \simeq t_a^* L_1$ or equivalently $t_{-a}^* h^* L_2 \simeq L_1$ for some $a \in A_1$. Fixing one such isomorphism ι we obtain an isomorphism in $\mathscr{T}_{g,d}$

$$h : A_1 \to A_2, \quad h \circ t_{-a} : P_1 \to P_2, \quad \iota : t_{-a}^* h^* L_2 \simeq L_1. \tag{5.1.4.6}$$

$\square$

The proof also shows the following:

Corollary 5.1.5 *The exact sequence 5.1.3.3 of fppf sheaves is also exact on the right.*

5.1.6. Proposition 5.1.4 is the starting point for our construction of a compactification of $\mathscr{A}_{g,d}$. Let us briefly outline the construction. We will first construct an open dense embedding $\mathscr{T}_{g,d} \subset \overline{\mathscr{T}}_{g,d}$, where $\overline{\mathscr{T}}_{g,d}$ classifies certain log schemes together with a line bundle and action of a semiabelian group scheme. The tautological theta group $\mathscr{G}$ over $\mathscr{T}_{g,d}$ extends canonically to $\overline{\mathscr{T}}_{g,d}$ and we construct $\overline{\mathscr{A}}_{g,d}$ as the rigidification of $\overline{\mathscr{T}}_{g,d}$ with respect to this extension of the theta group. We then show that $\overline{\mathscr{A}}_{g,d}$ is proper over $\mathbb{Z}$ with finite diagonal, and describe its local structure.

5.2 The Standard Construction

In this section we give a construction which will describe the geometric points of $\overline{\mathscr{T}}_{g,d}$ as well as the local rings of the boundary.

5.2.1. Let B be a base scheme and X a lattice of rank r dual to a torus T_X. We fix the following data:

(i) An inclusion $\phi : Y \to X$ of a sublattice with X/Y a finite group of order d_1.

(ii) An abelian scheme A of relative dimension $g - r$ over B together with an ample rigidified line bundle $\mathscr{M}$ defining a polarization of degree d_2 on A such that $d_1 d_2 = d$.

(iii) A homomorphism $c : X \to A^t$ corresponding as in 2.1.8 to an extension

$$0 \to T_X \to G \to A \to 0. \tag{5.2.1.1}$$

For $x \in X$ we write L_x for the rigidified line bundle on A corresponding to $c(x)$.

(iv) A homomorphism $c^t : Y \to A$ such that the diagram

$$
\begin{array}{ccc}
Y & \xrightarrow{\phi} & X \\
{\scriptstyle c^t}\downarrow & & \downarrow{\scriptstyle c} \\
A & \xrightarrow{\lambda_{\mathscr{M}}} & A^t
\end{array}
\tag{5.2.1.2}
$$

commutes.

(v) An integral regular X-periodic paving S of $X_{\mathbb{R}}$.

(vi) A trivialization τ of the biextension $(c^t \times c)^* \mathscr{B}^{-1}$ such that the induced trivialization of $(c^t \times c \circ \phi)^* \mathscr{B}^{-1}$ over $Y \times Y$ is compatible with the symmetric structure.

(vii) A trivialization ψ of $(c^t)^* \mathscr{M}^{-1}$ over Y inducing (as in 2.2.10) $\tau|_{Y \times Y}$.

Define $\widetilde{H}_S$ and H_S as in 4.1. Recall that by 4.1.6 the monoid H_S is finitely generated.

Lemma 5.2.2 *For any $y \in Y$ and $x \in X$ the sheaf $t^*_{c^t(y)}(\mathscr{M}^d \otimes L_x)$ is étale locally on B (non–canonically) isomorphic to $\mathscr{M}^d \otimes L_{x+d\phi(y)}$.*

Proof. By the theorem of the square, for $a, b \in A$ we have

$$t^*_a\mathscr{M} \otimes t^*_b\mathscr{M} \simeq t^*_{a+b}\mathscr{M} \otimes \mathscr{M},$$

after possibly replacing B by an étale covering. Using this and induction we have

$$t^*_{c^t(y)}(\mathscr{M}^d) \simeq (t^*_{dc^t(y)}\mathscr{M}) \otimes \mathscr{M}^{d-1}.$$

Since 5.2.1.2 commutes, we have

$$L_{x+d\phi(y)} = L_x \otimes L_{d\phi(y)} = L_x \otimes t^*_{dc^t(y)}\mathscr{M} \otimes \mathscr{M}^{-1}$$

so

$$\mathscr{M}^d \otimes L_{x+d\phi(y)} \simeq \mathscr{M}^{d-1} \otimes L_x \otimes t^*_{dc^t(y)}\mathscr{M}. \qquad (5.2.2.1)$$

Also

$$
\begin{aligned}
t^*_{c^t(y)}(\mathscr{M}^d \otimes L_x) &\simeq t^*_{c^t(y)}(\mathscr{M}^d) \otimes t^*_{c^t(y)}L_x \\
&\simeq t^*_{dc^t(y)}(\mathscr{M}) \otimes \mathscr{M}^{d-1} \otimes t^*_{c^t(y)}L_x.
\end{aligned}
$$

Since $t^*_{c^t(y)}L_x$ is isomorphic to L_x this gives the lemma. $\square$

5.2.3. To give an isomorphism

$$t^*_{c^t(y)}(\mathscr{M}^d \otimes L_x) \to \mathscr{M}^d \otimes L_{x+d\phi(y)}$$

is therefore equivalent to giving a section of

$$(\mathscr{M}^d \otimes L_x)(c^t(y)).$$

Therefore the section $\psi(y)^d\tau(y,x)$ defines a morphism, which we denote by the same letter,

$$\psi(y)^d\tau(y,x) : t^*_{c^t(y)}(\mathscr{M} \otimes L_x) \to \mathscr{M}^d \otimes L_{x+d\phi(y)}. \qquad (5.2.3.1)$$

5.2.4. Let $\beta : H_S \to \mathscr{O}_B$ be a morphism sending the nonzero elements of H_S to the nilradical of $\mathscr{O}_B$. Define P to be the integral points of $\mathrm{Cone}(1, X_{\mathbb{R}}) \subset \mathbb{R} \oplus X_{\mathbb{R}}$, and define

$$\mathcal{R} = \oplus_{(d,x)\in P}\mathscr{M}^d \otimes L_x$$

to be the quasi–coherent sheaf of algebras with algebra structure defined as in 4.1.10.2. We let $\widetilde{\mathscr{P}}$ denote the scheme $\underline{\mathrm{Proj}}_A(\mathcal{R})$. As in 4.1.10 there is a natural log structure $M_{\widetilde{\mathscr{P}}}$ so that we have a log smooth integral morphism

$$(\widetilde{\mathscr{P}}, M_{\widetilde{\mathscr{P}}}) \to (B, M_B), \tag{5.2.4.1}$$

where M_B denotes the log structure associated to β. As in 4.1.14, there is a natural action of G on $(\widetilde{\mathscr{P}}, M_{\widetilde{\mathscr{P}}})$ over (B, M_B) and the maps 5.2.3.1 define an action of Y on $\widetilde{\mathscr{P}}$ commuting with the G-action, together with a natural lifting of the Y-action to the line bundle $\mathscr{O}_{\widetilde{\mathscr{P}}}(1)$. Taking the quotient by the Y-action we obtain a log smooth integral morphism

$$(\mathscr{P}, M_{\mathscr{P}}) \to (B, M_B),$$

a G-action on $(\mathscr{P}, M_{\mathscr{P}})$ over (B, M_B), and a relatively ample invertible sheaf $L_{\mathscr{P}}$ on $\mathscr{P}$. We refer to this collection of data as the *standard family*.

Remark 5.2.5. As in 4.1.21, the actions of Y and T_X on the line bundle $\mathscr{O}_{\widetilde{\mathscr{P}}}(1)$ do not commute but rather satisfy the "Heisenberg relation": For sections $y \in X$ and $u \in T(B)$ with resulting automorphisms T_y and S_u of $(\widetilde{\mathscr{P}}, M_{\widetilde{\mathscr{P}}}, \mathscr{O}_{\widetilde{\mathscr{P}}}(1))$, we have

$$u(y) \circ S_u \circ T_y = T_y \circ S_u, \tag{5.2.5.1}$$

where $\mathbb{G}_m$ acts on $(\widetilde{\mathscr{P}}, M_{\widetilde{\mathscr{P}}}, \mathscr{O}_{\widetilde{\mathscr{P}}}(1))$ via multiplication on $\mathscr{O}_{\widetilde{\mathscr{P}}}(1)$.

Proposition 5.2.6. *Let $f : \mathscr{P} \to B$ denote the structural morphism. Then $f_* L_{\mathscr{P}}$ is locally free of rank d on B and the formation of $f_* L_{\mathscr{P}}$ commutes with arbitrary base change.*

Proof. We can without loss of generality assume that B is affine and look at global cohomology.

It follows from the construction that

$$H^0(\mathscr{P}, L_{\mathscr{P}}) \simeq H^0(\widetilde{\mathscr{P}}, \mathscr{O}_{\widetilde{\mathscr{P}}}(1))^Y. \tag{5.2.6.1}$$

If $\pi : \widetilde{\mathscr{P}} \to A$ is the projection, then by the same argument used in the proof of 4.3.6 the sheaf $\pi_* \mathscr{O}_{\widetilde{\mathscr{P}}}(1)$ is isomorphic to $\prod_{x \in X} \mathscr{M} \otimes L_x$. It follows that

$$H^0(\widetilde{\mathscr{P}}, \mathscr{O}_{\widetilde{\mathscr{P}}}(1))^Y = \left(\prod_{x \in X} H^0(A, \mathscr{M} \otimes L_x) \right)^Y. \tag{5.2.6.2}$$

Thus choosing liftings $x_i \in X$ of the elements of X/Y we see that $H^0(\mathscr{P}, L_{\mathscr{P}})$ is noncanonically isomorphic to

$$\prod_{x_i} H^0(A, \mathscr{M} \otimes L_{x_i}). \tag{5.2.6.3}$$

From this and standard base change properties of ample sheaves on abelian varieties the proposition follows. $\square$

Remark 5.2.7. The isomorphism

$$H^0(\mathscr{P}, L_\mathscr{P}) \simeq \left(\prod_{x \in X} H^0(A, \mathscr{M} \otimes L_x)\right)^Y \tag{5.2.7.1}$$

shows that in the case when B is a field, there is a Zariski dense open subset of $H^0(\mathscr{P}, L_\mathscr{P})$ consisting of sections whose zero loci do not contain any G-orbits (see [3, 5.2.7]). Furthermore if $\mathscr{M}$ is very ample on A then it also shows that the adjunction map $f^* f_* L_\mathscr{P} \to L_\mathscr{P}$ is surjective.

5.2.8. Fix positive integers g and d. Let $\overline{\mathscr{T}}_{g,d}$ be the stack over $\mathbb{Z}$ which to any scheme B associates the groupoid of data

$$(G, M_B, f : (\mathscr{P}, M_\mathscr{P}) \to (B, M_B), L_\mathscr{P})$$

such that:

   (i) G is a semi-abelian scheme of relative dimension g over B.
   (ii) M_B is a fine log structure on B.
   (iii) f is log smooth and G acts on $(\mathscr{P}, M_\mathscr{P})$ over (B, M_B).
   (iv) For every geometric point $\bar{b} \to B$, the fiber $(G_{\bar{b}}, \mathscr{P}_{\bar{b}}, M_{\mathscr{P}_{\bar{b}}}, L_{\mathscr{P}_{\bar{b}}})$ is isomorphic to the collection of data obtained from the saturation of the standard construction. Note in particular that this implies that $f_* L_\mathscr{P}$ is a locally free sheaf whose formation commutes with arbitrary base change on B.
   (v) The locally free sheaf $f_* L_\mathscr{P}$ has rank d.

In what follows we will show that a certain rigidification of $\overline{\mathscr{T}}_{g,d}$ provides a compactification of $\mathscr{A}_{g,d}$.

5.2.9. Already we can see that once we show that $\overline{\mathscr{T}}_{g,d}$ is algebraic then the substack $\mathscr{T}_{g,d} \subset \overline{\mathscr{T}}_{g,d}$ will be dense.

For this let R be a complete local ring and fix data as in 5.2.1 for $B = \mathrm{Spec}(R)$. Also fix a map $\beta : H_S^{\mathrm{sat}} \to R$ sending all nonzero elements to the maximal ideal. The map β defines a surjection $V[[H_S^{\mathrm{sat}}]] \to R$ where $V[[H_S^{\mathrm{sat}}]]$ is the completion of the monoid algebra $V[H_S^{\mathrm{sat}}]$ along the ideal defined by the map to $R/\mathfrak{m}_R$. We can then lift the data (i)-(vii) in 5.2.1 to $V[[H_S^{\mathrm{sat}}]]$ simply by pulling back along the retraction $V \to V[[H_S^{\mathrm{sat}}]]$. Applying the standard construction over $V[[H_S^{\mathrm{sat}}]]$ we obtain a compatible family of polarized projective schemes (P_n, L_n) with action of a semi-abelian group scheme G_n over the reductions of $V[[H_S^{\mathrm{sat}}]]$. By Grothendieck's existence theorem this family is uniquely algebraizable to a proper polarized scheme (P, L) over $V[[H_S^{\mathrm{sat}}]]$. Furthermore, by [13, Chapter III, section 4] the group schemes G_n are uniquely algebraizable to a semiabelian group scheme G over $V[[H_S^{\mathrm{sat}}]]$ which acts on P. The generic fiber of this polarized scheme with G-action then gives a point of $\mathscr{T}_{g,d}$ specializing to our given point of $\overline{\mathscr{T}}_{g,d}$.

5.3 Another Interpretation of $\widetilde{\mathscr{P}} \to \mathscr{P}$

5.3.1. Let B be a scheme, and fix data as in 5.2.1 defining the standard family

$$(\mathscr{P}, M_{\mathscr{P}}, L_{\mathscr{P}}) \to (B, M_B). \tag{5.3.1.1}$$

Let $f : \mathscr{P} \to B$ denote the projection. The morphism f is flat since it is the underlying morphism of an integral log smooth morphism 2.3.19. Also by the same argument proving 4.3.4 one sees that the map $\mathscr{O}_B \to f_*\mathscr{O}_{\mathscr{P}}$ is an isomorphism (and the same holds after arbitrary base change $B' \to B$). Let $\mathscr{P}ic(\mathscr{P})$ be the fibered category over B which to any $B' \to B$ associates the groupoid of line bundles on $\mathscr{P}_{B'}$. By [7, Appendix], the stack $\mathscr{P}ic(\mathscr{P})$ is an algebraic stack over B and in fact a $\mathbb{G}_m$–gerbe over the relative Picard functor $\underline{\mathrm{Pic}}_{\mathscr{P}/B}$. The stack $\mathscr{P}ic(\mathscr{P})$ is in fact a Picard stack (see [8, XVIII] for the notion of a Picard stack).

We also consider the stack $\mathrm{TORS}_{\mathscr{P}}(X)$ which to any scheme $B' \to B$ associates the groupoid of X–torsors over $\mathscr{P}_{B'}$. The stack $\mathrm{TORS}_{\mathscr{P}}(X)$ is also a Picard stack over B.

For any X–torsor $\widetilde{\mathscr{P}} \to \mathscr{P}$, we obtain a functor

$$F_{\widetilde{\mathscr{P}}} : T \to \mathscr{P}ic(\mathscr{P}), \tag{5.3.1.2}$$

by associating to any $u \in T = \mathrm{Hom}(X, \mathbb{G}_m)$ the $\mathbb{G}_m$-torsor

$$\widetilde{\mathscr{P}} \times^{X,u} \mathbb{G}_m. \tag{5.3.1.3}$$

In fact $F_{\widetilde{\mathscr{P}}}$ is naturally a morphism of Picard stacks (where T is viewed as a discrete Picard stack with no nontrivial automorphisms). In this way we obtain a morphism of Picard stacks

$$F : \mathrm{TORS}_{\mathscr{P}}(X) \to \mathrm{HOM}(T, \mathscr{P}ic(\mathscr{P})), \tag{5.3.1.4}$$

where the right side of 5.3.1.4 denotes the Picard stack of morphisms of Picard stacks $T \to \mathscr{P}ic(\mathscr{P})$ (see [8, XVIII.1.4.7] for the definition of a morphism of Picard stacks).

Theorem 5.3.2 *The morphism 5.3.1.4 is an equivalence of Picard stacks.*

Proof. The map 5.3.1.4 induces a morphism of fppf sheaves on B

$$R^1f_*X \to \mathscr{H}om(T, \underline{\mathrm{Pic}}_{\mathscr{P}/B}). \tag{5.3.2.1}$$

By [3, 4.2.4] this map is an isomorphism. From this it follows in particular that any homomorphism $T \to \underline{\mathrm{Pic}}_{\mathscr{P}/B}$ lifts fppf locally on B to a morphism of Picard stacks $T \to \mathscr{P}ic(\mathscr{P})$. Therefore $\mathrm{HOM}(T, \mathscr{P}ic(\mathscr{P}))$ is a $\mathrm{Hom}(T, \mathbb{G}_m) = X$–gerbe over $\mathscr{H}om(T, \underline{\mathrm{Pic}}_{\mathscr{P}/B})$.

Similarly, $\mathrm{TORS}_{\mathscr{P}}(X)$ is a X–gerbe over R^1f_*X, and the morphism 5.3.1.4 is a morphism of X-gerbes over the isomorphism 5.3.2.1. From this it follows that 5.3.1.4 is an isomorphism. $\square$

Proposition 5.3.3. *Let* $\pi : \widetilde{\mathcal{P}} \times^Y X \to \mathcal{P}$ *be the* X*-torsor obtained by pushing forward along* $Y \subset X$ *the* Y*-torsor* $\widetilde{\mathcal{P}} \to \mathcal{P}$ *defined in the standard construction. Then* $\widetilde{\mathcal{P}} \times^Y X$ *maps under* F *to the morphism of Picard stacks* $T \to \mathcal{P}ic(\mathcal{P})$ *sending* $u \in T$ *to* $u^* L_{\mathcal{P}} \otimes L_{\mathcal{P}}^{-1}$.

Proof. By construction the action of T on $\widetilde{\mathcal{P}}$ lifts to an action on $\pi^* L_{\mathcal{P}}$. This lifting of the action defines for any $u \in T$ a trivialization of $\pi^*(u^* L_{\mathcal{P}} \otimes L_{\mathcal{P}}^{-1})$. This defines an isomorphism of $\mathbb{G}_m$-torsors

$$\widetilde{\mathcal{P}} \times \mathbb{G}_m \to \widetilde{\mathcal{P}} \times_{\mathcal{P}} (u^* L_{\mathcal{P}} \otimes L_{\mathcal{P}}^{-1}). \tag{5.3.3.1}$$

If we let Y act on the left side by

$$y * (\tilde{p}, w) = (T_y(\tilde{p}), u(y) \cdot w), \tag{5.3.3.2}$$

then this isomorphism is even compatible with the Y-actions because of the Heisenberg relation 5.2.5. Taking the quotient by the Y-action we obtain an isomorphism

$$\widetilde{\mathcal{P}} \times^{Y,u} \mathbb{G}_m \simeq u^* L_{\mathcal{P}} \otimes L_{\mathcal{P}}^{-1}. \tag{5.3.3.3}$$

$\square$

5.3.4. As in 4.3.19, we can recover from

$$(\mathcal{P}, L_{\mathcal{P}}, G - \text{action}) \tag{5.3.4.1}$$

the paving S, the subgroup $Y \subset X$, and the maps $c : X \to A$ and $c^t : Y \to A^t$.
    From 5.3.2 and 5.3.3 we recover the X-torsor

$$\widetilde{\mathcal{Q}} := \widetilde{P} \times^Y X. \tag{5.3.4.2}$$

Let $\widetilde{\mathcal{Q}}^o$ denote a connected component of $\widetilde{\mathcal{Q}}$. We then recover $Y \subset X$ as the subgroup of elements fixing $\widetilde{\mathcal{Q}}^o$. We also obtain the G-action on $\widetilde{\mathcal{Q}}^o$ as in 4.3.19, as well as a lifting of the T_X-action to $L_{\widetilde{\mathcal{Q}}}^o$. Now observe that the Y-torsor $\widetilde{\mathcal{P}}$ is isomorphic to $\widetilde{\mathcal{Q}}^o$, and that any isomorphism is well-defined up to composition with translation by an element of Y. Because of the "Heisenberg relation" (which also holds by the construction in this context) there in fact exists a unique isomorphism of Y-torsors with line bundles

$$(\widetilde{\mathcal{P}}, L_{\widetilde{\mathcal{P}}}) \simeq (\widetilde{\mathcal{Q}}^o, L_{\widetilde{\mathcal{Q}}^o}) \tag{5.3.4.3}$$

compatible with the torus actions on the line bundles. In this way we recover $(\widetilde{\mathcal{P}}, L_{\widetilde{\mathcal{P}}})$ up to unique isomorphism.
    As in 4.3.19, we then also obtain a map

$$f : \widetilde{\mathcal{P}} \to A \tag{5.3.4.4}$$

well-defined up to translation by an element of A. Define $c^t : Y \to A$ by sending $y \in Y$ to the unique element $c^t(y) \in A(B)$ such that the diagram

$$
\begin{CD}
\widetilde{\mathscr{P}} @>t_y>> \widetilde{\mathscr{P}} \\
@VfVV @VVfV \\
A @>t_{c^t(y)}>> A
\end{CD}
\tag{5.3.4.5}
$$

commutes. As in 4.3.19 we also obtain line bundles $L^{(d,x)}$ on A as the x-eigenspaces in $f_*L^d_{\widetilde{\mathscr{P}}}$ and the resulting maps

$$L^{(d,x)} \otimes L^{(s,y)} \to L^{(d+s,x+y)} \tag{5.3.4.6}$$

determine the paving S. Finally we recover $c : X \to A^t$ by sending $x \in X$ to the class of the line bundle

$$L^{(1,x)} \otimes (L^{(1,0)})^{-1}. \tag{5.3.4.7}$$

5.4 The Theta Group

5.4.1. Let $(\mathscr{P}, M_{\mathscr{P}}) \to (B, M_B)$ be a log smooth morphism with polarization $L_{\mathscr{P}}$ and G–action obtained from the standard construction as in the preceding section. Let $\mathscr{G}_{(\mathscr{P}, M_{\mathscr{P}}, L_{\mathscr{P}})}$ denote the functor associating to any B–scheme T the group of automorphisms of the base change $(\mathscr{P}_T, M_{\mathscr{P}_T}, L_{\mathscr{P}_T})$ over $(T, M_B|_T)$ commuting with the G_T–action. Note that there is a natural inclusion $\mathbb{G}_m \hookrightarrow \mathscr{G}_{(\mathscr{P}, M_{\mathscr{P}}, L_{\mathscr{P}})}$ induced by the action of $\mathbb{G}_m$ on $L_{\mathscr{P}}$. The subgroup $\mathbb{G}_m \subset \mathscr{G}_{(\mathscr{P}, M_{\mathscr{P}}, L_{\mathscr{P}})}$ is central. We define

$$H(\mathscr{P}, M_{\mathscr{P}}, L_{\mathscr{P}}) := \mathscr{G}_{(\mathscr{P}, M_{\mathscr{P}}, L_{\mathscr{P}})}/\mathbb{G}_m.$$

The main result of this section is the following:

Theorem 5.4.2 (i) *The functor $\mathscr{G}_{(\mathscr{P}, M_{\mathscr{P}}, L_{\mathscr{P}})}$ is a group scheme.*
(ii) *There is a natural exact sequence*

$$1 \to \mathscr{G}_{(A, \mathscr{M})} \xrightarrow{\ a\ } \mathscr{G}_{(\mathscr{P}, M_{\mathscr{P}}, L_{\mathscr{P}})} \longrightarrow D(X/Y) \times X/Y \to 1, \tag{5.4.2.1}$$

where the map a sends $\mathbb{G}_m$ to $\mathbb{G}_m$ via the identity map.
(iii) *The group scheme $H(\mathscr{P}, M_{\mathscr{P}}, L_{\mathscr{P}})$ is a finite flat abelian group scheme of rank $d^2 = d_1^2 d_2^2$ over B (note that by 5.2.6 d^2 is also equal to the square of the rank of the vector bundle $f_*L_{\mathscr{P}}$ on B).*
(iv) *The pairing*

$$e : H(\mathscr{P}, M_{\mathscr{P}}, L_{\mathscr{P}}) \times H(\mathscr{P}, M_{\mathscr{P}}, L_{\mathscr{P}}) \to \mathbb{G}_m$$

sending $x, y \in H(\mathscr{P}, M_{\mathscr{P}}, L_{\mathscr{P}})$ to the commutator $\tilde{x}\tilde{y}\tilde{x}^{-1}\tilde{y}^{-1}$ is a perfect pairing. Here $\tilde{x}, \tilde{y} \in \mathscr{G}_{(\mathscr{P}, M_{\mathscr{P}}, L_{\mathscr{P}})}$ are liftings of x and y. Note that since $\mathbb{G}_m$ is central the expression $\tilde{x}\tilde{y}\tilde{x}^{-1}\tilde{y}^{-1}$ is independent of the choices.

The proof will be in several steps. In what follows we write just H for the functor $H(A, \mathscr{M})$ and $\mathscr{G}$ for $\mathscr{G}_{(A, \mathscr{M})}$. Let $j : H \hookrightarrow A$ be the inclusion.

Lemma 5.4.3 ([11, 4.3 (ii)]) *Let $E' \to H \times Y$ denote the pullback of the Poincare biextension over $A \times A^t$ by the morphism*

$$j \times \lambda_{\mathscr{M}} \circ c^t : H \times Y \to A \times A^t.$$

Then there is a canonical trivialization $\sigma : H \times Y \to E'$ whose formation is compatible with arbitrary base change on B.

Proof. Consider the commutative diagram

$$
\begin{array}{ccc}
A \times A^t & & A \times A^t \\
\nearrow \; \uparrow^{1 \times \lambda_{\mathscr{M}}} & & \uparrow^{1 \times \lambda_{\mathscr{M}}} \searrow \\
{}^{c^t \times 0}\Big(A \times A & \xleftarrow{\text{flip}} & A \times A \Big)^{j \times \lambda_{\mathscr{M}} \circ c^t} \\
\searrow \; \uparrow^{c^t \times j} & & \uparrow^{j \times c^t} \nearrow \\
Y \times H & \xleftarrow{\text{flip}} & H \times Y.
\end{array}
\tag{5.4.3.1}
$$

We then have isomorphisms of biextensions

$$
\begin{aligned}
E' &\simeq (j \times \lambda_{\mathscr{M}} \circ c^t)^* \mathscr{B} \\
&\simeq (j \times c^t)^* (1 \times \lambda_{\mathscr{M}})^* \mathscr{B} \\
&\simeq (j \times c^t)^* \text{flip}^* (1 \times \lambda_{\mathscr{M}})^* \mathscr{B} \quad \text{(symmetric structure on } \mathscr{B}) \\
&\simeq \text{flip}^* (c^t \times j)^* (1 \times \lambda_{\mathscr{M}})^* \mathscr{B} \\
&\simeq \text{flip}^* (c^t \times 0)^* \mathscr{B},
\end{aligned}
$$

and $(c^t \times 0)^* \mathscr{B}$ is canonically trivialized by 2.2.4. $\qquad\square$

Remark 5.4.4. The trivialization of E' can be described more explicitly as follows.

By descent theory it suffices to construct σ fppf locally on B.

Let $z \in H$ be a point and $y \in Y$ an element. Then the fiber of E' over (z, y) is equal to the fiber over z of the rigidified line bundle $t_y^* \mathscr{M} \otimes \mathscr{M}^{-1} \otimes \mathscr{M}(y)^{-1} \otimes \mathscr{M}(0)$. Equivalently $E'_{(z,y)}$ is isomorphic to the fiber over 0 of the line bundle

$$t_y^* t_z^* \mathscr{M} \otimes t_z^* \mathscr{M}^{-1} \otimes \mathscr{M}(y)^{-1} \otimes \mathscr{M}(0).$$

Since $z \in H$ there exists after replacing B by an fppf cover an isomorphism $t_z^* \mathscr{M} \to \mathscr{M}$. Choose one such isomorphism $\rho : t_z^* \mathscr{M} \to \mathscr{M}$. The resulting isomorphism

$$\rho \otimes \rho^t : t_y^* t_z^* \mathscr{M} \otimes t_z^* \mathscr{M}^{-1} \to t_y^* \mathscr{M} \otimes \mathscr{M}^{-1}$$

is then independent of the choice of ρ (any other choice of ρ differs by multiplication by a unit u and $(u\rho)^t = u^{-1}\rho^t$). This isomorphism therefore defines a canonical section of $E'_{(z,y)}$. We leave to the reader the verification that this trivialization agrees with the one defined in 5.4.3.

5.4.5. Let $E'' \to Y \times X$ denote the pullback of the Poincare biextension via the map $c^t \times c : Y \times X \to A \times A^t$, and let E''' denote the pullback of the Poincare biextension to $\{0\} \times Y$ via the map $c \circ \phi : Y \to A^t$. We then have a commutative diagram

$$
\begin{array}{ccccc}
E' & \longleftarrow & E''' & \longrightarrow & E'' \\
\downarrow & & \downarrow & & \downarrow \\
H \times Y & \longleftarrow & \{0\} \times Y & \longrightarrow & Y \times X,
\end{array}
$$

and hence we get by restriction two maps $\sigma|_Y, \tau|_Y : \{0\} \times Y \to E'''$.

Lemma 5.4.6 *These two morphisms* $\{0\} \times Y \to E'''$ *are equal.*

Proof. The fiber of E''' over $(0, y)$ is equal to $L_{\phi(y)}(0)$. Since $L_{\phi(y)}$ is rigidified there is a given section. It follows from the construction of σ that $\sigma|_Y(y)$ in fact is equal to this section. The same is true of $\tau|_Y$. This can be seen by noting that the restriction of τ to $Y \times Y$ is given by the section $\psi : Y \to \phi^* c^* \mathcal{M}^{-1}$ which shows that the section

$$
\tau|_Y(y) \in \mathcal{M}(y) \otimes \mathcal{M}(0) \otimes \mathcal{M}(y)^{-1} \otimes \mathcal{M}(0)^{-1}
$$

is equal to the section $\psi(y) \otimes 1 \otimes \psi(y)^{-1} \otimes 1$, where $1 \in \mathcal{M}(0)$ denotes any rigidification.   □

5.4.7. Let $E \to (H \times Y) \times X$ denote the biextension obtained by pulling back the Poincare biextension via the map

$$
(j \times c^t) \times c : (H \times Y) \times X \to A \times A^t, \quad (h, y, x) \mapsto (j(h) + c^t(y), c(x)).
$$

There is then a commutative diagram

$$
\begin{array}{ccccc}
E' & \longrightarrow & E & \longleftarrow & E'' \\
\downarrow & & \downarrow & & \downarrow \\
H \times Y & \xrightarrow{\ s\ } & (H \times Y) \times X & \xleftarrow{\ t\ } & Y \times X,
\end{array}
$$

where s is the map $(h, y) \mapsto (h, 0, \phi(y))$ and t is the map $(y, x) \mapsto (0, y, x)$.

Lemma 5.4.8 *There exists a unique trivialization* $\gamma : (H \times Y) \times X \to E$ *of the biextension E such that $s^* \gamma$ is equal to σ and $t^* \gamma$ is equal to τ.*

Proof. The square

$$
\begin{array}{ccc}
Y & \xrightarrow{\ v\ } & Y \times X \\
w \downarrow & & \downarrow t \\
H \times Y & \xrightarrow{\ s\ } & (H \times Y) \times X
\end{array}
$$

is both cartesian and cocartesian, where v is the map $y \mapsto (0, \phi(y))$ and w is the map $y \mapsto (0, y)$. It follows from this and 5.4.6 that there is a unique morphism of sheaves of sets

$$\gamma : (H \times Y) \times X \to E$$

with $s^*\gamma = \sigma$ and $t^*\gamma = \tau$. We have to check that γ is compatible with the symmetric biextension structure. This is equivalent to showing that the following diagrams commute

$$
\begin{array}{ccc}
(H \times Y) \times (H \times Y) \times X & \xrightarrow{+\times \mathrm{id}} & (H \times Y) \times X \\
{\scriptstyle p_{13}^*(\gamma) \wedge p_{23}^*(\gamma)} \downarrow & & \downarrow {\scriptstyle \gamma} \\
p_{13}^* E \wedge p_{23}^* E & \xrightarrow{\varphi} & E,
\end{array}
\qquad (5.4.8.1)
$$

where p_{13} and p_{23} are the two projections

$$(H \times Y) \times (H \times Y) \times X \to (H \times Y) \times X, \qquad (5.4.8.2)$$

and

$$
\begin{array}{ccc}
(H \times Y) \times X \times X & \xrightarrow{\mathrm{id} \times +} & (H \times Y) \times X \\
{\scriptstyle q_{12}^*(\gamma) \times q_{13}^*(\gamma)} \downarrow & & \downarrow {\scriptstyle \gamma} \\
q_{12}^* E \wedge p_{13}^* E & \xrightarrow{\psi} & E,
\end{array}
\qquad (5.4.8.3)
$$

where q_{12} and q_{13} are the two projections

$$(H \times Y) \times X \times X \to (H \times Y) \times X. \qquad (5.4.8.4)$$

This follows from the observation that the diagrams

$$
\begin{array}{ccc}
Y & \xrightarrow{0 \times 0 \times \phi} & Y \times Y \times X \\
{\scriptstyle 0 \times 0 \times \mathrm{id}} \downarrow & & \downarrow \\
(H \times H) \times Y & \longrightarrow & (H \times Y) \times (H \times Y) \times X
\end{array}
$$

and

$$
\begin{array}{ccc}
Y \times Y & \xrightarrow{0 \times \phi \times \phi} & Y \times X \times X \\
\downarrow & & \downarrow \\
H \times Y \times Y & \longrightarrow & (H \times Y) \times X \times X
\end{array}
$$

are cocartesian and the fact that $s^*\gamma = \sigma$ and $t^*\gamma = \tau$. $\square$

5.4.9. Let $\widetilde{E} \to (\mathscr{G} \times Y) \times X$ denote the pullback of E and let $\tilde{\gamma} : (\mathscr{G} \times Y \times X) \to \widetilde{E}$ be the pullback of γ. This trivialization $\tilde{\gamma}$ defines an action of $\mathscr{G} \times Y$ on $(\widetilde{\mathscr{P}}, M_{\widetilde{\mathscr{P}}}, \mathscr{O}_{\widetilde{\mathscr{P}}}(1))$ whose restriction to $\{1\} \times Y$ is the previously defined action.

This action of $\mathscr{G} \times Y$ is defined as follows. Let $g = (a, \iota : t_a^* \mathscr{M} \to \mathscr{M}) \in \mathscr{G}$, $y \in Y$ be sections. We then define for every $(d, x) \in P$ a morphism

$$\sigma_{g,y}^{(d,x)} : t_{a+y}^*(\mathscr{M}^d \otimes L_x) \to \mathscr{M}^d \otimes L_{x+dy}. \qquad (5.4.9.1)$$

For this note first that since $t_a^* \mathscr{M}$ is isomorphic to $\mathscr{M}$, we have noncanonical isomorphisms

$$t_{a+y}^*(\mathscr{M}^d \otimes L_x) \simeq t_y^*(\mathscr{M}^d \otimes L_x) \simeq \mathscr{M}^d \otimes L_{x+dy},$$

where the second isomorphism is by 5.2.2. Note also that ι induces an isomorphism

$$t_{a+y}^*(\mathscr{M}^d \otimes L_x) \simeq t_y^* \mathscr{M}^d \otimes t_{a+y}^* L_x.$$

Therefore giving the map 5.4.9.1 is equivalent to giving a section of

$$\mathscr{M}^d(y) \otimes L_x(a+y).$$

We take the section $\psi(y)^d \gamma((a, \iota), y, x)$.

Proposition 5.4.10. *The maps* $\sigma_{g,y}^{(d,x)}$ *define an action of* $\mathscr{G} \times Y$ *on the log scheme with line bundle* $(\widetilde{\mathscr{P}}, M_{\widetilde{\mathscr{P}}}, L_{\widetilde{\mathscr{P}}})$.

Remark 5.4.11. Passing to the quotient by the Y–action we obtain the inclusion

$$\mathscr{G} \hookrightarrow \mathscr{G}_{(\mathscr{P}, M_{\mathscr{P}}, L_{\mathscr{P}})}$$

in 5.4.2.1.

Proof. It suffices to show that for every $d, d' \geq 0$, $g = (a, \iota_a : t_a^* \mathscr{M} \to \mathscr{M})$, $g' = (a', \iota_{a'} : t_{a'}^* \mathscr{M} \to \mathscr{M}) \in \mathscr{G}$, $y, y' \in Y$, and $x, x' \in X$ the following two diagrams commute:

$$
\begin{array}{ccc}
t_{a+y}^*(\mathscr{M}^d \otimes L_x) \otimes t_{a+y}^*(\mathscr{M}^{d'} \otimes L_{x'}) & \xrightarrow{\sigma_{g,y}^{(d,x)} \otimes \sigma_{g,y}^{(d',x')}} & \mathscr{M}^d \otimes L_{x+dy} \otimes L_{x'+dy'} \\
\downarrow{\scriptstyle \text{can}} & & \downarrow{\scriptstyle \text{can}} \\
t_{a+y}^*(\mathscr{M}^{d+d'} \otimes L_{x+x'}) & \xrightarrow{\sigma_{g,y}^{(d+d',x+x')}} & \mathscr{M}^{d+d'} \otimes L_{x+x'+(d+d')y},
\end{array}
$$
$$(5.4.11.1)$$

$$
\begin{array}{ccc}
t_{a'+y'}^* t_{a+y}^*(\mathscr{M}^d \otimes L_x) & \xrightarrow{t_{a'+y'}^* \sigma_{g,y}^{(d,x)}} & t_{a'+y'}^*(\mathscr{M}^d \otimes L_{x+dy}) \\
\downarrow{\scriptstyle \text{can}} & & \downarrow{\scriptstyle \sigma_{g',y'}^{(d,x+dy)}} \\
t_{a+a'+y+y'}^*(\mathscr{M}^d \otimes L_x) & \xrightarrow{\sigma_{g+g',y+y'}^{(d,x)}} & \mathscr{M}^d \otimes L_{x+d(y+y')}.
\end{array}
$$
$$(5.4.11.2)$$

For the commutativity of 5.4.11.1 note first that by the definition of the maps $\sigma_{g,y}^{(d,x)}$ the result holds if $x = 0$ and $d' = 0$, $d = 0$ and $x' = 0$, or if

$x = x' = 0$. It follows that to prove the commutativity of 5.4.11.1 it suffices to consider the following two cases:

Case 1: $d = d' = 0$. In this case we are asking for the commutativity of the diagram

$$
\begin{array}{ccc}
(t^*_{a+y}L_x) \otimes (t^*_{a+y}L_{x'}) & \xrightarrow{\tilde{\gamma}(g,y,x) \otimes \tilde{\gamma}(g,y,x')} & L_x \otimes L'_x \\
\text{can} \downarrow & & \downarrow \text{can} \\
t^*_{a+y}L_{x+x'} & \xrightarrow{\tilde{\gamma}(g,y,x+x')} & L_{x+x'},
\end{array} \tag{5.4.11.3}
$$

which holds by the additivity of $\tilde{\gamma}$ in x.

Case 2: $x = x' = 0$, and $d = d' = 1$. In this case we are asking for the commutativity of the following diagram:

$$
\begin{array}{ccc}
t^*_{a+y}(\mathscr{M}) \otimes t^*_{a+y}(\mathscr{M}) & \xrightarrow{\iota_a \otimes \iota_a} & t^*_y \mathscr{M} \otimes t^*_y \mathscr{M} \\
\text{can} \downarrow & & \downarrow \psi(y) \otimes \psi(y) \\
t^*_{a+y}\mathscr{M}^2 & & \mathscr{M} \otimes L_y \otimes \mathscr{M} \otimes L_y \\
\iota_a^2 \downarrow & & \downarrow \text{can} \\
t^*_y \mathscr{M}^2 & \xrightarrow{\psi(y)^2} & \mathscr{M}^2 \otimes L_{2y},
\end{array} \tag{5.4.11.4}
$$

which is immediate.

For the commutativity of 5.4.11.2, consideration of the diagram

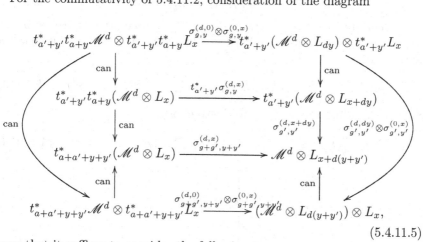

$$\tag{5.4.11.5}$$

shows that it suffices to consider the following two cases.

Case 1: $d = 0$. In this case we need to show that the diagram

$$
\begin{array}{ccc}
t^*_{a'+y'}t^*_{a+y}L_x & \xrightarrow{\tilde{\gamma}(g,y,x)} & t^*_{a'+y'}L_x \\
\downarrow{\scriptstyle\text{can}} & & \downarrow{\scriptstyle\tilde{\gamma}(g',y',x)} \\
t^*_{a+a'+y+y'}L_x & \xrightarrow{\tilde{\gamma}(g+g',y+y',x)} & L_x
\end{array}
\tag{5.4.11.6}
$$

commutes, which follows from the additivity in the first variable of $\tilde{\gamma}$.

Case 2: $d = 1$ and $x = 0$. In this case we need to show that the diagram

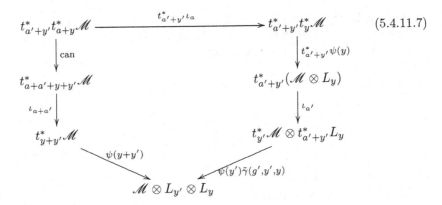

(5.4.11.7)

commutes.

The map

$$
\tilde{\gamma}(g',y',y) : t^*_{a'+y'}L_y \to L_y
\tag{5.4.11.8}
$$

can be described more explicitly as follows. By the definition of $\tilde{\gamma}(g',y',y)$ this map is the unique map of rigidified line bundles, where $t^*_{a'+y'}L_y$ is rigidified by the section

$$
\tau(y',y) \in L_{y'}(y) \simeq L_{y'}(y) \otimes L_{a'}(y) \simeq L_{y'+a'}(y) \simeq L_y(a'+y').
\tag{5.4.11.9}
$$

It follows that 5.4.11.8 is equal to the composite

$$
t^*_{a'+y'}L_y \simeq t^*_{y+y'}t^*_{a'}\mathcal{M} \otimes t^*_{y'}t^*_{a'}\mathcal{M}^{-1} \otimes \mathcal{M}(y)^{-1}
\tag{5.4.11.10}
$$

$$
\downarrow{\scriptstyle t^*_{y+y'}\iota_{a'}\otimes t^*_{y'}\iota_{a'}^{-1}\otimes 1}
$$

$$
t^*_{y+y'}\mathcal{M} \otimes t^*_{y'}\mathcal{M}^{-1} \otimes \mathcal{M}(y)^{-1}
$$

$$
\downarrow{\scriptstyle \tau(y',y)}
$$

$$
t^*_y\mathcal{M} \otimes \mathcal{M}^{-1} \otimes \mathcal{M}(y)^{-1} \simeq L_y.
$$

It follows that the diagram 5.4.11.7 can be identified with the outside octagon in the following Figure 1:

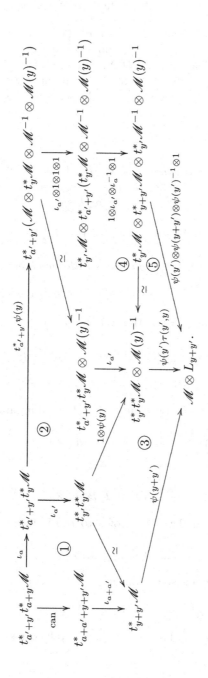

Figure 1

The inside diagrams labelled 1, 2, 4, and 5 all clearly commute, so to prove the commutativity of 5.4.11.7 it suffices to prove the commutativity of the inside diagram 3. This follows from the definition of $\tau(y', y)$. $\square$

5.4.12. Note that though the action of T on $\mathcal{O}_{\widetilde{\mathscr{P}}}(1)$ does not commute with the Y–action, the action of $D(X/Y) \subset T$ does commute with the Y–action. Therefore there is a canonical action of $D(X/Y)$ on $(\mathscr{P}, M_{\mathscr{P}}, L_{\mathscr{P}})$ defining an inclusion $D(X/Y) \hookrightarrow \mathscr{G}_{(\mathscr{P}, M_{\mathscr{P}}, L_{\mathscr{P}})}$. It follows from the preceding discussion that the elements of $D(X/Y)$ commute with the sections of $\mathscr{G}$, and hence we obtain an inclusion

$$\mathscr{G} \times D(X/Y) \hookrightarrow \mathscr{G}_{(\mathscr{P}, M_{\mathscr{P}}, L_{\mathscr{P}})}.$$

Lemma 5.4.13 *Any automorphism* $\alpha : (\mathscr{P}, M_{\mathscr{P}}, L_{\mathscr{P}}) \to (\mathscr{P}, M_{\mathscr{P}}, L_{\mathscr{P}})$ *such that the automorphism* $\alpha_{\mathscr{P}} : (\mathscr{P}, M_{\mathscr{P}}) \to (\mathscr{P}, M_{\mathscr{P}})$ *commutes with the* G–*action lifts fppf locally on* B *to an automorphism*

$$\tilde{\alpha} : (\widetilde{\mathscr{P}}, M_{\widetilde{\mathscr{P}}}, L_{\widetilde{\mathscr{P}}}) \to (\widetilde{\mathscr{P}}, M_{\widetilde{\mathscr{P}}}, L_{\widetilde{\mathscr{P}}})$$

where $\tilde{\alpha}_{\mathscr{P}} : (\widetilde{\mathscr{P}}, M_{\widetilde{\mathscr{P}}}) \to (\widetilde{\mathscr{P}}, M_{\widetilde{\mathscr{P}}})$ *commutes with the* G–*action.*

Proof. Recall from 5.3.2 that there is an equivalence of Picard categories

$$\text{TORS}_{\mathscr{P}}(X) \to \text{HOM}(T, \mathscr{P}ic(\mathscr{P})), \tag{5.4.13.1}$$

and that under this equivalence the X-torsor $\widetilde{\mathscr{P}} \times^Y X \to \mathscr{P}$ corresponds to the map

$$\rho : T \to \mathscr{P}ic(\mathscr{P}), \quad u \mapsto u^* L_{\mathscr{P}} \otimes L_{\mathscr{P}}^{-1}. \tag{5.4.13.2}$$

From the construction of the equivalence 5.4.13.1 the diagram

$$\begin{array}{ccc}
\text{TORS}_{\mathscr{P}}(X) & \xrightarrow{\text{5.4.13.1}} & \text{HOM}(T, \mathscr{P}ic(\mathscr{P})) \\
\alpha^* \downarrow & & \downarrow \alpha^* \\
\text{TORS}_{\mathscr{P}}(X) & \xrightarrow{\text{5.4.13.1}} & \text{HOM}(T, \mathscr{P}ic(\mathscr{P}))
\end{array} \tag{5.4.13.3}$$

is naturally 2–commutative. The lifting of α to $L_{\mathscr{P}}$ induces an isomorphism of functors

$$\alpha^* \circ (5.4.13.1) \simeq (5.4.13.1), \tag{5.4.13.4}$$

and hence α also induces an isomorphism $\alpha^*(\widetilde{\mathscr{P}} \times^Y X) \to \widetilde{\mathscr{P}} \times^Y X$ of X-torsors over $\mathscr{P}$. Let

$$\alpha^\dagger : \widetilde{\mathscr{P}} \times^Y X \to \widetilde{\mathscr{P}} \times^Y X$$

be the resulting lifting of α. For every scheme–valued section $u \in G$ the diagram

$$\widetilde{\mathscr{P}} \times^Y X \xrightarrow{\alpha^\dagger} \widetilde{\mathscr{P}} \times^Y X$$

$$\Big\downarrow u \qquad\qquad \Big\downarrow u$$

$$\widetilde{\mathscr{P}} \times^Y X \xrightarrow{\alpha^\dagger} \widetilde{\mathscr{P}} \times^Y X$$

commutes up to translation by an element $\rho(u) \in X$. This defines a morphism of schemes $\rho : G \to X$ which since G is has geometrically connected fibers is necessarily constant. Since $\rho(e) = 0$ we have $\rho(u) = 0$ for all u and hence $\alpha^\dagger$ commutes with the G–action.

Taking the quotient of $\widetilde{\mathscr{P}} \times^Y X$ by the Y–action we obtain an isomorphism

$$\mathscr{P} \times (X/Y) \to \mathscr{P} \times (X/Y)$$

which modulo the X/Y–action is equal to $\alpha_\mathscr{P}$. It follows that there exists an element $x \in X$ such that the composition $t_x \circ \alpha^\dagger$ (where t_x denotes the action of x on $\widetilde{\mathscr{P}} \times^Y X$) sends $\widetilde{\mathscr{P}}$ to itself. Note that the element $x \in X$ is unique up to adding an element of Y. Let

$$\tilde{\alpha}_\mathscr{P} : (\widetilde{\mathscr{P}}, M_{\widetilde{\mathscr{P}}}) \to (\widetilde{\mathscr{P}}, M_{\widetilde{\mathscr{P}}})$$

be the resulting lifting of $\alpha_\mathscr{P}$. The lifting of $\alpha_\mathscr{P}$ to $L_\mathscr{P}$ defines a lifting of $\tilde{\alpha}_\mathscr{P}$ to an isomorphism $\tilde{\beta} : \tilde{\alpha}_\mathscr{P}^* \mathscr{O}_{\widetilde{\mathscr{P}}}(1) \to \mathscr{O}_{\widetilde{\mathscr{P}}}(1)$. $\square$

5.4.14. The choice of the lifting $\tilde{\alpha}$ is unique up to composition with translation by an element of Y. If $\widetilde{L}$ denotes the scheme associated to the line bundle $\mathscr{O}_{\widetilde{\mathscr{P}}}(1)$, then for any local section $u \in T$ the diagram

$$\widetilde{L} \xrightarrow{\tilde{\alpha}} \widetilde{L}$$

$$\Big\downarrow u \qquad\qquad \Big\downarrow u$$

$$\widetilde{L} \xrightarrow{\tilde{\alpha}} \widetilde{L}$$

commutes up to multiplication by a scalar $\rho(u) \in \mathbb{G}_m$. This defines a homomorphism $\rho : T \to \mathbb{G}_m$, or equivalently a section $\rho \in X$. Changing $\tilde{\alpha}$ by $y \circ \tilde{\alpha}$ for some $y \in Y$ changes ρ to $\rho + \phi(y)$. We therefore obtain a canonical map

$$q : \mathscr{G}_{(\mathscr{P}, M_\mathscr{P}, L_\mathscr{P})} \to X/Y.$$

We will show that this map is surjective with kernel $\mathscr{G} \times D(X/Y)$. Note already that if α maps to $\bar{x} \in X/Y$ and $u \in D(X/Y)$ is a section, then $u\alpha u^{-1}$ is equal to $u(\bar{x}) \cdot \alpha$. Hence if we prove that q is surjective with kernel $\mathscr{G} \times D(X/Y)$, then all the statements in 5.4.2 follow.

Lemma 5.4.15 *Let $\alpha \in \mathscr{G}_{(\mathscr{P}, M_\mathscr{P}, L_\mathscr{P})}$ be an automorphism, and let $\tilde{\alpha}$ be a lifting to $(\widetilde{\mathscr{P}}, M_{\widetilde{\mathscr{P}}})$. Then there exists a unique automorphism $\bar{\alpha} : A \to A$ such that the diagram*

$$\widetilde{\mathscr{P}} \xrightarrow{\ \tilde{\alpha}\ } \widetilde{\mathscr{P}}$$

$$\pi \downarrow \qquad\qquad \downarrow \pi$$

$$A \xrightarrow{\ \bar{\alpha}\ } A$$

commutes. In fact, $\bar{\alpha}$ is equal to translation by an element mapping under $\lambda_{\mathscr{M}}$ to $c(\rho) \in A^t$.

Proof. The existence of the map $\bar{\alpha}$ follows from the same argument used in 4.2.3. Since $\bar{\alpha}$ commutes with the translation action it follows that $\bar{\alpha}$ is translation by some element $g \in A$. Looking at T–invariants of $\pi_* \mathscr{O}_{\widetilde{\mathscr{P}}}(1)$ one obtains isomorphisms

$$t_g^*(\mathscr{M} \otimes L_x) \to \mathscr{M} \otimes L_{x+\rho(\alpha)} \qquad (5.4.15.1)$$

for all x. From this it follows that $\lambda_{\mathscr{M}}(g) = c(\rho(\alpha))$. $\square$

5.4.16. The argument shows in particular that if $\alpha \in \mathscr{G}_{(\mathscr{P}, M_{\mathscr{P}}, L_{\mathscr{P}})}$ is in the kernel of q then $\bar{\alpha}$ is given by translation by an element $g \in H$ and the isomorphism 5.4.15.1 for $x = 0$ lifts this element to $\mathscr{G}$. This implies that $\mathrm{Ker}(q) = \mathscr{G} \times D(X/Y)$. Indeed changing α by the automorphism defined by this element of $\mathscr{G}$, we are reduced to considering the case when $g = e$ and 5.4.15.1 for $x = 0$ is the identity maps. In this case the maps 5.4.15.1 are given by compatible maps $L_x \to L_x$ which commute with the Y–action. We leave to the reader the task of verifying that such a map is given by an element of $D(X/Y)$.

5.4.17. For the surjectivity of q proceed as follows. Let $\rho \in X$ be an element. After replacing B by a finite flat cover we can choose an element $\alpha \in A$ with $\lambda_{\mathscr{M}}(\alpha) = L_\rho$. Then $t_\alpha^* \mathscr{M}$ is isomorphic to $\mathscr{M} \otimes L_\rho$. Fix one such isomorphism $\iota : t_\alpha^* \mathscr{M} \to \mathscr{M} \otimes L_\rho$. Define $\lambda(y) : t_\alpha^* L_y \to L_y$ to be the composite (recall that $\mathscr{M}$ is rigidified at 0)

$$t_\alpha^* L_y = t_\alpha^* t_y^* \mathscr{M} \otimes t_\alpha^* \mathscr{M}^{-1} \otimes \mathscr{M}(y)^{-1}$$

$$t_y^*(\iota) \otimes \iota^t \downarrow$$

$$t_y^*(\mathscr{M} \otimes L_\rho) \otimes (\mathscr{M}^{-1} \otimes L_\rho^{-1}) \otimes \mathscr{M}(y)^{-1}$$

$$1 \otimes \tau(y, \rho) \otimes 1 \otimes 1 \downarrow$$

$$t_y^* \mathscr{M} \otimes L_\rho \otimes \mathscr{M}^{-1} \otimes L_\rho^{-1} \otimes \mathscr{M}(y)^{-1} = L_y.$$

Note that $\lambda(y)$ does not depend on the choice of ι.

Lemma 5.4.18 *The diagram*

$$\begin{array}{ccc} t_y^* t_\alpha^* \mathscr{M} & \xrightarrow{\ \iota\ } & t_y^*(\mathscr{M} \otimes L_\rho) \\ \psi(y) \downarrow & & \downarrow \psi(y)\tau(y,\rho) \\ t_\alpha^*(\mathscr{M} \otimes L_y) & \xrightarrow{\iota \otimes \lambda(y)} & \mathscr{M} \otimes L_y \otimes L_\rho \end{array} \qquad (5.4.18.1)$$

commutes.

Proof. Locally on A we can choose bases for the line bundles involved as follows:

line bundle	basis
$t_\alpha^* \mathcal{M}$	u
L_ρ	l
$\mathcal{M}$	n
$t_y^* t_\alpha^* \mathcal{M}$	u'
$t_y^* \mathcal{M}$	n'
$t_y^* L_\rho$	l'.

We can furthermore make these choices so that

$$\iota(u) = n \otimes l, \quad \tau(y,\rho)(l') = l, \quad t_y^*(\iota)(u') = n' \otimes l'.$$

The commutativity of 5.4.18.1 is equivalent to the commutativity of the following diagram:

$$
\begin{array}{ccc}
t_y^* t_\alpha^* (\mathcal{M}) & \xrightarrow{\ t_y^* \iota\ } & t_y^* \mathcal{M} \otimes t_y^* L_\rho \\
\psi(y) \downarrow & & \downarrow \mathrm{id} \otimes \tau(y,\rho) \otimes \psi(y) \\
t_\alpha^*(\mathcal{M}) \otimes t_\alpha^* t_y^*(\mathcal{M}) \otimes t_\alpha^*(\mathcal{M}^{-1}) \otimes \mathcal{M}(y)^{-1} & & t_y^* \mathcal{M} \otimes L_\rho \otimes \mathcal{M}(y)^{-1} \\
\iota \otimes t_y^* \iota \otimes \iota^t \downarrow & & \| \\
(\mathcal{M} \otimes L_\rho) \otimes t_y^* \mathcal{M} \otimes t_y^* L_\rho \otimes \mathcal{M}^{-1} \otimes L_\rho^{-1} \otimes \mathcal{M}(y)^{-1} & \xrightarrow{\ \sigma\ } & L_\rho \otimes t_y^* \mathcal{M} \otimes \mathcal{M}(y)^{-1}.
\end{array}
$$

$$(5.4.18.2)$$

Here σ sends $m \otimes l \otimes m' \otimes l' \otimes n \otimes t \otimes s$ to $\langle l, t \rangle \cdot \langle m, n \rangle \cdot \tau(y,\rho)(l') \otimes m' \otimes s$. This diagram commutes for example by writing out what happens in terms of our basis:

$$
\begin{array}{ccc}
u' & \longrightarrow & n' \otimes l' \\
\downarrow & & \downarrow \\
u \otimes u' \otimes u^t \otimes \psi(y) & & n' \otimes l \otimes \psi(y) \\
\downarrow & & \| \\
(n \otimes l) \otimes (n' \otimes l') \otimes (n \otimes l)^t \otimes \psi(y) & \longrightarrow & l \otimes n' \otimes \psi(y).
\end{array}
$$

$\square$

Lemma 5.4.19 *For any $y, y' \in Y$ the diagram*

$$
\begin{array}{ccc}
t_\alpha(L_{y+y'}) & \xrightarrow{\ \lambda(y+y')\ } & L_{y+y'} \\
t_\alpha^* \mathrm{can} \downarrow & & \downarrow \mathrm{can} \\
t_\alpha^* L_y \otimes t_\alpha^* L_{y'} & \xrightarrow{\ \lambda(y) \otimes \lambda(y')\ } & L_y \otimes L_{y'}
\end{array}
$$

$$(5.4.19.1)$$

commutes.

Proof. The morphism $\lambda(y) : t_\alpha^* L_y \to L_y$ is determined by the induced isomorphism $t_\alpha^* L_y(0) \simeq L_y(\alpha) \to L_y(0) \simeq k(y)$, where the last isomorphism is given by the rigidification on L_y. In other words, $\lambda(y)$ is specified by a section of

$$L_y(\alpha) = \mathcal{M}(y + \alpha) \otimes \mathcal{M}(y)^{-1} \otimes \mathcal{M}(\alpha)^{-1} = \lambda_{\mathcal{M}}(\alpha)(y) = L_\rho(y).$$

By construction, the section corresponding to $\lambda(y)$ is the section $\tau(y, \rho) \in L_\rho(y)$. With this description, the commutativity of 5.4.19.1 is equivalent to the statement that under the isomorphism provided by the biextension structure

$$L_\rho(y) \otimes L_\rho(y') \to L_\rho(y + y'),$$

the element $\tau(y, \rho) \otimes \tau(y', \rho)$ maps to $\tau(y + y', \rho)$. This is true since τ is compatible with the biextension structure. $\square$

5.4.20. Let $\mathscr{P}ic_A$ denote the Picard stack of line bundles on A. Viewing X as a "discrete" Picard stack we have two morphisms of Picard stacks

$$F, t_\alpha^* F : X \to \mathscr{P}ic_A, \quad F(x) = L_x, \quad t_\alpha^* F(x) = t_\alpha^* L_x.$$

Here the isomorphisms

$$F(x + x') \simeq F(x) \otimes F(x'), \quad t_\alpha^* F(x + x') \simeq t_\alpha^* F(x) \otimes t_\alpha^* F(x')$$

are given by the unique isomorphisms $L_{x+x'} \simeq L_x \otimes L'_x$ which are compatible with the rigidication. The above elements $\lambda(y)$ define an isomorphism of functors $Y \to \mathscr{P}ic_A$

$$\lambda : t_\alpha^* F|_Y \to F|_Y.$$

Lemma 5.4.21 *After replacing B by an fppf cover, there exists an isomorphism $\tilde{\lambda} : t_\alpha^* F \to F$ of morphisms of Picard stacks whose restriction to Y is equal to λ.*

Proof. Since X is a free group and $t_\alpha^* L_x$ is non–canonically isomorphic to L_x, there exists by [8, XVIII.1.4.3] an isomorphism $\delta : t_\alpha^* F \to F$ of morphisms of Picard stacks. Moreover, one sees easily that δ is unique up to multiplication by a homomorphism $u : X \to \mathbb{G}_m$ (in other words, for such a homomorphism we can define a new morphism δ' by $\delta'(x) = u(x)\delta(x)$). The restriction of δ to Y may not equal λ, but will differ by a homomorphism $v : Y \to \mathbb{G}_m$. After replacing B by an fppf cover there exists a homomorphism $u : X \to \mathbb{G}_m$ extending v. Setting $\tilde{\lambda}$ equal to $u \cdot \delta$ we obtain the desired extension. $\square$

5.4.22. Fix one such isomorphism $\tilde{\lambda} : t_\alpha^* F \to F$ so we have for every $x \in X$ an isomorphism $\tilde{\lambda}(x) : t_\alpha^* L_x \to L_x$ such that for $x, x' \in X$ the diagram

$$
\begin{array}{ccc}
t_\alpha^* L_x \otimes t_\alpha^* L_{x'} & \xrightarrow{\tilde{\lambda}(x) \otimes \tilde{\lambda}(x')} & L_x \otimes L_{x'} \\
\simeq \downarrow & & \downarrow \simeq \\
t_\alpha^* (L_{x+x'}) & \xrightarrow{\tilde{\lambda}(x+x')} & L_{x+x'}
\end{array}
$$

commutes. Define

$$\beta(d,x): t_\alpha^*(\mathscr{M}^d \otimes L_x) \to \mathscr{M}^d \otimes L_\rho^d \otimes L_x \simeq \mathscr{M}^d \otimes L_{x+d\rho}$$

to be the map $\iota^{\otimes d} \otimes \tilde\lambda(x)$.

Lemma 5.4.23 *For any* $(d,x),(d',x') \in P$ *the diagram*

$$
\begin{array}{ccc}
t_\alpha^*(\mathscr{M}^d \otimes L_x) \otimes t_\alpha^*(\mathscr{M}^{d'} \otimes L_{x'}) & \xrightarrow{\beta(d,x) \otimes \beta(d',x')} & \mathscr{M}^d \otimes L_\rho^d \otimes L_x \otimes \mathscr{M}^{d'} \otimes L_\rho^{d'} \otimes L_{x'} \\
\simeq \downarrow & & \downarrow \simeq \\
t_\alpha^*(\mathscr{M}^{d+d'} \otimes L_{x+x'}) & \xrightarrow{\beta(d+d',x+x')} & \mathscr{M}^{d+d'} \otimes L_\rho^{d+d'} \otimes L_{x+x'}
\end{array}
$$

is commutative.

Proof. This is clear from the construction. $\square$

5.4.24. The maps $\beta(d,x)$ therefore define a morphism of graded algebras $t_\alpha^* \mathcal{R} \to \mathcal{R}$ on A. We thus obtain an automorphism $\tilde\alpha$ of the log scheme $(\widetilde{\mathscr{P}}, M_{\widetilde{\mathscr{P}}})$ together with a morphism $\tilde\beta : t_{\tilde\alpha}^* \mathscr{O}_{\widetilde{\mathscr{P}}}(1) \to \mathscr{O}_{\widetilde{\mathscr{P}}}(1)$. Note that the automorphism of $(\widetilde{\mathscr{P}}, M_{\widetilde{\mathscr{P}}})$ is equal to the automorphism provided by the action of G and the observation that the elements $\tilde\lambda(x)$ define a point $\tilde\alpha \in G$ over $\alpha \in A$. The following lemma now shows that $(\tilde\alpha, \tilde\beta)$ descends to a point of $\mathscr{G}_{(\mathscr{P}, M_{\mathscr{P}}, L_{\mathscr{P}})}$ mapping to the class of ρ in X/Y.

Lemma 5.4.25 *The map* $\beta : t_\alpha^* \mathcal{R} \to \mathcal{R}$ *commutes with the* Y*-action. Equivalently, for any* $y \in Y$ *and* $(d,x) \in P$ *the diagram*

$$
\begin{array}{ccc}
t_y^* t_\alpha^*(\mathscr{M}^d \otimes L_x) & \xrightarrow{\beta(d,x)} & t_y^*(\mathscr{M}^d \otimes L_{d\rho} \otimes L_x) \\
\psi(y)^d \tau(y,x) \downarrow & & \downarrow \psi(y)^d \tau(y,x+d\rho) \\
t_\alpha^*(\mathscr{M}^d \otimes L_{dy} \otimes L_x) & \xrightarrow{\beta(d,x+dy)} & \mathscr{M}^d \otimes L_{d\rho} \otimes L_{dy} \otimes L_x
\end{array}
\qquad (5.4.25.1)
$$

commutes.

Proof. The diagram 5.4.25.1 is obtained by tensoring the diagram

$$
\begin{array}{ccc}
t_y^* t_\alpha^* L_x & \xrightarrow{t_\alpha^* \tilde\lambda(x)} & t_y^* L_x \\
t_\alpha^* \tau(y,x) \downarrow & & \downarrow \tau(y,x) \\
t_\alpha^* L_x & \xrightarrow{\tilde\lambda(x)} & L_x
\end{array}
\qquad (5.4.25.2)
$$

with

$$
\begin{array}{ccc}
t_y^* t_\alpha^* (\mathscr{M}^d) & \xrightarrow{\ \iota^d\ } & t_y^* (\mathscr{M}^d \otimes L_\rho^d) \\
\psi(y)^d \downarrow & & \downarrow \psi(y)^d \tau(y, d\rho) \\
t_\alpha^* (\mathscr{M}^d \otimes L_{dy}) & \xrightarrow{\ \iota^d \otimes \lambda(y)^d\ } & \mathscr{M}^d \otimes L_{d\rho} \otimes L_{dy}.
\end{array}
\qquad (5.4.25.3)
$$

That 5.4.25.2 commutes follows from the construction (the maps determined by going around each way of the diagram are both determined by the section $\tau(y, x) \otimes \tau(y, \rho) \in L_x(y) \otimes L_x(\alpha) \simeq L_x(y + \alpha))$, and the commutativity of 5.4.25.3 follows from raising each vertex in 5.4.18.1 to the d–th power.   □

5.4.26. Note that if α is the element of $\mathscr{G}_{(\mathscr{P}, M_\mathscr{P}, L_\mathscr{P})}$ constructed above mapping to $\rho \in X/Y$ then for $u \in D(X/Y)$ we have

$$
u \circ \alpha = u(\rho)\alpha \circ u.
$$

In fact it follows from the constructions that the image of α in $H(\mathscr{P}, M_\mathscr{P}, L_\mathscr{P})$ is in the center (since the automorphism of $(\widetilde{\mathscr{P}}, M_{\widetilde{\mathscr{P}}})$ defined by α is given by the action of $\tilde{\alpha} \in G$), and hence $H(\mathscr{P}, M_\mathscr{P}, L_\mathscr{P})$ is abelian and the map e in (5.4.2 (iv)) makes sense. It follows that if $\gamma \in \mathscr{G}_{(\mathscr{P}, M_\mathscr{P}, L_\mathscr{P})}$ is an element with $e(\gamma, \tau) = 0$ for all τ then γ is in H. But by [36, Part I, Theorem I] the restriction of e to H is non–degenerate whence (iv). This completes the proof of 5.4.2.   □

Proposition 5.4.27. *If B is the spectrum of an algebraically closed field k, then there is a unique irreducible representation $\rho : \mathscr{G}_{(\mathscr{P}, M_\mathscr{P}, L_\mathscr{P})} \to GL(V)$ over k such that $\rho(\lambda)(v) = \lambda v$ for $\lambda \in \mathbb{G}_m$ and $v \in V$ (we say that ρ has weight 1). The dimension of V is equal to d, and if W is any weight one representation of $\mathscr{G}_{(\mathscr{P}, M_\mathscr{P}, L_\mathscr{P})}$ then W is isomorphic to a direct sum of copies of ρ.*

Proof. This is shown in [18, 8.32].   □

5.5 Deformation Theory

5.5.1. Let

$$
B' \to B \to B_0
$$

be a diagram of surjective morphisms of noetherian rings with nilpotent kernels. Assume that B_0 is an integral domain and $J = \mathrm{Ker}(B' \to B)$ is a square zero–ideal with J a B_0–module. We fix the following data:

(i) A semiabelian scheme G'/B' with reduction G (resp. G_0) to B (resp. B_0). We assume that G' is a split semiabelian scheme so that there is an extension

$$
0 \to T \to G' \to A' \to 0
$$

with T a split torus and A' an abelian scheme over B'. If X denotes $\operatorname{Hom}(T, \mathbb{G}_m)$ then this extension is given by a homomorphism $c' : X \to (A')^t$. Let $c : X \to A^t$ (resp. $c_0 : X \to A_0^t$) denote the reduction to B (resp. B_0).

(ii) Let $\phi : Y \hookrightarrow X$ be a subgroup of finite index and $c^t : Y \to A$ a homomorphism.

(iii) Let S be a regular paving of $X_{\mathbb{R}}$.

(iv) Let $\mathcal{M}$ be a rigidified ample line bundle on A such that the diagram

$$
\begin{array}{ccc}
Y & \xrightarrow{\phi} & X \\
{\scriptstyle c^t}\downarrow & & \downarrow{\scriptstyle c} \\
A & \xrightarrow{\lambda_{\mathcal{M}}} & A^t
\end{array}
$$

commutes.

(v) Fix trivializations $\tau : Y \times X \to (c^t \times c)^* \mathcal{B}^{-1}$ and $\psi : Y \to (c^t)^* \mathcal{M}^{-1}$ as in (5.2.1 (vi) and (vii)).

(vi) Let $\beta : H_S \to B'$ be a homomorphism sending all nonzero elements to the radical of B'. This homomorphism defines a log structure $M_{B'}$. Let M_B (resp. M_{B_0}) denote the pullback to $\operatorname{Spec}(B)$ (resp. $\operatorname{Spec}(B_0)$).

Let

$$(\mathscr{P}, M_{\mathscr{P}}, L_{\mathscr{P}}) \to (\operatorname{Spec}(B), M_B)$$

be the polarized log scheme with G–action obtained from the standard construction.

We define the notion of a *log smooth deformation of*

$$(\mathscr{P}, M_{\mathscr{P}}, L_{\mathscr{P}}, G - \text{action}) \tag{5.5.1.1}$$

as in 4.3.2.

Proposition 5.5.2. *Let $(\mathscr{P}', M_{\mathscr{P}'}, L_{\mathscr{P}'}, G' - \text{action})$ be a log smooth deformation of $(\mathscr{P}, M_{\mathscr{P}}, L_{\mathscr{P}}, G - \text{action})$ to $(\operatorname{Spec}(B'), M_{B'})$. Then there exists a lifting $\mathcal{M}'$ of $\mathcal{M}$ to A' and maps*

$$c'^t : Y \to A', \quad \psi' : Y \to c'^{t*}\mathcal{M}'^{-1}, \quad \tau' : Y \times X \to (c^t \times c)^* \mathcal{B}'^{-1}$$

as in (5.2.1 (vi) and (vii)) lifting c^t, c, and τ such that $(\mathscr{P}', M_{\mathscr{P}'}, L_{\mathscr{P}'}, G' - \text{action})$ is isomorphic to the polarized family with G'–action obtained from the standard construction using this data.

Proof. Since the argument is very similar to the one in 4.3.3 we sketch the proof leaving some of the details to the reader.

By the same argument used in 4.3.3 the deformation $(\mathscr{P}', M_{\mathscr{P}'}, L_{\mathscr{P}'}, G' - \text{action})$ defines a deformation

$$(\widetilde{\mathscr{P}}', M_{\widetilde{\mathscr{P}}'}, L_{\widetilde{\mathscr{P}}'}, Y - \text{action}, G' - \text{action})$$

of
$$(\widetilde{\mathscr{P}}, M_{\widetilde{\mathscr{P}}}, L_{\widetilde{\mathscr{P}}}, Y-\text{action}, G-\text{action}).$$

Let $f : \widetilde{P} \to A$ be the projection. Then as in 4.3.8 the scheme $Z' = (|A|, f_*\mathcal{O}_{\widetilde{\mathscr{P}}'})$ is a deformation of A to B', and the action of G' on $\widetilde{\mathscr{P}}'$ induces an action of A' on Z'. Choosing a point of Z' we obtain a morphism $f' : \widetilde{\mathscr{P}}' \to A'$ lifting f.

By the same argument used in the proof of 4.3.3 there exists a lifting $\mathscr{M}'$ of $\mathscr{M}$ to A' such that $(\widetilde{\mathscr{P}}', M_{\widetilde{\mathscr{P}}'}, L_{\widetilde{\mathscr{P}}'})$ is isomorphic to the log scheme obtained from the standard construction using $\mathscr{M}'$. Thus the only issue is how to describe the Y–action.

By the same argument used in 4.3.16 there exists a unique homomorphism $c'^t : Y \to A'$ such that for every $y \in Y$ the diagram

$$
\begin{array}{ccc}
\widetilde{\mathscr{P}}' & \xrightarrow{s_y} & \widetilde{\mathscr{P}}' \\
{\scriptstyle f'}\downarrow & & \downarrow{\scriptstyle f'} \\
A' & \xrightarrow{t_{c'^t(y)}} & A'
\end{array}
\qquad (5.5.2.1)
$$

commutes, where s_y denotes the action of y on $\widetilde{\mathscr{P}}$. By the same argument used in (loc. cit.) we have the relation $\lambda_{\mathscr{M}'}c'^t = c'\phi$. Thus the Y–action is given by isomorphisms

$$\iota_y^{(d,x)} : t^*_{c'^t(y)}(\mathscr{M}'^d \otimes L_{c'(x)}) \to \mathscr{M}'^d \otimes L_{c'(x+\phi(dy))}.$$

As in 4.3.17 giving such isomorphisms is equivalent to giving maps ψ' and τ' as in the proposition. □

5.6 Isomorphisms without Log Structures

5.6.1. Let B be a scheme and

$$(G, M_B, \mathscr{P}, L_{\mathscr{P}}, M_{\mathscr{P}}), (G', M'_B, \mathscr{P}', L_{\mathscr{P}'}, M_{\mathscr{P}'}) \in \overline{\mathscr{T}}_{g,d}(B)$$

two objects of $\overline{\mathscr{T}}_{g,d}$. Let I denote the functor on the category of B–schemes classifying triples of isomorphisms

$$\sigma : G \to G', \quad \gamma : \mathscr{P} \to \mathscr{P}', \quad \iota : \gamma^* L_{\mathscr{P}'} \to L_{\mathscr{P}}$$

such that the diagram

$$
\begin{array}{ccc}
G \times \mathscr{P} & \xrightarrow{\text{action}} & \mathscr{P} \\
{\scriptstyle \sigma \times \gamma}\downarrow & & \downarrow{\scriptstyle \gamma} \\
G' \times \mathscr{P}' & \xrightarrow{\text{action}} & \mathscr{P}'
\end{array}
$$

commutes. We will use write just $(G, \mathscr{P}, L_{\mathscr{P}}) \to (G', \mathscr{P}', L_{\mathscr{P}'})$ for such a collection of isomorphisms.

Proposition 5.6.2. *The functor I is a quasi–projective scheme.*

Proof. Let $I(\mathscr{P}, \mathscr{P}')$ denote the functor classifying isomorphisms $\mathscr{P} \to \mathscr{P}'$. Since $\mathscr{P}$ and $\mathscr{P}'$ are projective and flat over B, the functor $I(\mathscr{P}, \mathscr{P}')$ is a disjoint union of quasi–projective schemes (it is equal to an open subfunctor of the Hilbert scheme of $\mathscr{P} \times \mathscr{P}'$). Let $I((\mathscr{P}, L_{\mathscr{P}}), (\mathscr{P}', L_{\mathscr{P}'}))$ denote the functor of pairs $\gamma : \mathscr{P} \to \mathscr{P}'$ and $\iota : \gamma^* L_{\mathscr{P}'} \to L_{\mathscr{P}}$. For any $u \in \mathbb{G}_m(B)$ we can change the isomorphisms ι to $u \cdot \iota$, and therefore the is a natural action of $\mathbb{G}_m$ on $I((\mathscr{P}, L_{\mathscr{P}}), (\mathscr{P}', L_{\mathscr{P}'}))$. Let F denote the quotient sheaf (with respect to the fppf topology) $I((\mathscr{P}, L_{\mathscr{P}}), (\mathscr{P}', L_{\mathscr{P}'}))/\mathbb{G}_m$. There is a natural forgetful map $\pi : F \to I(\mathscr{P}, \mathscr{P}')$.

Lemma 5.6.3 *The morphism π identifies F with a closed subscheme of $I(\mathscr{P}, \mathscr{P}')$. In particular, since $I((\mathscr{P}, L_{\mathscr{P}}), (\mathscr{P}', L_{\mathscr{P}'}))$ is a $\mathbb{G}_m$–torsor over F, the functor $I((\mathscr{P}, L_{\mathscr{P}}), (\mathscr{P}', L_{\mathscr{P}'}))$ is a disjoint union of quasi–projective schemes.*

Proof. For any point $\sigma \in I(\mathscr{P}, \mathscr{P}')$ the fiber product $F \times_{I(\mathscr{P}, \mathscr{P}'),\sigma} B$ is represented by the fiber product of the diagram

$$
\begin{array}{ccc}
 & & \underline{\mathrm{Pic}}_{\mathscr{P}/B} \\
 & & \Delta \downarrow \\
B & \xrightarrow{[L_{\mathscr{P}}] \times [\sigma^* L_{\mathscr{P}'}]} & \underline{\mathrm{Pic}}_{\mathscr{P}/B} \times \underline{\mathrm{Pic}}_{\mathscr{P}/B}.
\end{array}
$$

Since $\underline{\mathrm{Pic}}_{\mathscr{P}/B}$ is a separated algebraic space by [5, 7.3] the result follows. $\square$

Next consider the forgetful map $I \to I((\mathscr{P}, L_{\mathscr{P}}), (\mathscr{P}', L_{\mathscr{P}'}))$. We claim that this map is representable by closed immersions. For this fix a pair of isomorphisms $\gamma : \mathscr{P} \to \mathscr{P}'$ and $\iota : \gamma^* L_{\mathscr{P}'} \to L_{\mathscr{P}}$. The fiber product

$$
\mathcal{F} := I \times_{I((\mathscr{P}, L_{\mathscr{P}}), (\mathscr{P}', L_{\mathscr{P}'})),(\gamma,\iota)} B
$$

is then equal to the functor which to any B–scheme B' associates the set of isomorphisms $\beta : G \to G'$ such that the diagram

$$
\begin{array}{ccc}
G_{B'} \times \mathscr{P}_{B'} & \xrightarrow{\text{action}} & \mathscr{P}_{B'} \\
\beta \times \gamma \downarrow & & \downarrow \gamma \\
G'_{B'} \times \mathscr{P}'_{B'} & \xrightarrow{\text{action}} & \mathscr{P}'_{B'}
\end{array}
\tag{5.6.3.1}
$$

commutes.

Lemma 5.6.4 *The functor $\mathcal{F}$ is a subfunctor of B.*

Proof. We have to show that for any $B' \to B$ as above there exists at most one isomorphism $\beta : G \to G'$ such that 5.6.3.1 commutes. For this it suffices to consider the case when B' is an Artinian local ring. In this case $\mathscr{P}'$ and $\mathscr{P}$ are given by the standard construction. It follows from the standard construction that the maximal open subschemes $\mathscr{P}'^{\mathrm{reg}} \subset \mathscr{P}'$ and $\mathscr{P}^{\mathrm{reg}} \subset \mathscr{P}$ where the structure morphism is smooth have the property that if we fix points $e' \in \mathscr{P}'^{\mathrm{reg}}$ and $e \in \mathscr{P}^{\mathrm{reg}}$ then the maps

$$G' \to \mathscr{P}'^{\mathrm{reg}}, \quad g \mapsto g(e')$$

and

$$G \to \mathscr{P}^{\mathrm{reg}}, \quad g \mapsto g(e)$$

are open immersions. If we choose $e' = \gamma(e)$ then the diagram

$$
\begin{array}{ccc}
G & \xrightarrow{\;g \mapsto g(e)\;} & \mathscr{P} \\
{\scriptstyle \beta}\big\downarrow & & \big\downarrow{\scriptstyle \gamma} \\
G' & \xrightarrow{\;g' \mapsto g'(e')\;} & \mathscr{P}'
\end{array}
$$

commutes and hence β is determined by γ. $\quad\square$

Lemma 5.6.5 *The structural map $\mathcal{F} \to B$ is represented by closed immersion.*

Proof. After replacing B by an étale cover, we can find a section $\theta \in f_* L_{\mathscr{P}}$ such that $(G, \mathscr{P}, L_{\mathscr{P}}, \theta)$ and $(G', \mathscr{P}', L_{\mathscr{P}'}, \iota(\theta))$ define two points s, s' of Alexeev's space $\mathscr{A}_{g,d}^{\mathrm{Alex}}(B)$. Since $\mathscr{A}_{g,d}^{\mathrm{Alex}}$ has finite diagonal the projection

$$B \times_{s, \mathscr{A}_{g,d}^{\mathrm{Alex}}, s'} B \to B$$

is a finite morphism, and by the above the map

$$B \times_{\mathscr{A}_{g,d}^{\mathrm{Alex}}} B \to I((\mathscr{P}, L_{\mathscr{P}}), (\mathscr{P}', L_{\mathscr{P}'}))$$

is a closed immersion. The desired closed subscheme of B is then the fiber product

$$(B \times_{\mathscr{A}_{g,d}^{\mathrm{Alex}}} B) \times_{I((\mathscr{P}, L_{\mathscr{P}}), (\mathscr{P}', L_{\mathscr{P}'})), (\gamma, \iota)} B \subset B.$$

$\square$

It follows that I is a disjoint union of quasi–projective schemes. To conclude the proof of 5.6.2 it therefore suffices to show that I is quasi–compact. For this let $\mathscr{U}$ (resp. $\mathscr{U}'$) denote the open subset of the total space of $f_* L_{\mathscr{P}}$ (resp. $f'_* L_{\mathscr{P}'}$) classifying sections giving $(G, \mathscr{P}, L_{\mathscr{P}})$ (resp. $(G', \mathscr{P}', L_{\mathscr{P}'})$) the structure of a point in $\mathscr{A}_{g,d}^{\mathrm{Alex}}$. If $s, s' \in \mathscr{A}_{g,d}^{\mathrm{Alex}}(\mathscr{U} \times_B \mathscr{U}')$ denote the tautological objects then we have a commutative diagram

$$
\begin{array}{ccc}
\underline{\mathrm{Isom}}_{\mathscr{A}_{g,d}^{\mathrm{Alex}}}(s, s') & \xrightarrow{\;\;\pi\;\;} & I \\
{\scriptstyle \text{q-compact}}\big\downarrow & & \big\downarrow \\
\mathscr{U} \times_B \mathscr{U}' & \xrightarrow{\;\text{q-compact}\;} & B,
\end{array}
$$

where π is surjective. It follows that I is also quasi–compact. This completes the proof of 5.6.2. $\quad\square$

5.7 Algebraization of Formal Log Structures

In this section we gather together some technical results about algebraization of formal log structures needed in what follows.

Proposition 5.7.1. *Let X be a noetherian scheme and F_1 and F_2 two constructible sheaves of sets on X. Suppose given for every point $x \in X$ a morphism $f_{\bar{x}} : F_{1,\bar{x}} \to F_{2,\bar{x}}$ of sets with continuous $\mathrm{Gal}(\bar{x}/x)$–action such that for every specialization $\eta \to x$ the diagram*

$$
\begin{array}{ccc}
F_{1,\bar{x}} & \xrightarrow{\ f_{\bar{x}}\ } & F_{2,\bar{x}} \\
{\scriptstyle \mathrm{sp}}\downarrow & & \downarrow{\scriptstyle \mathrm{sp}} \\
F_{1,\bar{\eta}} & \xrightarrow{\ f_{\bar{\eta}}\ } & F_{2,\bar{\eta}}
\end{array}
\qquad (5.7.1.1)
$$

commutes, where the arrows labelled sp are the specialization morphisms. Then there exists a unique morphism of sheaves $f : F_1 \to F_2$ inducing the maps $f_{\bar{x}}$ on stalks.

Proof. We proceed by noetherian induction. If X is zero–dimensional the result is clear. For the induction step, note that by proper descent for étale sheaves it suffices to consider the case when X is an integral normal scheme. Let $j : U \subset X$ be a dense open subset such that the restrictions $F_{i,U}$ of the sheaves F_i to U are locally constant, and let $i : Z \hookrightarrow X$ be the complement. If $\eta \in U$ is the generic point then the category of locally constant sheaves on U is equivalent to a full subcategory of the category of sets with continuous action of $\mathrm{Gal}(\bar{\eta}/\eta)$. The map $f_{\bar{\eta}} : F_{1,\bar{\eta}} \to F_{2,\bar{\eta}}$ therefore induces a morphism $f_U : F_{1,U} \to F_{2,U}$. By the induction hypothesis we also have a morphism $f_Z : F_{1,Z} \to F_{2,Z}$. Now recall that the category of sheaves on X is equivalent to the category of triples (M_U, M_Z, ϕ) where M_U is a sheaf on U, M_Z is a sheaf on Z, and $\phi : M_Z \to i^* j_* M_U$ is a morphism of sheaves on Z. To construct the morphism f it therefore suffices to show that the following diagram commutes

$$
\begin{array}{ccc}
F_{1,Z} & \longrightarrow & i^* j_* F_{1,U} \\
{\scriptstyle f_Z}\downarrow & & \downarrow{\scriptstyle f_U} \\
F_{2,Z} & \longrightarrow & i^* j_* F_{2,U},
\end{array}
$$

where the horizontal arrows are the natural maps $i^* F_i \to i^* j_* j^* F_i$. Looking at a point $x \in Z$ this amounts to the commutativity of the diagram

$$
\begin{array}{ccc}
F_{1,\bar{x}} & \xrightarrow{\ f_{\bar{x}}\ } & F_{2,\bar{x}} \\
{\scriptstyle \mathrm{sp}}\downarrow & & \downarrow{\scriptstyle \mathrm{sp}} \\
F_{1,\bar{\eta}} & \xrightarrow{\ f_{\bar{\eta}}\ } & F_{2,\bar{\eta}}
\end{array}
$$

which holds by assumption. $\square$

Theorem 5.7.2 *Let A be a complete noetherian local ring with maximal ideal $\mathfrak{a}$ and $f : X \to \operatorname{Spec}(A)$ a proper morphism of schemes. For $n \geq 0$ let X_n denote $X \otimes_A (A/\mathfrak{a}^{n+1})$. Then the natural functor between groupoids*

$$\text{(fine log structures on } X\text{)}$$

$$\downarrow \qquad\qquad (5.7.2.1)$$

$$\text{(compatible families of fine log structures on the } X_n\text{)}$$

is fully faithful.

Proof. Let M_1 and M_2 be two fine log structures on X and let $M_{i,n}$ denote the pullback of M_i to X_n. Assume given a compatible family of isomorphisms $\sigma_n : M_{1,n} \to M_{2,n}$. We have to show that there exists a unique isomorphism $\sigma : M_1 \to M_2$ inducing the σ_n.

We first construct the map $\bar{\sigma} : \overline{M}_1 \to \overline{M}_2$. Note that the map $\bar{\sigma}$ is necessarily unique. For if $\eta \in X$ is any point then since X is proper over A the closure of η contains a point $x \in X_0$. The resulting diagram

$$
\begin{array}{ccc}
\overline{M}_{1,\bar{x}} & \xrightarrow{\bar{\sigma}_{\bar{x}}} & \overline{M}_{2,\bar{x}} \\
{\scriptstyle\text{sp}}\downarrow & & \downarrow{\scriptstyle\text{sp}} \\
\overline{M}_{1,\bar{\eta}} & \xrightarrow{\bar{\sigma}_{\bar{\eta}}} & \overline{M}_{2,\bar{\eta}}
\end{array}
$$

must then commute, where the vertical arrows are the surjective specialization maps. Therefore $\bar{\sigma}_{\bar{x}}$ determines $\bar{\sigma}_{\bar{\eta}}$. On the other hand, the map $\bar{\sigma}_{\bar{x}}$ is equal to the map induced by the σ_n. By the uniqueness, we can therefore replace A by a finite extension. We choose a finite extension as follows. Let $\{Z_j \subset X\}$ be a stratification by irreducible locally closed subschemes such that the sheaves $\overline{M}_i$ are locally constant when restricted to Z_j, and let $\overline{Z}_j \subset X$ denote the scheme–theoretic closure of Z_j. After replacing A by a finite extension and refining our stratification we can assume that the closed fiber of each $\overline{Z}_j$ is connected.

We first define for every point $\eta \in X$ an isomorphism $f_{\bar{\eta}} : \overline{M}_{1,\bar{\eta}} \to \overline{M}_{2,\bar{\eta}}$. For this let Z denote the scheme–theoretic closure of η in X. Since X is proper over $\operatorname{Spec}(A)$ there exists a point $x \in Z \cap X_0$. We then have a diagram

$$
\begin{array}{ccc}
\overline{M}_{1,\bar{x}} & \xrightarrow{\bar{\sigma}_{\bar{x}}} & \overline{M}_{2,\bar{x}} \\
{\scriptstyle\pi_1}\downarrow & & \downarrow{\scriptstyle\pi_2} \\
\overline{M}_{1,\bar{\eta}} & & \overline{M}_{2,\bar{\eta}},
\end{array}
$$

where the vertical arrows are surjections. We claim that the morphism $\sigma_{\bar{x}}$ descends to a morphism $f_{\bar{\eta}} : \overline{M}_{1,\bar{\eta}} \to \overline{M}_{2,\bar{\eta}}$. For this note that if $F_i \subset \overline{M}_{i,\bar{x}}$ denotes the face of elements $f \in \overline{M}_{i,\bar{x}}$ such that for any lifting $\tilde{f} \in M_{i,\bar{x}}$ the image of $\alpha(f)$ in $\mathcal{O}_{Z,\bar{x}}$ is non-zero, then $\overline{M}_{i,\bar{\eta}}$ is the quotient of $\overline{M}_{i,\bar{x}}$ by F_i.

Lemma 5.7.3 *The map* $\bar{\sigma}_{\bar{x}} : \overline{M}_{1,\bar{x}} \to \overline{M}_{2,\bar{x}}$ *maps* F_1 *to* F_2.

Proof. Let $\widehat{\mathscr{O}}_{X,\bar{x}}$ denote the completion with respect to $\mathfrak{m}_A \cdot \mathscr{O}_{X,\bar{x}}$ of the local ring $\mathscr{O}_{X,\bar{x}}$ and let $\widehat{M}_i$ denote the pullback of M_i to $\mathrm{Spec}(\widehat{\mathscr{O}}_{X,\bar{x}})$. The isomorphisms σ_n then induce an isomorphism of log structures $\hat{\sigma}_{\bar{x}} : \widehat{M}_1 \to \widehat{M}_2$ inducing $\bar{\sigma}_{\bar{x}}$. The lemma then follows by noting that F_i is equal to the image in $\overline{M}_{i,\bar{x}}$ of the submonoid

$$\{m \in \widehat{M}_{i,\bar{x}} | \alpha(m) \neq 0 \text{ in } \widehat{\mathscr{O}}_{X,\bar{x}}\}. \tag{5.7.3.1}$$

$\square$

We therefore obtain a morphism $f^x_{\bar{\eta}} : \overline{M}_{1,\bar{\eta}} \to \overline{M}_{2,\bar{\eta}}$.

Lemma 5.7.4 *The map* $f^x_{\bar{\eta}}$ *is independent of the choice of* $x \in Z$.

Proof. Suppose η lies in a stratum Z_j, and let $\eta' \in Z_j$ be the generic point. Then x is also in the closure of η' and by the construction there is a commutative diagram

$$
\begin{array}{ccc}
\overline{M}_{1,\bar{x}} & \xrightarrow{\bar{\sigma}_{\bar{x}}} & \overline{M}_{2,\bar{x}} \\
\pi_1 \downarrow & & \downarrow \pi_2 \\
\overline{M}_{1,\bar{\eta}} & \xrightarrow{f^x_{\bar{\eta}}} & \overline{M}_{2,\bar{\eta}} \\
\pi'_1 \downarrow & & \downarrow \pi'_2 \\
\overline{M}_{1,\bar{\eta}'} & \xrightarrow{f^x_{\bar{\eta}'}} & \overline{M}_{2,\bar{\eta}'}.
\end{array}
$$

where the maps π'_i are isomorphisms since both η and η' lie in Z_j. It therefore suffices to consider the case when η is the generic point of Z_j. In this case the scheme Z has connected closed fiber and therefore it suffices to show that if $x' \in Z \cap X_0$ is a second point with x' specializing to x then the maps $f^x_{\bar{\eta}}$ and $f^{x'}_{\bar{\eta}}$ are equal. But this is clear for we have a commutative diagram

$$
\begin{array}{ccc}
\overline{M}_{1,\bar{x}} & \xrightarrow{\bar{\sigma}_{\bar{x}}} & \overline{M}_{2,\bar{x}} \\
\pi_1 \downarrow & & \downarrow \pi_2 \\
\overline{M}_{1,\bar{x}'} & \xrightarrow{\bar{\sigma}^x_{\bar{x}'}} & \overline{M}_{2,\bar{x}'} \\
\pi'_1 \downarrow & & \downarrow \pi'_2 \\
\overline{M}_{1,\bar{\eta}} & \xrightarrow{f^x_{\bar{\eta}}} & \overline{M}_{2,\bar{\eta}}.
\end{array}
$$

$\square$

In what follows we write just $f_{\bar{\eta}}$ for $f_{\bar{\eta}}^x$ (for any choice of x).

To construct $\bar{\sigma}$, it therefore suffices by 5.7.1 to show that for any two points $\eta, \eta' \in X$ with η' specializing to η the diagram

$$
\begin{array}{ccc}
\overline{M}_{1,\bar{\eta}} & \xrightarrow{f_{\bar{\eta}}} & \overline{M}_{2,\bar{\eta}} \\
\text{sp} \downarrow & & \downarrow \text{sp} \\
\overline{M}_{1,\bar{\eta}'} & \xrightarrow{f_{\bar{\eta}'}} & \overline{M}_{2,\bar{\eta}'}
\end{array}
$$

commutes. Let x be a point in the closed fiber of the closure of η. Then by the construction there is a commutative diagram

$$
\begin{array}{ccc}
\overline{M}_{1,\bar{x}} & \xrightarrow{\bar{\sigma}_{\bar{x}}} & \overline{M}_{2,\bar{x}} \\
\downarrow & & \downarrow \\
\overline{M}_{1,\bar{\eta}} & \xrightarrow{f_{\bar{\eta}}} & \overline{M}_{2,\bar{\eta}} \\
\downarrow & & \downarrow \\
\overline{M}_{1,\bar{\eta}'} & \xrightarrow{f_{\bar{\eta}'}} & \overline{M}_{2,\bar{\eta}'}
\end{array}
$$

where the vertical maps are surjections. This therefore completes the construction of $\bar{\sigma} : \overline{M}_1 \to \overline{M}_2$.

To lift the map (uniquely) to an isomorphism of log structures $\sigma : M_1 \to M_2$ proceed as follows. Let Q be the functor on X–schemes associating to any $f : T \to X$ the set of isomorphisms of log structures $f^*M_1 \to f^*M_2$ inducing $f^{-1}(\bar{\sigma}) : f^{-1}\overline{M}_1 \to f^{-1}\overline{M}_2$. Then as in 4.9.2 the functor Q is a separated algebraic space over X. The maps σ_n are then given by a compatible family of sections $s_n : X_n \to Q_n$ which since X is proper over A is induced by a unique section $s : X \to Q$ (using the Grothendieck existence theorem for algebraic spaces [43, 1.4]). This gives the desired isomorphism σ. This completes the proof of 5.7.2. $\square$

5.8 Description of the Group H_S^{gp}

5.8.1. Fix a lattice X as in 4.1.1, and let S be a regular paving of X defined by a quadratic form B_0. Define H_S as in 4.1.3.

Lemma 5.8.2 *The map*

$$
s : X \times X \to H_S^{gp}, \quad (x,y) \mapsto (1, x+y) * (1,0) - (1,x) * (1,y)
$$

is bilinear and symmetric.

Proof. The map s is clearly symmetric. It therefore suffices to show that for $x, y_1, y_2 \in X$ we have

$$s(x, y_1 + y_2) = s(x, y_1) + s(x, y_2).$$

For any element $(d, x) \in P$ write $\overline{(d, x)} \in \varinjlim_{\omega} P_{\omega}^{gp}$ for the image under the map 4.1.3.1. We compute

$$
\begin{aligned}
&s(x, y_1) + s(x, y_2) \\
&= (1, x + y_1) * (1, 0) - (1, x) * (1, y_1) + (1, x + y_2) * (1, 0) - (1, x) * (1, y_2) \\
&= (1, x + y_1 + y_2) * (1, y_2) - (1, x + y_2) * (1, y_1 + y_2) + (1, x + y_2) * (1, 0) \\
&\quad - (1, x) * (1, y_2) \\
&= \overline{(1, x + y_1 + y_2)} + \overline{(1, y_2)} - \overline{(2, x + y_1 + 2y_2)} - \overline{(1, x + y_2)} - \overline{(1, y_1 + y_2)} \\
&\quad + \overline{(2, x + y_1 + 2y_2)} + \overline{(1, x + y_2)} + \overline{(1, 0)} - \overline{(2, x + y_2)} - \overline{(1, x)} - \overline{(1, y_2)} \\
&\quad + \overline{(2, x + y_2)} \\
&= \overline{(1, x + y_1 + y_2)} + \overline{(1, 0)} - \overline{(2, x + y_1 + y_2)} - \overline{(1, x)} - \overline{(1, y_1 + y_2)} \\
&\quad + \overline{(2, x + y_1 + y_2)} \\
&= (1, x + y_1 + y_2) * (1, 0) - (1, x) * (1, y_1 + y_2) \\
&= s(x, y_1 + y_2).
\end{aligned}
$$

$\square$

We write also $s : S^2 X \to H_S^{gp}$ for the induced map from the second symmetric power.

5.8.3. The map s can also be described as follows. Let $\mathbb{Z}^{(X)} = \oplus_{x \in X} \mathbb{Z} \cdot x$ denote the free abelian group generated by X, with X-action induced by translation on X. For any abelian group N with X-action we then have a canonical isomorphism of groups

$$\mathrm{Hom}_X(\mathbb{Z}^{(X)}, N) \simeq N, \tag{5.8.3.1}$$

where the left side denotes morphisms of X-representations. In particular, $\mathbb{Z}^{(X)}$ is a projective X-representation. From this it also follows that

$$H_0(\mathbb{Z}^{(X)}) \simeq \mathbb{Z}. \tag{5.8.3.2}$$

As before let P^{gp} denote the X-representation with underlying abelian group $\mathbb{Z} \oplus X$ and X-action given by

$$y * (n, x) = (n, x + ny). \tag{5.8.3.3}$$

Let

$$\pi : \mathbb{Z}^{(X)} \to P^{gp} \tag{5.8.3.4}$$

be the surjection given by

$$(n_x)_x \mapsto \left(\sum n_x, \sum n_x \cdot x\right), \tag{5.8.3.5}$$

and let K denote the kernel of π.

Consider the exact sequence of X-representations

$$0 \longrightarrow X \xrightarrow{x \mapsto (0,x)} P^{gp} \xrightarrow{(n,x) \mapsto n} \mathbb{Z} \longrightarrow 0, \tag{5.8.3.6}$$

where X acts trivially on X and $\mathbb{Z}$. By direct calculation one sees that this induces an isomorphism

$$H_0(P^{gp}) \simeq H_0(\mathbb{Z}) \simeq \mathbb{Z}. \tag{5.8.3.7}$$

We therefore obtain an exact sequence

$$H_2(\mathbb{Z}) \to H_1(X) \to H_1(P^{gp}) \to H_1(\mathbb{Z}) \to X \to 0. \tag{5.8.3.8}$$

5.8.4. The homology of the trivial X-representation $\mathbb{Z}$ can be computed as follows. Consider the surjection

$$p : \mathbb{Z}^{(X)} \to \mathbb{Z} \tag{5.8.4.1}$$

sending $(n_x)_x$ to $\sum n_x$. The kernel V consists of elements $(n_x)_x$ such that $\sum n_x = 0$. Every such element can be written uniquely as

$$\sum_x n_x (1_x - 1_0), \tag{5.8.4.2}$$

where 1_x denotes the generator of the x-component in $\mathbb{Z}^{(X)}$. The kernel V of p therefore isomorphism to the free abelian group on the generators $\zeta_x := 1_x - 1_0$ with X-action given by

$$y * \zeta_x = \zeta_{x+y} - \zeta_y. \tag{5.8.4.3}$$

In particular

$$H_1(\mathbb{Z}) \simeq H_0(V) \simeq X, \tag{5.8.4.4}$$

from which it follows that the map $H_1(\mathbb{Z}) \to X$ in 5.8.3.8 is an isomorphism. We therefore have a short exact sequence

$$H_2(\mathbb{Z}) \to H_1(X) \to H_1(P^{gp}) \to 0. \tag{5.8.4.5}$$

Lemma 5.8.5 *The map*

$$H_1(X) \simeq H_1(\mathbb{Z}) \otimes X \simeq X \otimes X \to H_1(P^{gp}) \tag{5.8.5.1}$$

in 5.8.3.8 induces an isomorphism

$$\Gamma^2(X) = X \otimes X / \Lambda^2 X \to H_1(P^{gp}). \tag{5.8.5.2}$$

Proof. This is shown in [3, 5.1.4]. $\square$

5.8.6. The map

$$X \otimes X \to H_1(P^{gp}) \tag{5.8.6.1}$$

can be described explicitly as follows. From the commutative diagram

$$
\begin{array}{ccccccccc}
0 & \longrightarrow & V \otimes X & \longrightarrow & \mathbb{Z}^{(X)} \otimes X & \xrightarrow{p \otimes 1} & X & \longrightarrow & 0 \;, \\
& & \downarrow c & & \downarrow q & & \downarrow & & \\
0 & \longrightarrow & K & \longrightarrow & \mathbb{Z}^{(X)} & \xrightarrow{\pi} & P^{gp} & \longrightarrow & 0
\end{array}
\tag{5.8.6.2}
$$

where q is the map sending $1_x \otimes y$ to $1_{y+x} - 1_x$, we see that the map 5.8.6.1 sends $x \otimes y$ to the class in $H_0(K) \simeq H_1(P^{gp})$ of

$$1_{y+x} - 1_x - 1_y + 1_0 \in K \subset \mathbb{Z}^{(X)}. \tag{5.8.6.3}$$

Note that from this formula it is also clear that 5.8.6.1 descends to a map $\Gamma^2 X \to H_1(P^{gp})$.

Lemma 5.8.7 *The map c in 5.8.6.2 is surjective.*

Proof. We can expand the diagram 5.8.6.2 into a diagram

$$
\begin{array}{ccccccccc}
0 & \longrightarrow & V \otimes X & \longrightarrow & \mathbb{Z}^{(X)} \otimes X & \xrightarrow{p \otimes 1} & X & \longrightarrow & 0 \;, \\
& & \downarrow c & & \downarrow e & & \downarrow \text{id} & & \\
0 & \longrightarrow & K & \longrightarrow & V & \longrightarrow & X & \longrightarrow & 0 \\
& & \downarrow \text{id} & & \downarrow & & \downarrow & & \\
0 & \longrightarrow & K & \longrightarrow & \mathbb{Z}^{(X)} & \xrightarrow{\pi} & P^{gp} & \longrightarrow & 0
\end{array}
\tag{5.8.7.1}
$$

Now observe that

$$\zeta_x = e(1_0 \otimes x) \tag{5.8.7.2}$$

in V so the map e is surjective. From this it follows that c is also surjective.
$\square$

5.8.8. Now observe that there is a commutative diagram

$$
\begin{array}{ccccccccc}
0 & \longrightarrow & K & \longrightarrow & \mathbb{Z}^{(X)} & \xrightarrow{\pi} & P^{gp} & \longrightarrow & 0 \\
& & \downarrow & & \downarrow b & & \downarrow \simeq & & \\
0 & \longrightarrow & \widetilde{H}_S^{gp} & \longrightarrow & \varinjlim_\omega P_\omega^{gp} & \longrightarrow & P^{gp} & \longrightarrow & 0,
\end{array}
\tag{5.8.8.1}
$$

where b is the map sending 1_x to $(1, x)$, which upon taking homology induces a morphism

$$\Gamma^2 X \simeq H_0(K) \to H_0(\widetilde{H}_S^{gp}) = H_S^{gp}. \tag{5.8.8.2}$$

From the explicit description in 5.8.6 it follows that this map is equal to the map s.

Lemma 5.8.9 *Let* $g : P \rtimes H_S \to Q$ *be a morphism of monoids sending* H_S *to* $Q_{\geq 0}$, *and let* $A : \mathbb{Z}^{(X)} \to Q$ *denote the composite*

$$\mathbb{Z}^{(X)} \xrightarrow{\ b\ } (P \rtimes H_S)^{gp} \xrightarrow{\ g\ } Q. \tag{5.8.9.1}$$

Let $h : X_Q \to Q$ *be the function obtained as the lower envelope of the convex hull of the set*

$$G_A := \{(x, A(1_x)) | x \in X\} \subset X_Q \times Q. \tag{5.8.9.2}$$

Then for every $z \in X_Q$ *we have*

$$h(z) = \frac{1}{N} g(N, Nz), \tag{5.8.9.3}$$

where N *is any integer such that* $Nz \in X$.

Proof. By definition $h(z)$ is equal to the infimum of the numbers

$$\sum_i a_i g(1, x_i), \tag{5.8.9.4}$$

taken over all possible ways of writing $z = \sum_i a_i x_i$ with $a_i \in Q_{\geq 0}$, $\sum a_i = 1$, and $x_i \in X$. Let $\omega \in S$ denote a simplex containing z and write $z = \sum a_i y_i$ as above with $y_i \in \omega$. If N is an integer such that $Na_i \in \mathbb{Z}$ for all i we then obtain

$$\sum a_i g(1, y_i) = \frac{1}{N} [\sum (Na_i) g(1, y_i)]$$
$$= \frac{1}{N} (\sum g(Na_i, Na_i y_i))$$
$$= \frac{1}{N} g(\sum Na_i, \sum (Na_i) y_u) \quad \text{(since } g \text{ is linear on } P_\omega\text{)}$$
$$= \frac{1}{N} g(N, Nz).$$

It therefore suffices to show that for any other expression $z = \sum_{j=1}^r b_j x_j$ and positive integer N such that $Na_i, Nb_j \in \mathbb{Z}$ for all i, j we have

$$\frac{1}{N} g(N, Nz) \leq \sum_{j=1}^r b_j g(1, x_j), \tag{5.8.9.5}$$

or equivalently

$$\sum_{j=1}^r (Nb_j) g(1, x_j) \geq g(N, Nz). \tag{5.8.9.6}$$

For this note that

$$\sum_{j=1}^{r}(Nb_j)g(1,x_j) = \sum_{j=1}^{r}g(Nb_j, Nb_j x_j)$$

$$= \sum (N(\sum b_j), N\sum b_j x_j)$$

$$+ \sum_{j=1}^{r-1} g((Nb_{j+1}, Nb_{j+1}x_{j+1}) * (\sum_{i=1}^{j} Nb_i, \sum_{i=1}^{j} Nb_i x_i))$$

$$= g(N, Nz)$$

$$+ \sum_{j=1}^{r-1} g((Nb_{j+1}, Nb_{j+1}x_{j+1}) * (\sum_{i=1}^{j} Nb_i, \sum_{i=1}^{j} Nb_i x_i))$$

and

$$\sum_{j=1}^{r-1} g((Nb_{j+1}, Nb_{j+1}x_{j+1}) * (\sum_{i=1}^{j} Nb_i, \sum_{i=1}^{j} Nb_i x_i)) \geq 0 \qquad (5.8.9.7)$$

by assumption. □

Lemma 5.8.10 *Let $g : H_S \to \mathbb{Q}_{\geq 0}$ and let B denote the quadratic form*

$$S^2 X \xrightarrow{\ s\ } H_S^{gp} \xrightarrow{\ g\ } \mathbb{Q}. \qquad (5.8.10.1)$$

Then B is positive semidefinite and the paving S' associated to the quadratic function $a(x) := B(x,x)/2$ on X as in 4.1.1 is coarser than S, and the map g factors through $H_{S'}$.

Proof. That B is positive semidefinite is clear since $s(x \otimes x) \in H_S$.
Let

$$A : \mathbb{Z}^{(X)} \to \mathbb{Q} \qquad (5.8.10.2)$$

denote the function sending 1_x to $a(x)$. Note first of all that the induced map

$$A : K \to \mathbb{Q} \qquad (5.8.10.3)$$

descends to a map $H_0(K) \to \mathbb{Q}$. Indeed by 5.8.7 the map $c : V \otimes X \to K$ is surjective and

$$A(c(\zeta_x \otimes y)) = A(1_{y+x}) - A(1_y) - A(1_x) + A(1_0)$$
$$= \frac{1}{2}[B(x+y, x+y) - B(y,y) - B(x,x)]$$
$$= B(x,y).$$

On the other hand, for any $z \in X$ we also have

$$A(z \cdot c(\zeta_x \otimes y)) = A(1_{y+x+z}) - A(1_{y+z}) - A(1_{x+z}) + A(1_z) \qquad (5.8.10.4)$$

which an elementary calculation shows is also equal to $B(x,y)$.

Since $B(x,y) = g(s(x \otimes y))$ the above in fact shows that there is a commutative diagram

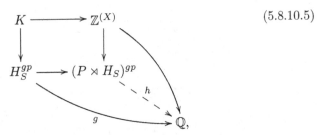

$$(5.8.10.5)$$

where the arrow h is obtained by noting that the left square in the commutative diagram

$$
\begin{array}{ccccccccc}
0 & \longrightarrow & K & \longrightarrow & \mathbb{Z}^{(X)} & \longrightarrow & P^{gp} & \longrightarrow & 0 \\
& & \downarrow & & \downarrow & & \downarrow & & \\
0 & \longrightarrow & H_S^{gp} & \longrightarrow & (P \rtimes H_S)^{gp} & \longrightarrow & P^{gp} & \longrightarrow & 0
\end{array}
$$

$$(5.8.10.6)$$

is cocartesian. Now by 5.8.9 the paving S' is equal to the domains of linearity of the function on $X_{\mathbb{Q}}$ given by

$$x \mapsto \frac{1}{N} h(N, Nx), \qquad (5.8.10.7)$$

where N is any integer such that $Nx \in X$. From this the proposition follows. $\square$

Corollary 5.8.11 *Let $\tilde{g} : \varinjlim P_\omega^{gp} \to \mathbb{Q}$ be a piecewise linear function such that the induced map $\tilde{H}_S^{gp} \to \mathbb{Q}$ descends to H_S^{gp}. Assume the quadratic form*

$$S^2 X \xrightarrow{\ s\ } H_S^{gp} \xrightarrow{\ \tilde{g}\ } \mathbb{Q} \qquad (5.8.11.1)$$

is trivial. Then $\tilde{g}$ is a linear map.

Proof. By 5.8.10 the map $\tilde{g}$ descends to a map $P^{gp} \to \mathbb{Q}$. $\square$

Corollary 5.8.12 *The map $s_{\mathbb{Q}} : S^2 X \otimes \mathbb{Q} \to H_S^{gp} \otimes \mathbb{Q}$ is surjective.*

Proof. It suffices to show that the map

$$\mathrm{Hom}(H_S^{gp}, \mathbb{Q}) \to \mathrm{Hom}(S^2 X, \mathbb{Q}) \qquad (5.8.12.1)$$

is injective, which follows from 5.8.11. $\square$

Lemma 5.8.13 *Let B_0 be a positive semidefinite quadratic form $S^2 X \to \mathbb{Q}$, and define $U(B_0) \subset \mathrm{Hom}(S^2 X, \mathbb{Q})$ to be the set of positive semidefinite quadratic forms B whose associated paving of $X_{\mathbb{R}}$ is coarser than the paving S_{B_0} defined by B_0. Then $U(B_0)$ is a cone in $\mathrm{Hom}(S^2 X, \mathbb{Q})$.*

Proof. This follows from the theory of the "second Voronoi decomposition" [39, Part I, §2]. □

In fact, as explained in [39, Part I, 2.5] the collection of the cones $U(B_0)$ form a $GL(X)$-invariant cone decomposition of the monoid of positive semi-definite real quadratic forms. In particular, any face of $U(B_0)$ is again of the form $U(B')$ for some B'.

If B_0 is a quadratic form defining the paving S, then by 5.8.10 the map $s_{\mathbb{Q}}$ induces a map

$$\bar{s}_{\mathbb{Q}} : \operatorname{Hom}(H_S^{gp}, \mathbb{Q}) \to U(B_0)^{gp},$$

which by 5.8.12 is injective.

Lemma 5.8.14 *The map $\bar{s}_{\mathbb{Q}}$ is surjective.*

Proof. It suffices to show that if $B : S^2 X \to \mathbb{Q}$ is a positive semidefinite quadratic form such that S is a refinement of the associated paving S_B, then the map B factors through H_S^{gp}.

Let us first recall how to associate to such a $B : S^2 X \to \mathbb{Z}$ a paving S_B of $X_{\mathbb{R}}$. Set $A(x) := B(x, x)/2$, and consider the set

$$G_A := \{(x, A(x)) | x \in X\} \subset X_{\mathbb{Q}} \times \mathbb{Q}.$$

The convex hull of this set is the graph of a piecewise linear function $g_B : X_{\mathbb{Q}} \to \mathbb{Q}$. The X–invariant paving S_B of $X_{\mathbb{R}}$ is defined to be the domains of linearity of g_B. Define $\tilde{g}_B : P_{\mathbb{Q}} \to \mathbb{Q}$ to be the function

$$\tilde{g}_B(x, d) := dg_B(x/d).$$

If (d, x) and (l, y) are in the same cone, we have

$$
\begin{aligned}
\tilde{g}_B(d + l, x + y) &= (d + l)g_B\left(\frac{x + y}{d + l}\right) \\
&= (d + l)g_B\left(\frac{d}{d + l}(x/d) + \frac{l}{d + l}(y/l)\right) \\
&= (d + l)\left[\frac{d}{d + l}g_B(x/d) + \frac{l}{d + l}g_B(y/l)\right] \\
&= dg_B(x/d) + lg_B(y/l) \\
&= \tilde{g}_B(d, x) + \tilde{g}_B(l, y).
\end{aligned}
$$

Therefore $\tilde{g}_B$ is piecewise linear on $P_{\mathbb{Q}}$ and hence defines a map

$$\tilde{h}_B : \widetilde{H}_{S_B, \mathbb{Q}}^{gp} \to \mathbb{Q} \tag{5.8.14.1}$$

by

$$(d, x) * (l, y) \mapsto \tilde{g}_B(d + l, x + y) - \tilde{g}_B(d, x) - \tilde{g}_B(l, y). \tag{5.8.14.2}$$

Next note that

$$\tilde{h}_B((1, x + y) * (1, 0)) - \tilde{h}_B((1, x) * (1, y))$$
$$= g(x + y) + g(0) - 2g_B(\frac{x + y}{2}) + 2g(\frac{x + y}{2}) - g(x) - g(y)$$
$$= A(x + y) - A(x) - A(y)$$
$$= B(x + y, x + y)/2 - B(x, x)/2 - B(y, y)/2$$
$$= B(x, y).$$

Fix $z \in X_{\mathbb{Q}}$. By the argument used in the proof of 4.1.2 there exists a linear function $L : X_{\mathbb{Q}} \to \mathbb{Q}$ such that for any $x \in X_{\mathbb{Q}}$ we have

$$g_B(x + z) = g_B(x) + L(x). \tag{5.8.14.3}$$

From this one deduces by an elementary calculation that for any element $z \in X_{\mathbb{Q}}$ and elements $(d, x), (l, y) \in P_{\mathbb{Q}}$ we have

$$\tilde{h}_B((d, x) * (l, y)) = \tilde{h}_B((d, x + dz) * (l, y + lz)), \tag{5.8.14.4}$$

and therefore $\tilde{h}_B$ descends to a map $h_B : H_{S_B, \mathbb{Q}} \to \mathbb{Q}$. Furthermore the composite

$$S^2 X_{\mathbb{Q}} \xrightarrow{s_{\mathbb{Q}}} H_{S_B, \mathbb{Q}}^{gp} \xrightarrow{h_B} \mathbb{Q}$$

is equal to B.

Now since S_B is coarser than S, there is a natural surjection $\pi : H_S \to H_{S_B}$ such that the diagram

$$H_{S, \mathbb{Q}}^{gp}$$
$$\nearrow \qquad \downarrow \pi$$
$$S^2 X_{\mathbb{Q}} \longrightarrow H_{S_B, \mathbb{Q}}^{gp}.$$

The composite

$$H_{S, \mathbb{Q}}^{gp} \longrightarrow H_{S_B, \mathbb{Q}}^{gp} \xrightarrow{h_B} \mathbb{Q}$$

therefore gives the desired factorization of B. □

Summarizing:

Proposition 5.8.15. *The map s induces an isomorphism*

$$\bar{s}_{\mathbb{Q}} : \text{Hom}(U(B_0), \mathbb{Q}) \to H_{S, \mathbb{Q}}^{gp}. \tag{5.8.15.1}$$

Lemma 5.8.16 *The isomorphism* $\text{Hom}(H_S^{gp}, \mathbb{Q}) \simeq U(B_0)^{gp}$ *identifies* $\text{Hom}(H_S, \mathbb{Q}_{\geq 0})$ *with the cone* $U(B_0)$ *(where B_0 is the quadratic form defining S and $U(B_0)$ is defined in 5.8.13).*

Proof. If $g \in \text{Hom}(H_S, \mathbb{Q}_{\geq 0})$ is an element with associated quadratic form B_g, then

$$B_g(x, x) = g((1, 2x) * (1, 0)) \geq 0. \tag{5.8.16.1}$$

It follows that B_g is positive semidefinite. Furthermore the paving defined by B_g is coarser than S by 5.8.10. This shows that $\text{Hom}(H_S, \mathbb{Q}_{\geq 0})$ is contained in $U(B_0)$. Conversely, the proof of 5.8.14 shows that any $B' \in U(B_0)$ is obtained from an element in $\text{Hom}(H_S, \mathbb{Q}_{\geq 0})$. $\square$

Proposition 5.8.17. *Let* $F \subset H_S^{\text{sat}}$ *be a face. Then the quotient* H_S^{sat}/F *is isomorphic to* $H_{S'}^{\text{sat}}/(\text{torsion})$ *for some regular paving* S' *such that* S *refines* S'.

Proof. Set

$$F^* = \{g \in \text{Hom}(H_S, \mathbb{Q}_{\geq 0}) | g(F) = 0\} \subset \text{Hom}(H_S, \mathbb{Q}_{\geq 0}) = U(B_0). \tag{5.8.17.1}$$

The F^* is a face of $U(B_0)$ and therefore equal to $U(B')$ for some B'. Let S' be the paving defined by B' (which is coarser than S), and let $\pi : H_S \to H_{S'}$ be the natural surjection. We claim that π induces an isomorphism $\pi^{\text{sat}} : H_S^{\text{sat}}/F \to H_{S'}^{\text{sat}}/(\text{torsion})$.

To check that π descends to H_S^{sat}/F and that the induced map is injective, it suffices to show that the corresponding map on duals

$$\text{Hom}(H_{S'}, \mathbb{Q}_{\geq 0}) \to \text{Hom}(H_S^{\text{sat}}, \mathbb{Q}_{\geq 0}) \tag{5.8.17.2}$$

descends to a surjection

$$\text{Hom}(H_{S'}, \mathbb{Q}_{\geq 0}) \to \text{Hom}(H_S^{\text{sat}}/F, \mathbb{Q}_{\geq 0}). \tag{5.8.17.3}$$

This we have already shown as both sides in fact are identified with $U(B')$.

Thus it remains to see that $H_S^{\text{sat}} \to H_{S'}^{\text{sat}}/(\text{torsion})$ is surjective. For this note that the projection $\pi : H_S \to H_{S'}$ induces a morphism $\bar{\pi} : H_{S'} \to H_S^{\text{sat}}/F$. Since H_S^{sat}/F is saturated (being the quotient of a saturated monoid by a face), the map $\bar{\pi}$ extends uniquely to a map $H_{S'}^{\text{sat}} \to H_S^{\text{sat}}/F$. This map is surjective since the composite $H_S^{\text{sat}} \to H_{S'}^{\text{sat}} \to H_S^{\text{sat}}/F$ is surjective, and induces an isomorphism on the associated groups tensored with $\mathbb{Q}$. It follows that the induced map

$$H_{S'}^{\text{sat}}/(\text{torsion}) \to H_S^{\text{sat}}/F \tag{5.8.17.4}$$

is an isomorphism. $\square$

Lemma 5.8.18 *Let* $B : S^2 X \to \mathbb{Q}$ *be a positive semi-definite quadratic form and let* $X_1 \subset X$ *be a subgroup such that* $B(x, y) = 0$ *for all* $x, y \in X_1$. *Then* B *factors through* $S^2 \overline{X}$, *where* $\overline{X} := X/X_1$.

Proof. Assume there exists $x \in X$ and $y \in X_1$ such that $B(x,y) \neq 0$. After possibly replacing y by $-y$ we may assume that $B(x,y) > 0$. Then for any integer r we have

$$B(x - ry, x - ry) = B(x,x) - 2rB(x,y). \tag{5.8.18.1}$$

Choosing r sufficiently big this will be negative which is a contradiction.   □

5.8.19. Let $B : S^2X \to \mathbb{Q}$ be a quadratic form defining S' and let $X_1 \subset X$ be the subspace of elements $x \in X$ for which $B(x,-)$ is the zero map.

Let $\overline{X}$ denote the quotient X/X_1. Note that $\overline{X}$ is a torsion free abelian group. Then B descends to a non-degenerate quadratic form $\overline{B} : S^2\overline{X} \to \mathbb{Q}$ and S' is equal to the inverse image in X of the paving $\overline{S}'$ defined by $\overline{B}$ on $\overline{X}$.

Let $\overline{P}$ denote the integral points of $\mathrm{Cone}(1, \overline{X}_\mathbb{R}) \subset \mathbb{R} \times X_\mathbb{R}$,

Lemma 5.8.20 *The projection $H_{S'} \to H_{\overline{S}'}$ is an isomorphism.*

Proof. For $\bar{\omega} \in \overline{S}'$, let $P_{\bar{\omega}} \subset P$ denote the submonoid corresponding to the inverse image of $\bar{\omega}$ under the bijection $S' \to \overline{S}'$, and let $\overline{P}_{\bar{\omega}} \subset \overline{P}$ be the submonoid defined by $\bar{\omega}$. For every $\bar{\omega} \in \overline{S}'$, the kernel of the map

$$P_{\bar{\omega}}^{gp} \to \overline{P}_{\bar{\omega}}^{gp} \tag{5.8.20.1}$$

is canonically isomorphic to X_1. Indeed there is a canonical inclusion

$$X_1 \hookrightarrow P_{\bar{\omega}}^{gp}, \quad x \mapsto (d, y + x) - (d, y), \tag{5.8.20.2}$$

where $(d,y) \in P_{\bar{\omega}}$ is any element (we leave to the reader the verification that this is independent of the choice of (d,y) and is a homomorphism). If $(d,y) \in P_{\bar{\omega}}^{gp} \subset P^{gp} \simeq \mathbb{Z} \oplus X$ is an element mapping to zero in $\overline{P}^{gp} \simeq \mathbb{Z} \oplus \overline{X}$, then we must have $d = 0$ and $y \in X_1$.

Since the map $H_{S'} \to H_{\overline{S}'}$ is clearly surjective, it suffices to prove that the induced map on groups

$$\xi : H_{S'}^{gp} \to H_{\overline{S}'}^{gp} \tag{5.8.20.3}$$

is injective. For this consider the commutative diagram

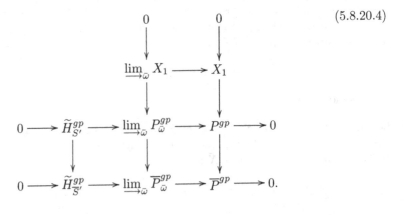

$$\tag{5.8.20.4}$$

Since $\overline{X}_{\mathbb{R}}$ is connected the map

$$\varinjlim_{\omega} X_1 \to X_1 \qquad (5.8.20.5)$$

is an isomorphism. It follows that $\widetilde{H}_{S'}^{gp} \to \widetilde{H}_{\overline{S}'}^{gp}$ is injective and hence an isomorphism. Applying $H_0(-)$ we deduce that ξ is an isomorphism. $\square$

Lemma 5.8.21 *Let M be a sharp saturated monoid and $\rho : (P \rtimes H_S) \oplus_{H_S} H_{\overline{S}} \to M$ a morphism sending at least one positive degree element q of P to zero. Then ρ factors through $\overline{P} \rtimes H_{\overline{S}}$.*

Proof. Consider the map $\rho^{gp} : (P \rtimes H_S) \oplus_{H_S} H_{\overline{S}} \to M^{gp}$ defined by ρ. The kernel of the map

$$((P \rtimes H_S) \oplus_{H_S} H_{\overline{S}})^{gp} \to (\overline{P} \rtimes H_{\overline{S}})^{gp} \qquad (5.8.21.1)$$

is canonically isomorphic to X_1, by consideration of the commutative diagram

$$
\begin{array}{ccccccccc}
0 & \longrightarrow & H_S^{gp} & \longrightarrow & (P \rtimes H_S)^{gp} & \longrightarrow & P^{gp} & \longrightarrow & 0 \\
& & \downarrow & & \downarrow & & \downarrow & & \\
0 & \longrightarrow & H_{\overline{S}}^{gp} & \longrightarrow & (\overline{P} \rtimes H_{\overline{S}})^{gp} & \longrightarrow & \overline{P}^{gp} & \longrightarrow & 0
\end{array}
\qquad (5.8.21.2)
$$

and noting that the kernel of $P^{gp} \to \overline{P}^{gp}$ is isomorphic to X_1. It follows that the obstruction to ρ descending to a map from $\overline{P} \rtimes H_{\overline{S}}$ is a map $o : X_1 \to M^{gp}$.

This map o can be described as follows. Let $q \in P$ be an element with $\rho(q) = 0$ in M, and let $\omega \in S$ be a simplex such that $q \in P_\omega$. The restriction of ρ to P_ω then factors through a morphism

$$\rho_\omega : P_{\omega,q} \to M. \qquad (5.8.21.3)$$

Now observe that the inclusion $X_1 \subset P_\omega^{gp}$ lifts to an inclusion $X_1 \subset P_{\omega,q}$ by writing $q = (d, y)$ and sending $x \in X_1$ to

$$(d, y + x_1) - (d, y) \in P_{\omega,q}. \qquad (5.8.21.4)$$

It follows that $X_1 \subset P_{\omega,q}^*$ and that the map o is equal to the map obtained by restricting ρ_ω to X_1. Since M is sharp the map

$$P_{\omega,q}^* \to M \qquad (5.8.21.5)$$

induced by $\rho_{\omega,q}$ is zero, which implies that o is also zero. $\square$

5.9 Specialization

Theorem 5.9.1 *Let* $f : (\mathscr{P}, M_{\mathscr{P}}) \to (B, M_B)$ *be a log smooth proper morphism of fine saturated log schemes, let* L *be a relatively ample line bundle on* $\mathscr{P}$, *and let* G/B *be semiabelian group scheme over* B *acting on* $(\mathscr{P}, M_{\mathscr{P}})$ *over* (B, M_B). *Then the condition that the geometric fiber of* $(\mathscr{P}, M_{\mathscr{P}}, L)/(B, M_B)$ *is isomorphic to the standard construction is represented by an open subset of* B.

The proof will be in several steps.

5.9.2. We first consider the following special case. Let $B = \mathrm{Spec}(V)$ be a complete discrete valuation ring with generic point $\eta \in B$ and closed point $s \in B$. Assume that the torus of G_η is split so that we have an extension

$$0 \to D(\overline{X}) \to G_\eta \to G_{1,\eta} \to 0,$$

where $\overline{X}$ is a finitely generated free abelian group and $G_{1,\eta}$ is an abelian variety. Since V is normal, the torus $D(\overline{X})$ extends uniquely to a torus $D(\overline{X}) \subset G$ over V [13, I.2.9]. Let G_1/V denote the quotient $G/D(\overline{X})$. Then G_1 is a semi-abelian scheme over V. We further assume that the torus of the closed fiber G_s is split. The reductions of G then define an extension over $\mathrm{Spf}(V)$

$$0 \to D(X) \to \widetilde{G} \to A \to 0,$$

and there is a natural inclusion $D(\overline{X}) \subset D(X)$ defining a surjection $X \to \overline{X}$. Let $X_1 \subset X$ be the kernel. Assume that the closed fiber $(\mathscr{P}_s, M_{\mathscr{P}_s}, L_s)$ is obtained from the standard construction. Then by 5.5.2 we know that $(\mathscr{P}, M_{\mathscr{P}}, L_{\mathscr{P}})$ is obtained by algebraizing the standard construction over the reductions.

In particular, we obtain a paving S of $X_{\mathbb{R}}$, maps c, c^t, ψ, and τ over V, and a chart $\beta : H_S^{\mathrm{sat}} \to M_B$ such that $(\mathscr{P}, M_{\mathscr{P}}, L)$ is obtained from the standard construction using this data.

5.9.3. Recall [3, 4.2.4] that there is an isomorphism of fppf sheaves on $\mathrm{Spec}(V)$.

$$R^1 f_* X \to \mathscr{H}om_{\mathrm{Gp}}(T_X, \underline{\mathrm{Pic}}(\mathscr{P})),$$

and similarly

$$R^1 f_* \overline{X} \simeq \mathscr{H}om_{\mathrm{Gp}}(T_{\overline{X}}, \underline{\mathrm{Pic}}(\mathscr{P})). \tag{5.9.3.1}$$

Now we have a map

$$\rho : T_{\overline{X}} \to \underline{\mathrm{Pic}}(\mathscr{P}), \quad u \mapsto u^* L \otimes L^{-1}.$$

which therefore defines a global section of $R^1 f_* \overline{X}$. After replacing V by a finite flat covering we can assume this class is obtained from a class in $H^1(\mathscr{P}, \overline{X})$. Let $\widetilde{Q}_1 \to \mathscr{P}$ denote the corresponding $\overline{X}$-torsor.

Lemma 5.9.4 *The G-action on $\mathscr{P}$ lifts uniquely to a G-action on $\widetilde{Q}_1$.*

Proof. The uniqueness of the lifting can be seen as follows. If ρ_1 and ρ_2 are two liftings then we get a map $G \to \overline{X}$ sending g to $\rho_1(g)\rho_2(g)^{-1} \in \overline{X}$ (since $\widetilde{Q}_1$ is an $\overline{X}$–torsor over $\mathscr{P}$ and the two actions on $\mathscr{P}$ coincide). Such a map must necessarily be constant since G is geometrically connected

Let $f : T_{\overline{X}} \to \underline{\mathrm{Pic}}(\mathscr{P})$ denote the morphism corresponding to $\widetilde{Q}_1$ under the isomorphism 5.9.3.1 (so $f(u) = [u^*L \otimes L^{-1}]$). By the theorem of the square [3, 4.1.6 and 4.1.7], for any section $g \in G$ we have

$$g^*(u^*L \otimes L^{-1}) \simeq u^*L \otimes L^{-1}. \tag{5.9.4.1}$$

Therefore $g^*\widetilde{Q}_1$ is isomorphic to $\widetilde{Q}_1$.

For $g \in G$ let W_g denote the $\overline{X}$-torsor of isomorphisms

$$\tilde{g} : \widetilde{Q}_1 \to \widetilde{Q}_1 \tag{5.9.4.2}$$

of $\overline{X}$-torsors over $g : \mathscr{P} \to \mathscr{P}$. For two sections $g, g' \in G$ the map

$$W_g \times W_{g'} \to W_{gg'} \tag{5.9.4.3}$$

sending $(\tilde{g}, \tilde{g}')$ to the composite

$$\widetilde{Q}_1 \xrightarrow{\ \tilde{g}'\ } \widetilde{Q}_1 \xrightarrow{\ \tilde{g}\ } \widetilde{Q}_1 \tag{5.9.4.4}$$

defines an isomorphism

$$W_g \wedge W_{g'} \to W_{gg'}. \tag{5.9.4.5}$$

In this way we obtain a morphism of Picard stacks

$$W : G \to \mathrm{TORS}(\overline{X}). \tag{5.9.4.6}$$

We want to show that this morphism is trivial. By [8, XVIII.1.4.5] the isomorphism class of W is classified by an element in the group

$$\mathrm{Ext}^0(G, \overline{X} \to 0) \simeq \mathrm{Ext}^1(G, \overline{X}), \tag{5.9.4.7}$$

where the complex $\overline{X} \to 0$ has $\overline{X}$ in degree -1. The result therefore follows from the fact that

$$\mathrm{Ext}^1(G, \overline{X}) = 0 \tag{5.9.4.8}$$

which follows from noting that if

$$0 \to \overline{X} \to E \to G \to 0 \tag{5.9.4.9}$$

is an extension, then the connected component of E maps isomorphically to G and hence the sequence is split. $\square$

5.9.5. Let $L_{\widetilde{Q}_1}$ denote the pullback of L to $\widetilde{Q}_1$. By the construction of the functor F in 5.3.1.4, the action of $D(\overline{X})$ lifts to an action on $L_{\widetilde{Q}_1}$. This lifting is unique up to twisting by a character $\chi : D(\overline{X}) \to \mathbb{G}_m$.

5.9.6. Let $\widetilde{Q}_{1,n}$ denote the reduction modulo $\mathfrak{m}_V^{n+1}$. Then $\widetilde{Q}_{1,n}$ can be described as follows. The surjection $X \to \overline{X}$ induces a commutative diagram

$$
\begin{array}{ccc}
H^1(\mathscr{P}_n, X) & \longrightarrow & \mathrm{Hom}(D(X), \underline{\mathrm{Pic}}(\mathscr{P}_n)) \\
\downarrow & & \downarrow \\
H^1(\mathscr{P}_n, \overline{X}) & \longrightarrow & \mathrm{Hom}(D(\overline{X}), \underline{\mathrm{Pic}}(\mathscr{P}_n)).
\end{array}
\tag{5.9.6.1}
$$

Let $\widetilde{\mathscr{P}}_n \to \mathscr{P}_n$ denote the space 5.2.4.1 used in the standard construction, and let $\widetilde{R}_n$ denote the X-torsor $\widetilde{\mathscr{P}}_n \times^Y X$. Then it follows from the commutativity of 5.9.6.1 that $(\widetilde{Q}_{1,n}, L_{\widetilde{Q}_{1,n}})$ is isomorphic to $(\widetilde{R}_n/X_1, L_{\widetilde{R}_n}/X_1)$. Moreover, by the uniqueness of the lifting of the $D(\overline{X})$–action we can choose such an isomorphism compatibly with the $D(\overline{X})$–actions.

5.9.7. By the same argument used in 5.2.9 we can lift $(G, \mathscr{P}, M_{\mathscr{P}}, L)$ to $\mathrm{Spec}(V[[H_S^{\mathrm{sat}}]])$ with the log structure defined by the natural map $H_S^{\mathrm{sat}} \to V[[H_S^{\mathrm{sat}}]]$. Let $\mathscr{G} \to \mathrm{Spec}(V[[H_S]])$ be the resulting semiabelian scheme. The generic fiber of G is abelian and has a polarization λ. By the construction in [13, Chapter II §2] we then have the dual semiabelian scheme $\mathscr{G}^t$ over $V[[H_S^{\mathrm{sat}}]]$ and an isogeny $\mathscr{G} \to \mathscr{G}^t$ inducing the polarization on the generic fiber. Let G^t be the pullback to V (along the morphism $V[[H_S^{\mathrm{sat}}]] \to V$) of $\mathscr{G}^t$ and $\lambda : G \to G^t$ the induced isogeny. Since the map $G \to G^t$ is an isogeny the toric part of G_η^t is a quotient of $D(\overline{X})$ and in particular is split. Let $D(\overline{Y}) \subset G^t$ denote the closure of the toric part of the generic fiber, and let G_1^t denote the quotient $G^t/D(\overline{Y})$. The surjection $D(\overline{X}) \to D(\overline{Y})$ dualizes to an inclusion $\overline{Y} \subset \overline{X}$. If $D(Y)$ denote the toric part of G_s^t then the inclusion $Y \subset X$ is the inclusion used in the standard construction, and by the same reasoning as above we have $Y \to \overline{Y}$ such that the diagram

$$
\begin{array}{ccc}
Y & \longrightarrow & X \\
\downarrow & & \downarrow \\
\overline{Y} & \longrightarrow & \overline{X}
\end{array}
$$

commutes. In particular, $\overline{Y}$ is just the image of Y in $\overline{X}$. Let $Y_1 \subset Y$ denote the kernel of $Y \to \overline{Y}$.

Lemma 5.9.8 *The chart $\beta : H_S^{\mathrm{sat}} \to M_B$ induces an isomorphism*

$$
H_{\overline{S}}^{\mathrm{sat}}/(torsion) \to \overline{M}_{B,\eta}
\tag{5.9.8.1}
$$

for some regular paving $\overline{S}$ of $\overline{X}_{\mathbb{R}}$.

Proof. Consider the deformation $(\mathscr{G}, \mathscr{P}^\dagger, L^\dagger)$ of $(G, \mathscr{P}, L)$ to $V[[H_S^{\mathrm{sat}}]]$ provided by the argument used in 5.2.9. Choose a map

$$\mathrm{Spec}(W) \to \mathrm{Spec}(V[[H_S^{\mathrm{sat}}]]) \qquad (5.9.8.2)$$

with W a complete discrete valuation ring sending the closed point of $\mathrm{Spec}(W)$ to η and the generic point to the generic point of $\mathrm{Spec}(V[[H_S^{\mathrm{sat}}]])$. Let ν be a valuation on W. We then obtain a quadratic form

$$\overline{B} : S^2 X \xrightarrow{s} H_S^{gp} \xrightarrow{\beta} \mathrm{Frac}(W)^* \xrightarrow{\nu} \mathbb{Z}. \qquad (5.9.8.3)$$

This quadratic form is positive semidefinite as the image of H_S in W has non-negative valuation. By [13, III.10.2 and the discussion preceding this theorem] this quadratic form in fact descends to a positive definite quadratic form (which we denote by the same letter)

$$\overline{B} : S^2 \overline{X} \to \mathbb{Z}. \qquad (5.9.8.4)$$

Let $\overline{S}$ denote the paving on $\overline{X}_{\mathbb{R}}$ defined by this quadratic form. By 5.8.10 and 5.8.20 the map $H_S \to \mathbb{Q}_{\geq 0}$ defined by ν factors through $H_{\overline{S}}$. On the other hand, since $\overline{B}$ is positive definite the map $H_{\overline{S}}^{\mathrm{sat}}/(torsion) \to \mathbb{Q}$ sends all nonzero elements to positive elements of $\mathbb{Z}$ by 4.1.8. $\qquad \square$

Lemma 5.9.9 *There exists a unique $\overline{Y}$-torsor $\widetilde{\mathscr{P}}_1 \to \mathscr{P}$ with an isomorphism $\widetilde{\mathscr{P}}_1 \times^{\overline{Y}} X \simeq \widetilde{Q}_1$.*

Proof. From the short exact sequence

$$0 \to \overline{Y} \to \overline{X} \to \overline{X}/\overline{Y} \to 0$$

we obtain a short exact sequence

$$0 \to H^1(\mathscr{P}, \overline{Y}) \to H^1(\mathscr{P}, \overline{X}) \to H^1(\mathscr{P}, \overline{X}/\overline{Y}).$$

Let $o \in H^1(\mathscr{P}, \overline{X}/\overline{Y})$ denote the image of the class defined by $\widetilde{Q}_1$. We have to show that $o = 0$. By the proper base change theorem [8, XII.5.1] it suffices to show this after reducing modulo the maximal ideal of V (note that $\overline{X}/\overline{Y}$ is a torsion constructible sheaf). But in this case we have $\widetilde{R}_0 \simeq \widetilde{\mathscr{P}}_0 \times^Y X$ whence

$$\widetilde{Q}_{1,0} \simeq (\widetilde{\mathscr{P}}_0 \times^Y X)/X_1 \simeq \widetilde{\mathscr{P}}_0/Y_1 \times^Y \overline{X},$$

where Y_1 denotes the kernel of $Y \to \overline{Y}$. $\qquad \square$

Lemma 5.9.10 *The G-action on $\widetilde{Q}_1$ restricts to a G-action on $\widetilde{\mathscr{P}}_1$.*

Proof. This can be verified over the closed fiber, where it follows from the fact that the G-action on $\widetilde{Q}_n$ is induced by a G-action on $\widetilde{\mathscr{P}}_n$. $\qquad \square$

5.9.11. Let us describe the torsor $\widetilde{\mathscr{P}}_{1,\eta}$ in more detail.

Denote by P_1 the intersection of P with $\mathrm{Cone}(1, X_{1,\mathbb{R}}) \subset \mathrm{Cone}(1, X_{\mathbb{R}})$. We then obtain a graded subalgebra

$$\oplus_{(d,x)\in P_1} \mathscr{M}^d \otimes L_x \subset \oplus_{(d,x)\in P} \mathscr{M}^d \otimes L_x. \tag{5.9.11.1}$$

Define

$$\widetilde{W} := \underline{\mathrm{Proj}}_A(\oplus_{(d,x)\in P_1} \mathscr{M}^d \otimes L_x) \tag{5.9.11.2}$$

The restrictions of ψ and τ to X_1 and $Y_1 \times X_1$ respectively define an action of Y_1 on $\widetilde{W}$ together with its tautological line bundle $L_{\widetilde{W}}$. Also by the same argument as in 4.1.14 we obtain an action of the semiabelian group scheme G_1 on $\widetilde{W}$ and a lifting to $L_{\widetilde{W}}$ of the restriction of this action to T_{X_1}.

The scheme $\widetilde{W}$ has a natural log structure $M_{\widetilde{W}}$ and the structure morphism extends naturally to a morphism

$$(\widetilde{W}, M_{\widetilde{W}}) \to (\mathrm{Spec}(V), M_V). \tag{5.9.11.3}$$

The log structure $M_{\widetilde{W}}$ defined as in 4.1.10 by locally on A choosing a trivialization of $\mathscr{M}$ and compatible trivializations of the L_x's and then noting that $\widetilde{W}$ is isomorphic to

$$\underline{\mathrm{Proj}}_A(\mathscr{O}_A \otimes_{\mathbb{Z}[H_S]} \mathbb{Z}[P_1 \rtimes H_S]). \tag{5.9.11.4}$$

The inclusion 5.9.11.1 defines a rational map

$$f : \widetilde{\mathscr{P}} \dashrightarrow \widetilde{W} \tag{5.9.11.5}$$

compatible with the G and G_1-actions. Note that the maximal open set $\widetilde{\mathscr{U}} \subset \widetilde{\mathscr{P}}$ where f is defined is dense in the closed fiber of $\widetilde{\mathscr{P}}$, Y_1-invariant, and f extends to a map of log schemes

$$f : (\widetilde{\mathscr{U}}, M_{\widetilde{\mathscr{U}}}) \to (\widetilde{W}, M_{\widetilde{W}}). \tag{5.9.11.6}$$

Lemma 5.9.12 *After possibly replacing V by a finite extension, the generic fiber $\widetilde{W}_\eta$ with its $D(X_1)$-action is isomorphic to*

$$\underline{\mathrm{Spec}}_A(\mathscr{O}_A[X_1]) \tag{5.9.12.1}$$

with the standard action of $D(X_1)$.

Proof. Let K denote the field of fractions of V, and let $F \subset H_S$ be the face of elements mapping to units in K. After possibly replacing V by a finite extension, we can find a homomorphism $\rho : P_1 \rtimes H_S \to K^*$ such that the restriction to F is equal to the composite

$$F \longrightarrow H_S \xrightarrow{\ \beta\ } K^*. \tag{5.9.12.2}$$

The map of algebras over A_η

$$\rho^* : \oplus_{(d,x)\in P_1 \rtimes H_S} \mathscr{M}_\eta^d \otimes L_{x,\eta} \to \oplus_{(d,x)\in P_1 \rtimes H_S} \mathscr{M}_\eta^d \otimes L_{x,\eta} \qquad (5.9.12.3)$$

which in degree (d,x) is equal to multiplication by $\rho(d,x)$ then extends to a commutative diagram of log schemes

$$
\begin{array}{ccc}
(\widetilde{W}_\eta, M_{\widetilde{W}_\eta}) & \longrightarrow & (\widetilde{W}_\eta, M_{\widetilde{W}_\eta}) \\
\downarrow & & \downarrow \\
\mathrm{Spec}(H_S \xrightarrow{\beta} K) & \xrightarrow{a} & \mathrm{Spec}(H_S \xrightarrow{\beta'} K),
\end{array}
\qquad (5.9.12.4)
$$

where β sends all elements of F to 1 and a is the morphism which is the identity on K and given on log structures by the map

$$H_S \to K^* \oplus H_S, \quad h \mapsto (\rho(h)^{-1}, h). \qquad (5.9.12.5)$$

From this it follows that it suffices to consider the case when $H_S \to K$ factors through $H_{\overline{S}}$.

From the isomorphism

$$(P_1 \rtimes H_S) \oplus_{H_S} H_{\overline{S}} \simeq P_1 \oplus H_{\overline{S}} \qquad (5.9.12.6)$$

we see that $\widetilde{W}_\eta$ is isomorphic to

$$\underline{\mathrm{Proj}}_A(\mathscr{O}_A[P_1]). \qquad (5.9.12.7)$$

Next observe that for any nonzero element $(d,x) \in P_1$, the localized monoid $P_{1,(d,x)}$ is isomorphic to P_1^{gp}. Indeed if $(l,y) \in P_1$ then in $P_{1,(d,x)}$ we have

$$d(l,y) + d(l, 2lx - y) = (2dl, 2dlx) \qquad (5.9.12.8)$$

and hence (l,y) is invertible in $P_{1,(d,x)}$.

Therefore

$$\underline{\mathrm{Proj}}_A(\mathscr{O}_A[P_1]) \simeq \underline{\mathrm{Proj}}_A(\mathscr{O}_A[P_1^{gp}]) \simeq \underline{\mathrm{Spec}}_A(\mathscr{O}_A[X_1]). \qquad (5.9.12.9)$$

□

Lemma 5.9.13 *(i) The rational map on the generic fiber*

$$f_\eta : \widetilde{\mathscr{P}}_\eta \dashrightarrow \widetilde{W}_\eta \qquad (5.9.13.1)$$

is a morphism.

(ii) The log scheme with $D(\overline{X})$-action $(\widetilde{\mathscr{P}}_\eta, M_{\widetilde{\mathscr{P}}_\eta})$ over $\widetilde{W}_\eta$ is étale locally on $\widetilde{W}_\eta$ isomorphic to

$$\underline{\mathrm{Proj}}_{\widetilde{W}_\eta}(\mathscr{O}_{\widetilde{W}_\eta} \otimes_{\mathbb{Z}[H_{\overline{S}}]} \mathbb{Z}[\overline{P} \rtimes H_{\overline{S}}]), \qquad (5.9.13.2)$$

where the map $H_{\overline{S}} \to \mathscr{O}_{\widetilde{W}_\eta}$ sends all nonzero elements to 0.

Proof. Locally on A we can make identifications

$$\widetilde{W} \simeq \underline{\mathrm{Proj}}_A(\mathcal{O}_A \otimes_{\mathbb{Z}[H_S]} \mathbb{Z}[P_1 \rtimes H_S]), \qquad (5.9.13.3)$$

and

$$\widetilde{\mathcal{P}}_\eta \simeq \underline{\mathrm{Proj}}_A(\mathcal{O}_A \otimes_{\mathbb{Z}[H_S]} \mathbb{Z}[P \rtimes H_S]) \qquad (5.9.13.4)$$

such that the map f is induced by the map

$$\mathbb{Z}[P_1 \rtimes H_S] \to \mathbb{Z}[P \rtimes H_S]. \qquad (5.9.13.5)$$

Over the generic fiber, we can after possibly changing the chart β so that $H_S \to K$ factors through $H_{\overline{S}}$ as in the proof of 5.9.12, identify the map f_η with the map

$$\underline{\mathrm{Proj}}_{A_\eta}(\mathcal{O}_{A_\eta} \otimes_{\mathbb{Z}[H_{\overline{S}}]} \mathbb{Z}[(P \rtimes H_S) \oplus_{H_S} H_{\overline{S}}]) - - \blacktriangleright \underline{\mathrm{Spec}}_{A_\eta}(\mathcal{O}_{A_\eta}[X_1])$$

$$(5.9.13.6)$$

induced by the inclusion

$$P_1 \oplus H_{\overline{S}} \simeq (P_1 \rtimes H_S) \oplus_{H_S} H_{\overline{S}} \hookrightarrow (P \rtimes H_S) \oplus_{H_S} H_{\overline{S}} \qquad (5.9.13.7)$$

and the natural identification of X_1 with the degree 0 elements in any localization of P_1 (see again the argument in 5.9.12).

To prove (i), it suffices to show that if $(d, x) \in P$ is an element then the image of P_1 in the localization

$$((P \rtimes H_S) \oplus_{H_S} H_{\overline{S}})_{(d,x)} \qquad (5.9.13.8)$$

is contained in the units. This follows from noting that for any element $(1, y)) \in P_1$ we have

$$(d - 2, x) + (1, y) + (1, -y) = (d, x). \qquad (5.9.13.9)$$

This proves (i).

For (ii) note that there is an isomorphism of (log) schemes over the scheme $\underline{\mathrm{Spec}}_{A_\eta}(\mathcal{O}_{A_\eta}[X_1])$

$$\underline{\mathrm{Proj}}_{A_\eta}(\mathcal{O}_{A_\eta} \otimes_{\mathbb{Z}[H_{\overline{S}}]} \mathbb{Z}[(P \rtimes H_S) \oplus_{H_S} H_{\overline{S}}])$$

$$\simeq \underline{\mathrm{Proj}}_{A_\eta}(\mathcal{O}_{A_\eta} \otimes_{\mathbb{Z}[H_{\overline{S}}]} \mathbb{Z}[P_1^{gp} \oplus_{P_1} ((P \rtimes H_S) \oplus_{H_S} H_{\overline{S}})]).$$

From the commutative diagram

$$(5.9.13.10)$$

$$
\begin{array}{ccc}
P_1^{gp} & \!\!=\!\!=\!\!=\!\!=\!\!= & P_1^{gp} \\
\downarrow & & \downarrow \\
0 \longrightarrow H_{\overline{S}}^{gp} \longrightarrow (P \rtimes H_S)^{gp} \oplus_{H_{\overline{S}}^{gp}} H_{\overline{S}}^{gp} \longrightarrow P^{gp} \longrightarrow 0 \\
\| \qquad\qquad\qquad \downarrow \qquad\qquad\qquad \downarrow \\
0 \longrightarrow H_{\overline{S}}^{gp} \longrightarrow (\overline{P} \rtimes H_{\overline{S}})^{gp} \longrightarrow \overline{P}^{gp} \longrightarrow 0
\end{array}
$$

it follows that the projection

$$P_1^{gp} \oplus_{P_1} ((P \rtimes H_S) \oplus_{H_S} H_{\overline{S}}) \to \overline{P} \rtimes H_{\overline{S}} \tag{5.9.13.11}$$

identifies $\overline{P} \rtimes H_{\overline{S}}$ with the quotient of

$$P_1^{gp} \oplus_{P_1} ((P \rtimes H_S) \oplus_{H_S} H_{\overline{S}}) \tag{5.9.13.12}$$

by the subgroup P_1^{gp}.

We claim that the projection 5.9.13.11 admits a section inducing a decomposition

$$P_1^{gp} \oplus_{P_1} ((P \rtimes H_S) \oplus_{H_S} H_{\overline{S}}) \simeq (\overline{P} \rtimes H_{\overline{S}}) \oplus P_1^{gp}. \tag{5.9.13.13}$$

For this it suffices to construct a section of the map on groups

$$(P \rtimes H_S)^{gp} \oplus_{H_{\overline{S}}^{gp}} H_{\overline{S}}^{gp} \to (\overline{P} \rtimes H_{\overline{S}})^{gp}. \tag{5.9.13.14}$$

Such a section exists since this projection induces an isomorphism on torsion subgroups (both torsion subgroups being equal to the torsion subgroup of $H_{\overline{S}}^{gp}$). Fixing one such decomposition 5.9.13.13 we obtain an isomorphism

$$\widetilde{\mathscr{P}}_\eta \simeq \widetilde{W}_\eta \times_{A_n} \underline{\mathrm{Proj}}_{A_n} (\mathscr{O}_{A_n} \otimes_{\mathbb{Z}[H_{\overline{S}}]} \mathbb{Z}[\overline{P} \rtimes H_{\overline{S}}]) \tag{5.9.13.15}$$

as desired.   □

5.9.14. Let $p : X_{\mathbb{R}} \to \overline{X}_{\mathbb{R}}$ denote the projection. For $\bar{\omega} \in \overline{S}$ let $P_{\bar{\omega}} \subset P$ denote the integral points of the cone $\mathrm{Cone}(1, p^{-1}(\bar{\omega})) \subset \mathbb{R} \oplus X_{\mathbb{R}}$. Define

$$\widetilde{\mathscr{P}}_{\bar{\omega}} \subset \widetilde{\mathscr{P}} \tag{5.9.14.1}$$

to be the closed subscheme

$$\underline{\mathrm{Proj}}_A (\mathscr{O}_A \otimes_{\mathbb{Z}[H_S]} \mathbb{Z}[P_{\bar{\omega}} \rtimes H_S]) \subset \underline{\mathrm{Proj}}_A (\mathscr{O}_A \otimes_{\mathbb{Z}[H_S]} \mathbb{Z}[P \rtimes H_S]) \tag{5.9.14.2}$$

induced by the projection

$$\mathscr{O}_A \otimes_{\mathbb{Z}[H_S]} \mathbb{Z}[P \rtimes H_S] \to \mathscr{O}_A \otimes_{\mathbb{Z}[H_S]} \mathbb{Z}[P_{\bar{\omega}} \rtimes H_S] \tag{5.9.14.3}$$

sending an element $p \in P$ to 0 unless $p \in P_{\bar{\omega}}$ and to $p \in P_{\bar{\omega}}$ otherwise (note that this gives a map of algebras for if $p, q \in P$ are two elements whose images in $\overline{P}$ lie in distinct cones then $p*q$ maps to zero in $\mathscr{O}_A$). The subscheme $\widetilde{\mathscr{P}}_{\bar{\omega}}$ is Y_1-invariant, and hence we obtain a compatible collection of closed subspaces

$$\mathscr{P}_{\bar{\omega},n} := \widetilde{\mathscr{P}}_{\bar{\omega},n}/Y_1 \subset \widetilde{\mathscr{P}}_n/Y_1 = \mathscr{P}_{1,n}. \tag{5.9.14.4}$$

Lemma 5.9.15 *For every n the scheme $\mathscr{P}_{\bar{\omega},n}$ is projective over B_n, with ample line bundle $L_{\mathscr{P}_{\bar{\omega},n}}$ the sheaf obtained by descent from the Y_1-linearized invertible sheaf $\mathscr{O}_{\widetilde{\mathscr{P}}_n}(1)$.*

Proof. Since $\widetilde{\mathscr{P}}_n$ is locally of finite type over B_n, the space $\mathscr{P}_{\bar{\omega},n}$ is also locally of finite type over B_n. It therefore suffices to prove that $\mathscr{P}_{\bar{\omega},n}$ is quasi-compact. For this we may as well assume that $n = 0$ in which case $\mathscr{P}_{\bar{\omega},0}$ is a scheme over $k := V/\mathfrak{m}_V$.

For each $\omega \in S$ let $\mathscr{P}_{\omega,0} \subset \widetilde{\mathscr{P}}_0$ denote the closed subscheme

$$\underline{\operatorname{Proj}}_{A_0}(\mathscr{O}_{A_0}[P_\omega]) \subset \widetilde{\mathscr{P}}_0. \tag{5.9.15.1}$$

Then $\widetilde{\mathscr{P}}_{\bar{\omega},0}$ is the union of those $\mathscr{P}_{\omega,0}$ with ω mapping to $\bar{\omega}$ in $\overline{S}$. Let $q : S \to \overline{S}$ be the projection. Then Y_1 acts on $q^{-1}(\bar{\omega})$ such that the induced action of $y \in Y_1$ on $\widetilde{\mathscr{P}}_{\bar{\omega},0}$ sends $\mathscr{P}_{\omega,0}$ to $\mathscr{P}_{y(\omega),0}$. It therefore suffices to show that the set

$$q^{-1}(\bar{\omega})/Y_1 \tag{5.9.15.2}$$

is finite. This is clear as the elements of this set form a covering of the compact set

$$p^{-1}(\bar{\omega})/Y_1 \subset X_\mathbb{R}/Y_1. \tag{5.9.15.3}$$

The statement that $L_{\mathscr{P}_{\bar{\omega},n}}$ is ample follows from the Nakai-Moishezon criterion [28, Theorem 2], and the fact that $L_{\mathscr{P}_{\bar{\omega},n}}$ pulls back to an ample sheaf on each $\widetilde{\mathscr{P}}_{\omega,n}$. $\square$

5.9.16. It follows that the closed subscheme $\mathscr{P}_{\bar{\omega},n} \subset \widetilde{\mathscr{P}}_{1,n}$ are uniquely algebraizable to closed subschemes

$$\mathscr{P}_{\bar{\omega}} \subset \widetilde{\mathscr{P}}_1 \tag{5.9.16.1}$$

which are of finite type and cover $\widetilde{\mathscr{P}}_1$.

Note also that if $\bar{\omega}, \bar{\omega}' \in \overline{S}$ are two elements, then the intersection

$$\mathscr{P}_{\bar{\omega}} \cap \mathscr{P}_{\bar{\omega}'} \subset \widetilde{\mathscr{P}}_1 \tag{5.9.16.2}$$

is equal to $\mathscr{P}_{\bar{\omega} \cap \bar{\omega}'}$.

5.9.17. By a similar argument to the above, for every integer n the action of Y_1 on $\widetilde{W}_n$ is properly discontinuous, and the line bundle $\mathscr{O}_{\widetilde{W}}(1)$ descends to an ample line bundle L_{W_n} on the quotient

$$W_n := \widetilde{W}_n/Y_1. \tag{5.9.17.1}$$

By the Grothendieck existence theorem the compatible family

$$\{W_n, L_{W_n}\} \tag{5.9.17.2}$$

is therefore induced by a unique projective scheme with ample line bundle $(W, L_W)/V$.

5.9.18. Let

$$\widetilde{\Gamma} \subset \widetilde{\mathscr{P}} \times_B \widetilde{W} \tag{5.9.18.1}$$

denote the closure of

$$\Gamma_f : \mathscr{U} \to \widetilde{\mathscr{P}} \times_B \widetilde{W}, \tag{5.9.18.2}$$

and for $\bar{\omega} \in \overline{S}$ let $\widetilde{\Gamma}_{\bar{\omega}}$ denote the closure of the restriction of f to $\mathscr{U} \cap \widetilde{\mathscr{P}}_{\bar{\omega}}$.

5.9.19. The scheme $\widetilde{\Gamma}$ can be described as follows. Set

$$\mathscr{C}_{\widetilde{\mathscr{P}}} := \underline{\mathrm{Spec}}_A(\oplus_{(d,x)\in P} \mathscr{M}^d \otimes L_x), \tag{5.9.19.1}$$

and

$$\mathscr{C}_{\widetilde{W}} := \underline{\mathrm{Spec}}_A(\oplus_{(d,x)\in P_1} \mathscr{M}^d \otimes L_x), \tag{5.9.19.2}$$

and let

$$f_{\mathscr{C}} : \mathscr{C}_{\widetilde{\mathscr{P}}} \to \mathscr{C}_{\widetilde{W}} \tag{5.9.19.3}$$

by the map of schemes defined by the map of sheaves of algebras 5.9.11.1.
 The $\mathbb{G}_m$-action on $\mathscr{C}_{\widetilde{W}}$ defines a map

$$\mathscr{C}_{\widetilde{\mathscr{P}}} \times \mathbb{G}_m \to \mathscr{C}_{\widetilde{\mathscr{P}}} \times_B \mathscr{C}_{\widetilde{W}}, \quad (p,u) \mapsto (p, u * f_{\mathscr{C}}(p)), \tag{5.9.19.4}$$

which since the $\mathbb{G}_m$-action on $\mathscr{C}_{\widetilde{W}}$ is faithful is an immersion. Let

$$Z \subset \mathscr{C}_{\widetilde{P}} \times_B \mathscr{C}_{\widetilde{W}} \tag{5.9.19.5}$$

denote the closure of 5.9.19.4. If $\mathscr{C}_{\widetilde{\mathscr{P}}}^{\circ}$ (resp. $\mathscr{C}_{\widetilde{W}}^{\circ}$) denotes the complement of the vertex in $\mathscr{C}_{\widetilde{\mathscr{P}}}$ (resp. $\mathscr{C}_{\widetilde{W}}$) then the inverse image of $\widetilde{\Gamma}$ in $\mathscr{C}_{\widetilde{\mathscr{P}}}^{\circ} \times_B \mathscr{C}_{\widetilde{W}}^{\circ}$ is equal to the intersection

$$Z \cap (\mathscr{C}_{\widetilde{\mathscr{P}}}^{\circ} \times_B \mathscr{C}_{\widetilde{W}}^{\circ}). \tag{5.9.19.6}$$

Now the scheme Z can be described as follows. If we locally on A trivialize $\mathscr{M}$ and the L_x's, then the map 5.9.19.4 is induced by the map of algebras

$$\mathscr{O}_A \otimes_{\mathbb{Z}[H_S]} \mathbb{Z}[(P \rtimes H_S) \oplus_{H_S} (P_1 \rtimes H_S)] \to \mathscr{O}_A \otimes_{\mathbb{Z}[H_S]} \mathbb{Z}[(P \rtimes H_S) \oplus \mathbb{Z}] \tag{5.9.19.7}$$

induced by the natural inclusion into the first factor

$$P \rtimes H_S \hookrightarrow (P \rtimes H_S) \oplus \mathbb{Z}, \tag{5.9.19.8}$$

and the map

$$P_1 \rtimes H_S \to (P \rtimes H_S) \oplus \mathbb{Z} \tag{5.9.19.9}$$

induced by the natural inclusion

$$P_1 \rtimes H_S \hookrightarrow P \rtimes H_S \tag{5.9.19.10}$$

and the degree map

$$\deg : P_1 \rtimes H_S \to \mathbb{Z}. \tag{5.9.19.11}$$

Now observe that the map 5.9.19.7 factors as a surjection

$$\mathscr{O}_A \otimes_{\mathbb{Z}[H_S]} \mathbb{Z}[(P \rtimes H_S) \oplus_{H_S} (P_1 \rtimes H_S)] \to \mathscr{O}_A \otimes_{\mathbb{Z}[H_S]} \mathbb{Z}[(P \rtimes H_S) \oplus \mathbb{N}] \quad (5.9.19.12)$$

followed by an inclusion

$$\mathscr{O}_A \otimes_{\mathbb{Z}[H_S]} \mathbb{Z}[(P \rtimes H_S) \oplus \mathbb{N}] \hookrightarrow \mathscr{O}_A \otimes_{\mathbb{Z}[H_S]} \mathbb{Z}[(P \rtimes H_S) \oplus \mathbb{Z}]. \quad (5.9.19.13)$$

It follows that in this local description the scheme Z is equal to

$$\underline{\mathrm{Spec}}_A (\mathscr{O}_A \otimes_{\mathbb{Z}[H_S]} \mathbb{Z}[(P \rtimes H_S) \oplus \mathbb{N}]). \quad (5.9.19.14)$$

Note that this description of Z also holds after arbitrary base change $B' \to B$.

Corollary 5.9.20 *The projection map $\widetilde{\Gamma} \to \widetilde{\mathscr{P}}$ is of finite type, and for every integer n the reduction $\widetilde{\Gamma}_n \subset \widetilde{\mathscr{P}}_n \times_{B_n} \widetilde{W}_n$ is equal to the scheme-theoretic closure of the graph of $f_n : \mathscr{U}_n \to \widetilde{W}_n$.*

By a similar analysis one also obtains the following:

Corollary 5.9.21 *For every $\bar{\omega} \in \overline{S}$, the projection map $\widetilde{\Gamma}_{\bar{\omega}} \to \widetilde{\mathscr{P}}_{\bar{\omega}}$ is of finite type, and for every integer n the reduction $\widetilde{\Gamma}_{\bar{\omega},n} \subset \widetilde{\mathscr{P}}_{\bar{\omega},n} \times_{B_n} \widetilde{W}_n$ is equal to the scheme-theoretic closure of the graph of $f_n : \mathscr{U}_n \cap \widetilde{\mathscr{P}}_{\bar{\omega}} \to \widetilde{W}_n$.*

5.9.22. Let $\widetilde{\Gamma}_{\bar{\omega},n}$ denote the reduction of $\widetilde{\Gamma}_{\bar{\omega},n}$ modulo $\mathfrak{m}^{n+1}$. Note that

$$\widetilde{\Gamma} = \cup_{\bar{\omega} \in \overline{S}} \widetilde{\Gamma}_{\bar{\omega}}. \quad (5.9.22.1)$$

Since f is compatible with the Y_1-actions, the subscheme $\widetilde{\Gamma}_{\bar{\omega}} \subset \widetilde{\mathscr{P}} \times_B \widetilde{W}$ is a Y_1-invariant, where Y_1 acts diagonally on $\widetilde{\mathscr{P}} \times_B \widetilde{W}$. Since the action of Y_1 on $\widetilde{\mathscr{P}}_n \times_{B_n} \widetilde{W}_n$ is properly discontinuous, the action of Y_1 on $\widetilde{\Gamma}_{\bar{\omega},n}$ is also properly discontinuous.

5.9.23. Define

$$\Gamma_{\bar{\omega},n} := \widetilde{\Gamma}_{\bar{\omega},n}/Y_1 \quad (5.9.23.1)$$

to be the quotient. The space $\Gamma_{\bar{\omega},n}$ comes equipped with an invertible sheaf $L_{\Gamma_{\bar{\omega},n}}$ induced by descent from the Y_1-linearized invertible sheaf

$$p_1^* \mathscr{O}_{\widetilde{\mathscr{P}}}(1) \otimes p_2^* \mathscr{O}_{\widetilde{W}}(1) \quad (5.9.23.2)$$

on $\widetilde{\mathscr{P}} \times_B \widetilde{W}$.

Lemma 5.9.24 *For every n and $\bar{\omega}$, the space $\Gamma_{\bar{\omega},n}$ is proper over B_n and $L_{\Gamma_{\bar{\omega},n}}$ is an ample invertible sheaf on $\Gamma_{\bar{\omega},n}$.*

Proof. The proof is very similar to the proof of 5.9.15.

First note that it suffices to consider the case when $n = 0$. Choose a set of representatives $\omega_1, \ldots, \omega_r$ for the finite set $q^{-1}(\bar{\omega})/Y_1$ the map

$$\coprod_{i=1}^{r} \Gamma_{\omega,0} \to \Gamma_{\bar{\omega},0} \qquad (5.9.24.1)$$

is surjective. The ampleness of $L_{\Gamma_{\bar{\omega}},0}$ follows as in 5.9.15 from the fact that this sheaf pulls back to an ample sheaf on each $\Gamma_{\omega,0}$. □

5.9.25. The projective system $(\Gamma_{\bar{\omega},n}, L_{\Gamma_{\bar{\omega}},n})$ is therefore uniquely algebraizable to a projective scheme with ample line bundle $(\Gamma_{\bar{\omega}}, L_{\Gamma_{\bar{\omega}}})$ over B. Furthermore the diagrams

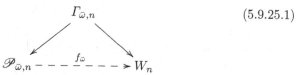

$$(5.9.25.1)$$

induce a diagram

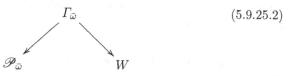

$$(5.9.25.2)$$

by algebraizing the closed subschemes

$$\Gamma_{\bar{\omega},n} \subset \mathscr{P}_{\bar{\omega},n} \times_{B_n} W_n. \qquad (5.9.25.3)$$

Let

$$\Gamma \subset \widetilde{\mathscr{P}}_1 \times_B W \qquad (5.9.25.4)$$

be the scheme-theoretic union of the $\mathscr{P}_{\bar{\omega}}$ (note that Γ is reduced since each $\Gamma_{\bar{\omega}}$ is reduced), so we have

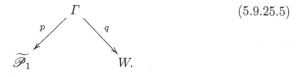

$$(5.9.25.5)$$

The map p is birational since this is true after reducing modulo $\mathfrak{m}$. We therefore obtain a rational map (which we again denote by f)

$$f : \widetilde{\mathscr{P}}_1 \dashrightarrow W. \qquad (5.9.25.6)$$

Let L_Γ denote the tensor product $p^* L_{\mathscr{P}_1} \otimes q^* L_W$, so L_Γ is an ample invertible sheaf on Γ whose restriction to $\Gamma_{\bar{\omega}}$ is equal to $L_{\Gamma_{\bar{\omega}}}$.

Lemma 5.9.26 *The map on generic fibers* $p_\eta : \Gamma_\eta \to \widetilde{\mathscr{P}}_{1,\eta}$ *is an isomorphism. In particular, the rational map 5.9.25.6 induces a morphism*

$$f_\eta : \mathscr{P}_{1,\eta} \to W_\eta. \tag{5.9.26.1}$$

Proof. Since the diagrams

$$\begin{array}{ccc}
\Gamma_{\tilde\omega} & \longrightarrow & \Gamma \\
\downarrow & & \downarrow \\
\mathscr{P}_{\tilde\omega} & \longrightarrow & \widetilde{\mathscr{P}}_1
\end{array} \tag{5.9.26.2}$$

are cartesian, it suffices to show that the projections

$$p_\eta : \Gamma_{\tilde\omega,\eta} \to \mathscr{P}_{\tilde\omega,\eta} \tag{5.9.26.3}$$

are isomorphisms.

Let $p \in \mathscr{P}_{\tilde\omega}$ be a point of the closed fiber, and let T denote the $\mathscr{P}_{\tilde\omega}$-scheme

$$\rho : T := \mathrm{Spec}(\widehat{\mathscr{O}}_{\mathscr{P}_{\tilde\omega},p}) \to \mathscr{P}_{\tilde\omega}. \tag{5.9.26.4}$$

Let $\Gamma_{\tilde\omega,T}$ denote the base change of $\Gamma_{\tilde\omega}$ to T. Then it suffices to show that the map on generic fibers

$$\Gamma_{\tilde\omega,T_\eta} \to T_\eta \tag{5.9.26.5}$$

is an isomorphism. Furthermore, to verify this we may replace T by a finite flat extension if necessary.

If $t \in T$ denotes the closed point, we may therefore assume that there exists a morphism

$$\tilde t : \mathrm{Spec}(k(t)) \to \widetilde{\mathscr{P}}_{\tilde\omega,0} \tag{5.9.26.6}$$

lifting the map to $\mathscr{P}_{\tilde\omega,0}$. Since the maps

$$\widetilde{\mathscr{P}}_{\tilde\omega,n} \to \mathscr{P}_{\tilde\omega,n} \tag{5.9.26.7}$$

are étale and T is the spectrum of a complete local ring, this map $\tilde t$ lifts uniquely to a morphism

$$\tilde\rho : T \to \widetilde{\mathscr{P}}_{\tilde\omega} \tag{5.9.26.8}$$

such that for every n the composite map on reductions

$$T_n \xrightarrow{\tilde\rho} \widetilde{\mathscr{P}}_{\tilde\omega,n} \longrightarrow \mathscr{P}_{\tilde\omega,n} \tag{5.9.26.9}$$

is equal to the given map ρ. Let $\Gamma'_{\tilde\omega,T}$ denote the fiber product of the diagram

$$\begin{array}{c}
\widetilde{\Gamma}_{\tilde\omega} \\
\downarrow \\
T \xrightarrow{\ \tilde\rho\ } \widetilde{\mathscr{P}}_{\tilde\omega}.
\end{array} \tag{5.9.26.10}$$

For every n, the two reductions $\Gamma'_{\bar{\omega},T_n}$ and $\Gamma_{\bar{\omega},T_n}$ are canonically isomorphic. By the Grothendieck existence theorem it follows that there is a canonical isomorphism over T

$$\Gamma_{\bar{\omega},T} \simeq \Gamma'_{\bar{\omega},T}. \tag{5.9.26.11}$$

This implies the lemma for by 5.9.13 the morphism on generic fibers

$$\Gamma'_{\bar{\omega},T_\eta} \to T_\eta \tag{5.9.26.12}$$

is an isomorphism. $\square$

5.9.27. The same argument used in the proof of 5.9.26 enables us to describe the local structure of the map

$$f_\eta : \widetilde{\mathscr{P}}_{1,\eta} \to W_\eta. \tag{5.9.27.1}$$

Let $w \in W$ be a point of the closed fiber and let Z denote the spectrum of $\hat{\mathscr{O}}_{W,\bar{w}}$. Let

$$z : Z \to W \tag{5.9.27.2}$$

be the natural map. Then as in the proof of 5.9.26 we can choose a map $\tilde{z} : Z \to \widetilde{W}$ such that the fiber product

$$\Gamma \times_W Z \tag{5.9.27.3}$$

is isomorphic to $\widetilde{\Gamma} \times_{\widetilde{W}} Z$. By 5.9.13 it follows that there is a non-canonical isomorphism

$$\widetilde{\mathscr{P}}_{1,\eta} \times_{W_\eta} Z_\eta \simeq \underline{\mathrm{Proj}}_{Z_\eta} (\mathscr{O}_{Z_\eta} \otimes_{\mathbb{Z}[H_{\overline{S}}]} \mathbb{Z}[\overline{P} \rtimes H_{\overline{S}}]). \tag{5.9.27.4}$$

5.9.28. There is a map of invertible sheaves

$$f_\eta^* L_{W_\eta} \to L_{\widetilde{\mathscr{P}}_{1,\eta}} \tag{5.9.28.1}$$

defined as follows. First note that to give such a map it suffices to define a compatible collection of morphisms

$$j_{\bar{\omega}}^* f_\eta^* L_{W_\eta} \to L_{\mathscr{P}_{\bar{\omega},\eta}} \tag{5.9.28.2}$$

for each $\bar{\omega} \in \overline{S}$, where $j_{\bar{\omega}} : \mathscr{P}_{\bar{\omega}} \hookrightarrow \widetilde{\mathscr{P}}_1$ is the inclusion. Indeed by the above local description of $\widetilde{\mathscr{P}}_{1,\eta}$ and the argument of [4, proof of 4.3], for any invertible sheaf $\mathscr{L}$ on $\widetilde{\mathscr{P}}_{1,\eta}$ the sequence

$$\mathscr{L} \to \prod_i j_{\bar{\omega}_i *} j_{\bar{\omega}_i}^* \mathscr{L} \rightrightarrows \prod_{i,k} j_{\bar{\omega}_i \cap \bar{\omega}_k *} j_{\bar{\omega}_i \cap \bar{\omega}_k}^* \mathscr{L} \tag{5.9.28.3}$$

is exact, where we have ordered the top-dimensional elements $\{\bar{\omega}_i\}$ of $\overline{S}$ in some way.

To give the map 5.9.28.2 it in turn suffices to define a section of

$$L_{\Gamma_{\bar\omega}} \otimes q^* L_W^{-2}(N\Gamma_0), \tag{5.9.28.4}$$

where Γ_0 denotes the closed fiber of Γ (a Cartier divisor on Γ) and N is an integer.

Lemma 5.9.29 *Let* $L_{\widetilde{\Gamma}_{\bar\omega}}$ *denote the invertible sheaf on* $\widetilde{\Gamma}_{\bar\omega}$ *obtained by restricting the sheaf 5.9.23.2 to* $\widetilde{\Gamma}_{\bar\omega}$. *Then there exists an integer N such that the section*

$$s_{\mathscr{U}_{\bar\omega}} \in \Gamma(\mathscr{U} \cap \widetilde{\Gamma}_{\bar\omega}, L_{\widetilde{\Gamma}_{\bar\omega}} \otimes q^* L_W^{-2}) \tag{5.9.29.1}$$

defined by the natural map $f|_{\mathscr{U}}^* \mathscr{O}_{\widetilde{W}}(1) \to \mathscr{O}_{\widetilde{\mathscr{P}}}(1)|_{\mathscr{U}}$ *extends to a section of*

$$L_{\widetilde{\Gamma}_{\bar\omega}} \otimes p_2^* \mathscr{O}_{\widetilde{W}}(1)^{\otimes -2}(N\widetilde{\Gamma}_{\bar\omega,0}). \tag{5.9.29.2}$$

Proof. Let $\mathscr{V} \subset \widetilde{\Gamma}_{\bar\omega}$ be a quasi-compact open subset such that the Y_1-translates of $\mathscr{V}$ cover $\widetilde{\Gamma}_{\bar\omega}$ (for example take the inverse image of such an open subset in $\widetilde{\mathscr{P}}_{\bar\omega}$). Let N be an integer such that the restriction of $s_{\mathscr{U}_{\bar\omega}}$ to $\mathscr{V}$ extends to an element

$$s \in \Gamma(\mathscr{V}, L_{\widetilde{\Gamma}_{\bar\omega}} \otimes p_2^* \mathscr{O}_{\widetilde{W}}(1)^{\otimes -2}(N\widetilde{\Gamma}_{\bar\omega,0})). \tag{5.9.29.3}$$

Then for every $y \in Y_1$ the pullback

$$y^*(s) \in \Gamma(y(\mathscr{V}), L_{\widetilde{\Gamma}_{\bar\omega}} \otimes p_2^* \mathscr{O}_{\widetilde{W}}(1)^{\otimes -2}(N\widetilde{\Gamma}_{\bar\omega,0})) \tag{5.9.29.4}$$

is a section whose restriction to $\mathscr{U}_{\bar\omega} \cap y(\mathscr{V})$ is equal to the restriction of $s_{\mathscr{U}}$. Since $\widetilde{\Gamma}_{\bar\omega}$ is integral this implies that these locally defined section $y^*(s)$ agree on the overlaps and hence proves the lemma. $\square$

5.9.30. This section

$$s \in \Gamma(\widetilde{\Gamma}_{\bar\omega}, L_{\widetilde{\Gamma}_{\bar\omega}} \otimes p_2^* \mathscr{O}_{\widetilde{W}}(1)^{\otimes -2}(N\widetilde{\Gamma}_{\bar\omega,0})) \tag{5.9.30.1}$$

defines for every n a Y_1-invariant section of

$$\Gamma(\widetilde{\Gamma}_{\bar\omega,n}, L_{\widetilde{\Gamma}_{\bar\omega,n}} \otimes p_2^* \mathscr{O}_{\widetilde{W}}(1)^{\otimes -2}(N\widetilde{\Gamma}_{\bar\omega,0})) \tag{5.9.30.2}$$

and by passing to the quotient by the Y_1-action and algebraizing a section

$$s \in \Gamma(\Gamma_{\bar\omega}, L_{\Gamma_{\bar\omega}} \otimes p_2^* L_W^{\otimes -2}(N\Gamma_{\bar\omega,0})). \tag{5.9.30.3}$$

On the generic fiber this defines the morphism 5.9.28.1.

Lemma 5.9.31 *For* $\bar x \in \overline{X}$ *and* $d \geq 1$, *let* $(f_{\eta*} L_{\widetilde{\mathscr{P}}_{1,\eta}}^d)_{\bar x}$ *denote the $\bar x$-eigenspace of the sheaf with $D(\overline{X})$-action $f_{\eta*} L_{\widetilde{\mathscr{P}}_{1,\eta}}^d$.*

   (i) For every $(d,\bar x)$ the sheaf $(f_{\eta*} L_{\widetilde{\mathscr{P}}_{1,\eta}}^d)_{\bar x}$ is locally free of rank 1 on W_η.

(ii) The map 5.9.28.1 induces an isomorphism

$$L_{W_\eta} \rightarrow (f_{\eta*} L_{\widetilde{\mathscr{P}}_{1,\eta}})_0. \tag{5.9.31.1}$$

(iii) The natural rational map over W_η

$$\widetilde{\mathscr{P}}_{1,\eta} \dashrightarrow \underline{\mathrm{Proj}}_{W_\eta} (\oplus_{(d,\bar{x})\in \overline{P}} (f_{\eta*} L^d_{\widetilde{\mathscr{P}}_{1,\eta}})_{\bar{x}}) \tag{5.9.31.2}$$

is a morphism, and in fact an isomorphism of schemes.

Proof. All of the assertions can be verified after base-changing to the completion of W_η at a point. In this case the result follows from the local descriptioin 5.9.27.4 (and the verification, left to the reader, that this local description is also compatible with the $D(\overline{X})$-action and the map on line bundles). □

5.9.32. In what follows for $(d, \bar{x}) \in \overline{P}$ we write

$$L^{(d,\bar{x})} := (f_{\eta*} L^d_{\widetilde{\mathscr{P}}_{1,\eta}})_{\bar{x}}. \tag{5.9.32.1}$$

Lemma 5.9.33 *There exists a unique G_1-action on W reducing to the action on W_n described in 5.9.11 for all n. The generic fiber W_η is a torsor under $G_{1,\eta}$ with respect to this action.*

Proof. This follows from the observation that $\widetilde{W}$ is a relatively complete model in the sense of [13, III.3.1], and the results of [13, III, §4]. □

Lemma 5.9.34 *Let $\pi : G \rightarrow G_1$ be the projection. Then the diagram*

$$
\begin{array}{ccc}
G_\eta \times \mathscr{P}_{1,\eta} & \xrightarrow{\text{action}} & \mathscr{P}_{1,\eta} \\
\downarrow{\scriptstyle \pi \times f_\eta} & & \downarrow{\scriptstyle f_\eta} \\
G_{1,\eta} \times W_\eta & \xrightarrow{\text{action}} & W_\eta
\end{array}
\tag{5.9.34.1}
$$

commutes.

Proof. Note that by the construction for any finite extension V'/V and section $g \in G(V')$ the diagram over $K' = \mathrm{Frac}(V')$

$$
\begin{array}{ccc}
\mathscr{P}_{1,\eta} & \xrightarrow{g} & \mathscr{P}_{1,\eta} \\
\downarrow{\scriptstyle f_\eta} & & \downarrow{\scriptstyle f_\eta} \\
W_\eta & \xrightarrow{\pi(g)} & W_\eta
\end{array}
\tag{5.9.34.2}
$$

commutes. On the other hand, the set of points of $G_{\bar{\eta}}$ for which the diagram 5.9.34.2 commutes is a closed subgroup scheme of $G_{\bar{\eta}}$. If H_η denotes the quotient of G_η by this subgroup scheme, and if H denotes the connected component of the Neron model of H_η, then the projection map $G_\eta \to H_\eta$ extends uniquely to a map $G \to H$. After replacing V by a finite extension we may assume that H is a semiabelian scheme over V. The map $G \to H$ is then surjective. On the other hand, for any V'-valued point $g \in G(V')$ as above, the image of g in $H(V')$ is the identity. It follows that H, and hence also H_η, is trivial. $\quad\square$

5.9.35. The line bundles $L^{(d,\bar{x})}$ admit the following alternate characterization.

Let $M_{\widetilde{\mathscr{P}}_1}$ denote the pullback to $\widetilde{\mathscr{P}}_1$ of the log structure $M_{\mathscr{P}}$, and similarly let $M_{\widetilde{\mathscr{P}}}$ denote the pullback of $M_{\mathscr{P}^\wedge}$ to $\widetilde{\mathscr{P}}^\wedge$ (the formal scheme over the completion of $\mathscr{P}$ along the maximal ideal of V used in the standard construction 5.2.4.1). On $\widetilde{\mathscr{P}}^\wedge$ we have by the construction a surjection $P \rtimes H_S \to \overline{M}_{\widetilde{\mathscr{P}}}$ such that for any point $x \in \tilde{P}^\wedge$ and $y \in Y$ the diagram

$$
\begin{array}{ccc}
P \rtimes H_S & \longrightarrow & \overline{M}_{\widetilde{\mathscr{P}}^\wedge,\bar{x}} \\
\lambda_y \downarrow & & \downarrow \text{action} \\
P \rtimes H_S & \longrightarrow & \overline{M}_{\widetilde{\mathscr{P}}^\wedge,y(\bar{x})}
\end{array}
$$

commutes where λ_y is the translation action. Consider the sheaf of sets $\mathscr{H}om(P \rtimes H_S, \overline{M}_{\widetilde{\mathscr{P}}_1})/Y_1$ where $\mathscr{H}om(P \rtimes H_S, \overline{M}_{\widetilde{\mathscr{P}}_1})$ denotes the sheaf of morphisms of monoids, and the Y_1–action is given by the translation action on $P \rtimes H_S$. The sheaf $\mathscr{H}om(P \rtimes H_S, \overline{M}_{\widetilde{\mathscr{P}}_1})/Y_1$ is a constructible sheaf of sets on $\widetilde{\mathscr{P}}_1$. The map $P \rtimes H_S \to \overline{M}_{\widetilde{\mathscr{P}}^\wedge}$ defines a global section

$$
\gamma_0 \in H^0(\widetilde{\mathscr{P}}_{1,0}, \mathscr{H}om(P \rtimes H_S, \overline{M}_{\widetilde{\mathscr{P}}_{1,0}})/Y_1).
$$

Lemma 5.9.36 *For any constructible sheaf of sets $\mathcal{F}$ on $\widetilde{\mathscr{P}}_1$ the natural map*

$$
H^0(\widetilde{\mathscr{P}}_1, \mathcal{F}) \to H^0(\widetilde{\mathscr{P}}_{1,0}, \mathcal{F}_1)
$$

is an isomorphism.

Proof. The scheme $\widetilde{\mathscr{P}}_1$ is obtained by gluing proper V–schemes along closed subschemes (the schemes $\mathscr{P}_{\bar{\omega}}$). It follows that to give a global section of $\mathcal{F}$ is equivalent to giving a compatible collection of sections over these proper subschemes, and the global sections over $\widetilde{\mathscr{P}}_{1,s}$ can be described similarly. The result therefore follows from the proper base change theorem [8, XII.5.1]. $\quad\square$

We therefore have a global section $\gamma \in H^0(\widetilde{\mathscr{P}}_1, \mathscr{H}om(P \rtimes H_S, \overline{M}_{\widetilde{\mathscr{P}}_1})/Y_1)$ inducing γ_0.

Lemma 5.9.37 *The restriction of γ to $\widetilde{\mathscr{P}}_{1,\eta}$ is contained in the image of the inclusion*

$$
\mathscr{H}om(\overline{P} \rtimes H_{\overline{S}}, \overline{M}_{\widetilde{\mathscr{P}}_1}) \hookrightarrow \mathscr{H}om(P \rtimes H_S, \overline{M}_{\widetilde{\mathscr{P}}_1})/Y_1.
$$

Proof. This follows from 5.8.21 and 5.9.8. □

On $\widetilde{\mathscr{P}}_{1,\eta}$ the section γ therefore induces a map $\bar{\gamma} : \overline{P} \rtimes H_{\overline{S}}^{\mathrm{sat}} \to \overline{M}_{\widetilde{\mathscr{P}}_{1,\eta}}$.

Lemma 5.9.38 *Let X be a projective normal toric variety with torus T and for every $x \in X$ let T_x denote the stabilizer of the point x. Let L be a T-linearized line bundle on X such that for every $x \in X$ the character $T_x \to \mathrm{Aut}(L(x)) \simeq \mathbb{G}_m$ is trivial. Then L is trivial as a T-linearized line bundle.*

Proof. Choose a birational morphism of toric varieties $\pi : X' \to X$ with center of codimension ≥ 2 in X with X' smooth. Let L' denote $\pi^* L$. Then since X is normal we have $H^0(X, L) = H^0(X', L')$. If the lemma holds for (X', L') then this implies that $H^0(X, L)$ is 1–dimensional and that the map of invertible sheaves

$$H^0(X, L) \otimes_k \mathscr{O}_X \to L$$

becomes an isomorphism after applying π^*, hence is an isomorphism. It therefore suffices to consider the case when X is smooth. In this case it is shown in [40] that the underlying line bundle of L is trivial, and the statement that the action is trivial follows immediately from the assumptions. □

5.9.39. Let $f_\eta : \widetilde{P}_{1,\eta} \to W_\eta$ be the projection defined in 5.9.25.6. For $(d, \bar{x}) \in \overline{P}$, let $N^{(d,\bar{x})}$ denote the line bundle on $\widetilde{\mathscr{P}}_{1,\eta}$ corresponding to the $\mathbb{G}_m$-torsor of liftings of $\bar{\gamma}(d, \bar{x})$ to $M_{\widetilde{\mathscr{P}}_{1,\eta}}$. The line bundle $N^{(d,\bar{x})}$ comes equipped with a morphism $N^{(d,\bar{x})} \to \mathscr{O}_{\widetilde{P}_{1,\eta}}$, and hence

$$U^{(d,x)} := N^{(d,\bar{x})} \otimes L_{\widetilde{\mathscr{P}}_{1,\eta}}^d \tag{5.9.39.1}$$

has a natural morphism $U^{(d,x)} \to L_{\widetilde{\mathscr{P}}_{1,\eta}}^d$.

Lemma 5.9.40 *Let $\zeta \in \widetilde{\mathscr{P}}_{1,\eta}$ be a point and let $T_\zeta \subset D(\overline{X})$ be the stabilizer of ζ. Then the character $T_\zeta \to \mathrm{Aut}(U^{(d,\bar{x})}(\zeta)) = \mathbb{G}_m$ is equal to the character*

$$T_\zeta \longrightarrow D(\overline{X}) \xrightarrow{\ \bar{x}\ } \mathbb{G}_m.$$

Proof. Let $\bar{\zeta} \subset \widetilde{\mathscr{P}}_1$ denote the closure of ζ, and let $z \in \bar{\zeta}$ be a point in the closed fiber. Let $w \in \widetilde{\mathscr{P}}$ denote a point over z. Then there is a commutative diagram

$$
\begin{array}{ccccccccc}
0 & \longrightarrow & \widehat{\mathscr{O}}^*_{\widetilde{\mathscr{P}}, \bar{w}} & \longrightarrow & (M_{\widetilde{\mathscr{P}}}|_{\widehat{\mathscr{O}}_{\widetilde{\mathscr{P}}, \bar{w}}})_{\bar{w}} & \longrightarrow & \overline{M}_{\widetilde{\mathscr{P}}, \bar{w}} & \longrightarrow & 0 \\
& & \| & & \| & & \| & & \\
0 & \longrightarrow & \widehat{\mathscr{O}}^*_{\widetilde{\mathscr{P}}_1, \bar{z}} & \longrightarrow & (M_{\widetilde{\mathscr{P}}_1}|_{\widehat{\mathscr{O}}_{\widetilde{\mathscr{P}}_1, \bar{z}}})_{\bar{z}} & \longrightarrow & \overline{M}_{\widetilde{\mathscr{P}}_1, \bar{z}} & \longrightarrow & 0 \\
& & \downarrow & & \downarrow & & \downarrow & & \\
0 & \longrightarrow & \widehat{\mathscr{O}}_{\widetilde{\mathscr{P}}_{1,\eta}, \bar{\zeta}} & \longrightarrow & (M_{\widetilde{\mathscr{P}}_{1,\eta}}|_{\widehat{\mathscr{O}}_{\widetilde{\mathscr{P}}_1, \bar{\zeta}}})_{\bar{\zeta}} & \longrightarrow & \overline{M}_{\widetilde{\mathscr{P}}_1, \bar{\zeta}} & \longrightarrow & 0,
\end{array}
$$

and by the construction the map $\bar\gamma : \overline{P} \rtimes H_{\overline{S}} \to \overline{M}_{\widetilde{\mathscr{P}}_{1,\zeta}}$ is induced by the map $P \rtimes H_S \to \overline{M}_{\widetilde{\mathscr{P}},\bar{w}}$. Choose any $(d,x) \in P$ liftings $(d,\bar{x})$. Since the stabilizer T_z in $D(\overline{X})$ of z contains the torus T_ζ, it suffices to show that the action of T_z on the tensor product of $L^d_{\widetilde{\mathscr{P}}}$ with the torsor of liftings to $(M_{\widetilde{\mathscr{P}}}|_{\hat{\mathscr{O}}_{\widetilde{\mathscr{P}},\bar{w}}})_{\bar{w}}$ of the image of (d,x) in $\overline{M}_{\widetilde{\mathscr{P}},\bar{w}}$ is given by the character (d,x). This follows from the standard construction. $\square$

Lemma 5.9.41 *Let $S_{(d,\bar{x})}$ denote the trivial line bundle on $\widetilde{\mathscr{P}}_{1,\eta}$ with $D(\overline{X})$-linearization given by the character $\bar{x}$. Then the line bundle $U^{(d,\bar{x})}$ is isomorphic as a $D(\overline{X})$-linearized line bundle to $f_\eta^* f_{\eta *} U^{(d,\bar{x})} \otimes S_{(d,\bar{x})}$.*

Proof. By 5.9.38 applied to $U^{(d,\bar{x})} \otimes S^{-1}_{(d,\bar{x})}$ the restriction of $U^{(d,\bar{x})}$ to each $\mathscr{P}_{\bar{\omega},\eta}$ is the pullback of a line bundle on W_η. Let $w \in W_\eta$ be a point, and let T denote the spectrum of $\hat{\mathscr{O}}_{W_\eta,\bar{w}}$. To prove that the map

$$f_\eta^* f_{\eta *} U^{(d,\bar{x})} \otimes S_{(d,\bar{x})} \to U^{(d,\bar{x})} \qquad (5.9.41.1)$$

is an isomorphism, it suffices to show that this is the case after base change to T. By 5.9.27.4 this base change to T is isomorphic to

$$\underline{\mathrm{Proj}}_T(\mathscr{O}_T \otimes_{\mathbb{Z}[H_{\overline{S}}]} \mathbb{Z}[\overline{P} \rtimes H_{\overline{S}}]). \qquad (5.9.41.2)$$

Using the argument of [4, proof of 4.3] one sees that in this case a line bundle on 5.9.41.2 whose restriction to each

$$\underline{\mathrm{Proj}}_T(\mathscr{O}_T \otimes_{\mathbb{Z}[H_{\overline{S}}]} \mathbb{Z}[\overline{P}_{\bar\omega} \rtimes H_{\overline{S}}]) \subset \underline{\mathrm{Proj}}_T(\mathscr{O}_T \otimes_{\mathbb{Z}[H_{\overline{S}}]} \mathbb{Z}[\overline{P} \rtimes H_{\overline{S}}]) \quad (5.9.41.3)$$

is trivial is classified by an element of

$$H^1(\overline{X}_{\mathbb{R}}, \mathscr{O}_T^*) = 0. \qquad (5.9.41.4)$$

The statement about the action follows from the preceding lemma. $\square$

Corollary 5.9.42 *The map $U^{(d,\bar{x})} \to L_{\widetilde{\mathscr{P}}_{1,\eta}}$ induces an isomorphism*

$$(f_{\eta *} U^{(d,\bar{x})})_{\bar{x}} \to L^{(d,\bar{x})}. \qquad (5.9.42.1)$$

5.9.43. Note that by construction there is for any $(d,\bar{x}), (l,\bar{y}) \in \overline{P}$ a canonical isomorphism of $\mathbb{G}_m$-torsors with maps to $M_{\widetilde{\mathscr{P}}_{1,\eta}}$ between

$$N^{(d,\bar{x})} \otimes N^{(l,\bar{y})} \qquad (5.9.43.1)$$

and the torsor of liftings of

$$\bar\gamma(d+l, \bar{x}+\bar{y}, (d,\bar{x}) * (l,\bar{y})) \in \overline{M}_{\widetilde{\mathscr{P}}_{1,\eta}}. \qquad (5.9.43.2)$$

On the other hand, translation by the image of $(d, \bar{x}) * (l, \bar{y}))$ in $M_{\widetilde{\mathscr{P}}_{1,\eta}}$ defines an isomorphism between $N^{(d+l, \bar{x}+\bar{y})}$ and this torsor of liftings of 5.9.43.2. We therefore obtain a canonical isomorphism

$$N^{(d,\bar{x})} \otimes N^{(l,\bar{y})} \longrightarrow N^{(d+l,\bar{x}+\bar{y})}. \tag{5.9.43.3}$$

Twisting this isomorphism by $L_{\widetilde{\mathscr{P}}_{1,\eta}}^{d+l}$ and pushing down to W_η we obtain a canonical isomorphism

$$\mathrm{can} : L^{(d,\bar{x})} \otimes L^{(l,\bar{y})} \longrightarrow L^{(d+l,\bar{x}+\bar{y})}. \tag{5.9.43.4}$$

Now define for any $\bar{x} \in \overline{X}$ the line bundle

$$L_{\bar{x}} := L^{(d,\bar{x})} \otimes L_{W_\eta}^{-d}, \tag{5.9.43.5}$$

where d is an integer ≥ 1. Up to canonical isomorphism this is independent of the choice of d as can be seen as follows.

First note that the isomorphism 5.9.43.4 induces in particular a canonical isomorphism

$$L^{(d,\bar{x})} \otimes L^{(d,-\bar{x})} \simeq L_{W_\eta}^{2d}. \tag{5.9.43.6}$$

Therefore for any other integer $d' \geq 1$ we have a canonical isomorphism

$$L^{(d,\bar{x})} \otimes L_{W_\eta}^{-d} \otimes (L^{(d',\bar{x})})^{-1} \otimes L_{W_\eta}^{d'}$$
$$\simeq L^{(d,\bar{x})} \otimes L_{W_\eta}^{-d} \otimes L^{(d',-\bar{x})} \otimes L_{W_\eta}^{-2d'} \otimes L_{W_\eta}^{d'}$$
$$\simeq L^{(d+d',\bar{x}-\bar{x})} \otimes L_{W_\eta}^{-d-d'}$$
$$\simeq \mathcal{O}_{\widetilde{\mathscr{P}}_{1,\eta}}.$$

This implies the independence on d. From 5.9.43.4 we also obtain for every $\bar{x}, \bar{y} \in \overline{X}$ a canonical isomorphism

$$L_{\bar{x}} \otimes L_{\bar{y}} \longrightarrow L_{\bar{x}+\bar{y}}. \tag{5.9.43.7}$$

In particular we obtain a homomorphism

$$c : \overline{X} \longrightarrow \underline{\mathrm{Pic}}(W_\eta). \tag{5.9.43.8}$$

We will show below that in fact this homomorphism has image in $\underline{\mathrm{Pic}}^0(W_\eta)$.

5.9.44. There is an action of $\overline{Y}$ on W defined as follows.

First note that there is an action of Y on $\widetilde{W}$. An element $y \in Y$ acts by choosing an isomorphism

$$\rho_y : t_y^* \mathcal{M} \longrightarrow \mathcal{M} \otimes L_y \tag{5.9.44.1}$$

which induces an isomorphism

$$\rho : t_y^*(\oplus_{(d,x)\in P_1} \mathscr{M}^d \otimes L_x) \to \oplus_{(d,x)\in P_1} \mathscr{M}^d \otimes L_x \otimes L_y^d. \qquad (5.9.44.2)$$

This map in turn induces a morphism of schemes

$$\underline{\mathrm{Proj}}_A(\oplus_{(d,x)\in P_1} \mathscr{M}^d \otimes L_x) \xrightarrow{\;\simeq\;} \underline{\mathrm{Proj}}_A(\oplus_{(d,x)\in P_1} \mathscr{M}^d \otimes L_x \otimes L_y^d)$$

$$\searrow^{t_y} \qquad\qquad \downarrow$$

$$\underline{\mathrm{Proj}}_A(\oplus_{(d,x)\in P_1} \mathscr{M}^d \otimes L_x)$$

$$(5.9.44.3)$$

over the translation action of y on A.

For every n, this action of Y on $\widetilde{W}$ induces an action of Y on $\widetilde{W}_n$ which commutes with the Y_1-action, and hence induces an action of $\overline{Y}$ on W_n. For every $y \in \overline{Y}$ the action of y on W_n is given by its graph in $W_n \times_{B_n} W_n$. By the Grothendieck existence theorem applied to $W_n \times_{B_n} W_n$ the action of y can therefore be uniquely algebraized to an action on W. In this way we obtain an action of $\overline{Y}$ on W.

Note that the rational map

$$f : \widetilde{\mathscr{P}} \dashrightarrow \widetilde{W} \qquad\qquad (5.9.44.4)$$

is compatible with the Y-action, and hence Γ also admits an action of $\overline{Y}$ compatible with the actions on $\widetilde{\mathscr{P}}_1$ and W. It follows that the morphism on the generic fiber

$$f_\eta : \widetilde{P}_{1,\eta} \to W_\eta \qquad\qquad (5.9.44.5)$$

is compatible with the $\overline{Y}$-actions.

Note also that by construction the $\overline{Y}$-action on each W_n commutes with the G_1-action, and by an argument similar to the one used in 5.9.34 the action of $\overline{Y}$ on W_n commutes with the G_1-action. It follows that for every $y \in \overline{Y}$ there exists a unique element $c^t(y) \in G_{1,\eta}$ such that the diagram

$$
\begin{array}{ccc}
\widetilde{\mathscr{P}}_{1,\eta} & \xrightarrow{\;t_y\;} & \widetilde{\mathscr{P}}_{1,\eta} \\
{\scriptstyle f_\eta}\downarrow & & \downarrow{\scriptstyle f_\eta} \\
W_\eta & \xrightarrow{\;t_{c^t(y)}\;} & W_\eta
\end{array}
\qquad\qquad (5.9.44.6)
$$

commutes. In this way we obtain a homomorphism

$$c^t : \overline{Y} \to G_{1,\eta}. \qquad\qquad (5.9.44.7)$$

Lemma 5.9.45 *The diagram*

$$
\begin{array}{ccc}
\overline{Y} & \longrightarrow & \overline{X} \\
{\scriptstyle c^t}\downarrow & & \downarrow{\scriptstyle c} \\
G_{1,\eta} & \xrightarrow{\;\lambda_{L_{W_\eta}}\;} & \underline{\mathrm{Pic}}(W_\eta)
\end{array}
\qquad\qquad (5.9.45.1)
$$

commutes.

Proof. By construction $t_y^* U^{(1,x)} = U^{(1,x+y)}$ for every $x \in \overline{X}$ and $y \in \overline{Y}$. From this it follows that

$$L_{c(y)} = L_{W_\eta}^{-1} \otimes t_y^* L_{W_\eta} = \lambda_{L_{W_\eta}}(c^t(y)). \qquad (5.9.45.2)$$

□

Corollary 5.9.46 *The image of c is contained in $\underline{\mathrm{Pic}}^0(W_\eta)$.*

Proof. By [35, Cor. 2, p. 178] the quotient $\underline{\mathrm{Pic}}(W_\eta)/\underline{\mathrm{Pic}}^0(W_\eta)$ is torsion free. Since the image of $\overline{Y}$ is contained in $\underline{\mathrm{Pic}}^0(W_\eta)$ by the commutativity of 5.9.45.1 and $\overline{Y} \subset \overline{X}$ has finite index this implies the result.   □

Lemma 5.9.47 *The homomorphism $c : \overline{X} \to G_{1,\eta}^t$ is equal to the homomorphism defined by the extension*

$$0 \to D(\overline{X}) \to G_\eta \to G_{1,\eta} \to 0. \qquad (5.9.47.1)$$

Proof. It suffices to verify this after making a base change on B. We may therefore assume that there exists a point of $\widetilde{\mathscr{P}}_{1,\eta}$ contained in the maximal dense open subset where G_η acts freely. This defines an inclusion $j : G_\eta \hookrightarrow \widetilde{\mathscr{P}}_{1,\eta}$ sitting in a commutative diagram

$$
\begin{array}{ccc}
G_\eta & \xrightarrow{\ j\ } & \widetilde{\mathscr{P}}_{1,\eta} \\
 & \searrow{\scriptstyle \pi} & \downarrow{\scriptstyle f_\eta} \\
 & & G_{1,\eta}.
\end{array}
\qquad (5.9.47.2)
$$

Furthermore, using the local description in 5.9.27.4 one sees that the map

$$\pi^* L_{W_\eta} \simeq j^* f_\eta^* L_{W_\eta} \to j^* L_{\widetilde{\mathscr{P}}_{1,\eta}} \qquad (5.9.47.3)$$

is an isomorphism, and induces for every $x \in \overline{X}$ an isomorphism

$$(f_{\eta *} L_{\widetilde{\mathscr{P}}_{1,\eta}})_x \to (\pi_*(\mathscr{O}_{G_\eta})) \otimes L_{W_\eta}^{-1}. \qquad (5.9.47.4)$$

This isomorphism induces for every $x \in \overline{X}$ an isomorphism

$$L_x \simeq (\pi_* \mathscr{O}_{G_\eta})_x. \qquad (5.9.47.5)$$

This implies the lemma.   □

5.9.48. Fix a point $w \in W_\eta$ giving an identification $W_\eta \simeq G_1$. Also fix a rigidification of L_{W_η} at the origin. Summarizing: we have constructed data as below such that the log scheme $\widetilde{\mathscr{P}}_{1,\eta}$ over $G_{1,\eta}$ is obtained by applying the construction 5.2.4 using this data.

(i) An inclusion $\phi : \overline{Y} \hookrightarrow \overline{X}$.
(ii) A rigidified ample invertible sheaf $L_{G_{1,\eta}}$ on $G_{1,\eta}$ (the sheaf obtained from L_{W_η}).
(iii) A homomorphism $c : \overline{X} \to G^t_{1,\eta}$ defining the extension

$$0 \to D(\overline{X}) \to G_\eta \to G_{1,\eta} \to 0. \tag{5.9.48.1}$$

(iv) A homomorphism $c^t : \overline{Y} \to G_{1,\eta}$ such that the diagram

$$
\begin{array}{ccc}
\overline{Y} & \xrightarrow{\phi} & \overline{X} \\
{\scriptstyle c^t}\downarrow & & \downarrow{\scriptstyle c} \\
G_{1,\eta} & \xrightarrow{\lambda_{L_{G_{1,\eta}}}} & G^t_{1,\eta}
\end{array}
\tag{5.9.48.2}
$$

commutes.
(v) An integral regular paving $\overline{S}$ of $\overline{X}_\mathbb{R}$, and a morphism $\beta : H_{\overline{S}} \to K$ sending all nonzero elements to zero.

5.9.49. To complete the proof of 5.9.1 in the case when B is the spectrum of a complete discrete valuation ring it remains to describe the $\overline{Y}$-action on $\widetilde{\mathscr{P}}_{1,\eta}$ in terms of maps ψ and τ as in (5.2.1, (vi) and (vii)).

The action of $\overline{Y}$ on $\widetilde{\mathscr{P}}_{1,\eta}$ defines for every $\bar{y} \in \overline{Y}$ an isomorphism $t^*_{\bar{y}} L_{G_{1,\eta}} \to L_{G_{1,\eta}} \otimes L_{\bar{y}}$. Equivalently a section $\bar{\psi} : \overline{Y} \to c^{t*} L^{-1}_{G_{1,\eta}}$. Furthermore the isomorphisms

$$t^*_{\bar{y}}(L^d_{G_{1,\eta}} \otimes L_{\bar{x}}) \to L^d_{G_{1,\eta}} \otimes L_{\bar{x}} \otimes L_{\bar{y}}$$

can be written uniquely as $\bar{\psi}(\bar{y})^d$ multiplied by an isomorphism $\tau(\bar{y}, \bar{x})$: $t^*_{\bar{y}} L_{\bar{x}} \to L_{\bar{x}}$. These isomorphisms define a trivialization

$$\tau : \overline{Y} \times \overline{X} \to (c^t \times c)^* \mathscr{B}^{-1}.$$

As in 4.1.17 the compatibility with the algebra structure precisely means that this restricts on $\overline{Y} \times \overline{Y}$ to a trivialization as a symmetric biextension. Summarizing all of this we see that $(\mathscr{P}_\eta, M_{\mathscr{P}_\eta}, L_{P_\eta})$ is obtained from the standard construction.

5.9.50. To complete the proof of 5.9.1 it now suffices to show that the set of points of B where the fiber is isomorphic to the standard construction is constructible. For this we may assume B integral. Furthermore, it suffices to show that if the generic fiber is given by the standard construction, then there exists a dense open subset of B where all the fibers are obtained by the standard construction. In fact, by a standard "spreading out" argument one sees that there exists a dominant morphism $B' \to B$ such that the pullback to B' is given by the standard construction. We leave the details to the reader.

5.10 Isomorphisms in $\overline{\mathscr{T}}_{g,d}$

5.10.1. Let B be a scheme and $\mathscr{S}_i := (G_i, M_B^i, P_i, M_{P_i}, L_{P_i}) \in \overline{\mathscr{T}}_{g,d}(B)$ $(i = 1,2)$ be two objects. Let I denote the quasi–projective scheme of isomorphisms $(G_1, P_1, L_{P_1}) \to (G_2, P_2, L_{P_2})$ of polarized schemes with semi–abelian group action 5.6.2. Denote by I' the functor

$$\underline{\mathrm{Isom}}_{\overline{\mathscr{T}}_{g,d}}(\mathscr{S}_1, \mathscr{S}_2).$$

Proposition 5.10.2. *The natural morphism of functors $I' \to I$ is representable by closed immersions. In particular, I' is a quasi–projective B–scheme.*

Proof. In the case when B is the spectrum of an artinian local ring the result can be seen as follows. It suffices to prove the result after making a flat base change (note that I' is a sheaf with respect to the fppf topology), and therefore we can assume that the families $\mathscr{S}_i$ are obtain by saturating the standard construction. Let $(G_i, N_B^i, P_i, N_{P_i}, L_{P_i})$ $(i = 1,2)$ be the data obtained by the standard construction such that $\mathscr{S}_i$ is obtained from this family by saturating. Note that the data $Y \subset X$, S, and the maps c, c^t must be the same in both constructions by 5.3.4. We may further assume that $\overline{N}_{P_i}$ are constant sheaves. Then by the same argument used in 4.4 an isomorphism $\iota : (G_1, P_1, L_{P_1}) \to (G_2, P_2, L_{P_2})$ extends to a unique isomorphism $\iota : (G_1, N_B^1, P_1, N_{P_1}, L_{P_1}) \to (G_2, N_B^2, P_2, N_{P_2}, L_{P_2})$. To extend this isomorphism to an element of I' it suffices to find an isomorphism $\tilde{\iota} : M_B^1 \to M_B^2$ such that the diagram

$$
\begin{array}{ccc}
N_B^1 & \longrightarrow & M_B^1 \\
\iota \downarrow & & \downarrow \tilde{\iota} \\
N_B^2 & \longrightarrow & M_B^2
\end{array}
$$

commutes. For then we get an isomorphism

$$M_{P_1} \simeq N_{P_1} \oplus_{N_B^1} M_B^1 \to N_{P_2} \oplus_{N_B^2} M_B^2 = M_{P_2}$$

by pushout.

Let $T_i \subset \overline{N}_B^{i,gp}$ be the torsion subgroup. It is a constant finite group by assumption. From the commutative diagram

$$
\begin{array}{ccccccccc}
0 & \longrightarrow & \mathscr{O}_B^* & \longrightarrow & N_B^{i,gp} & \longrightarrow & \overline{N}_B^{i,gp} & \longrightarrow & 0 \\
& & \| & & \downarrow & & \downarrow & & \\
0 & \longrightarrow & \mathscr{O}_B^* & \longrightarrow & M_B^{i,gp} & \longrightarrow & \overline{M}_B^{i,gp} & \longrightarrow & 0
\end{array}
$$

and the snake lemma it follows that the kernel of $N_B^{i,gp} \to M_B^{i,gp}$ is canonically isomorphic to T_i. For any element $t \in T_1$ the image $\iota(t) \in N_B^{2,gp}$ is equal to

$\lambda(u) + t'$ for unique $t' \in T_2$ and $u \in \mathscr{O}_B^*$. This defines a homomorphism $T_1 \to \mathbb{G}_m$ and the condition that the isomorphism $N_B^1 \to N_B^2$ extends to an isomorphism $M_B^1 \to M_B^2$ is precisely the condition that this homomorphism is trivial. From this the case when B is the spectrum of an artinian local ring follows.

For the general case, note first that by the same argument used in the proof of 4.9.1 any isomorphism $\iota : (G_1, P_1, L_{P_1}) \to (G_2, P_2, L_{P_2})$ induces isomorphisms $\bar{\sigma} : \overline{M}_{P_1} \to \overline{M}_{P_2}$ and $\bar{\delta} : \overline{M}_B^1 \to \overline{M}_B^2$ agreeing with the morphisms obtained over artinian local rings by the preceding paragraph. Let $I_1 \to B$ be the scheme classifying isomorphisms $M_B^1 \to M_B^2$ inducing $\bar{\delta}$ as in the proof of 4.9.1. By the same argument as in section 4.9, there is over I_1 a scheme $I_2 \to I_1$ classifying isomorphisms $M_{P_1} \to M_{P_2}$ extending ι compatible with the G_i–actions over the tautological isomorphism $M_B^1 \to M_B^2$. The scheme I_2 then represents I'. That $I' \to I$ is a closed immersion follows from the case of an artinian local ring which shows that $I' \to I$ is a monomorphism and surjective, hence a closed immersion. $\square$

Corollary 5.10.3 *The stack $\overline{\mathscr{T}}_{g,d}$ is a quasi–compact algebraic stack with separated diagonal.*

5.11 Rigidification

5.11.1. Let $\mathscr{I} \to \overline{\mathscr{T}}_{g,d}$ denote the inertia stack of $\overline{\mathscr{T}}_{g,d}$. Recall that the stack $\mathscr{I}$ associates to a scheme B the groupoid of pairs $(\mathscr{S}, \alpha)$, where $\mathscr{S} \in \overline{\mathscr{T}}_{g,d}(B)$ and α is an automorphism of $\mathscr{S}$. In other words, $\mathscr{I}$ is the automorphism group scheme of the universal family over $\overline{\mathscr{T}}_{g,d}$.

Lemma 5.11.2 *Let B be a scheme and G/B a semi-abelian scheme over B. Then for any automorphism $\alpha : G \to G$, the condition that $\alpha = \mathrm{id}$ is represented by a closed subscheme of B.*

Proof. By a standard limit argument, it suffices to consider the case when B is noetherian. Let E_i be the locally free sheaf on B underlying the coordinate ring of the i–th infinitesimal neighborhood of the identity in G. Then α induces an automorphism $\alpha_i : E_i \to E_i$ of vector bundles and α is equal to the identity if and only if each α_i is the identity. Let $Z_i \subset B$ denote the closed subscheme representing the condition that α_i is the identity. We then obtain a decreasing sequence of closed subschemes

$$\cdots \subset Z_i \subset Z_{i-1} \subset \cdots \subset Z_0 \subset B.$$

Since B is noetherian this sequence is eventually constant and so $\cap Z_i \subset B$ represents the condition that α is the identity. $\square$

5.11.3. It follows that there is a closed substack $\mathcal{G} \subset \mathcal{I}$ classifying pairs $(\mathscr{S}, \alpha)$ where α is the identity on the underlying semi-abelian scheme of $\mathscr{S}$. In fact, $\mathcal{G} \subset \mathcal{I}$ is a normal (relative over $\overline{\mathscr{T}}_{g,d}$) subgroup scheme of $\mathcal{I}$. There is also a natural inclusion $\mathbb{G}_m \subset \mathcal{G}$ given by the action on the tautological line bundle.

Lemma 5.11.4 *The subgroup scheme $\mathbb{G}_m \subset \mathcal{G}$ is a normal closed subgroup scheme and the quotient $\mathcal{H} := \mathcal{G}/\mathbb{G}_m$ is a finite flat commutative group scheme of rank d^2 over $\overline{\mathscr{T}}_{g,d}$. In particular, $\mathcal{G}$ is flat over $\overline{\mathscr{T}}_{g,d}$.*

Proof. Note first that $\mathbb{G}_m \subset \mathcal{G}$ is a normal subgroup as this can be checked over artinian local rings. The quotient $\mathcal{H} = \mathcal{G}/\mathbb{G}_m$ therefore exists as an algebraic space. This group space is quasi-finite over $\overline{\mathscr{T}}_{g,d}$ and therefore is a relative scheme. Let R be a complete noetherian local ring and $\mathrm{Spec}(R) \to \overline{\mathscr{T}}_{g,d}$ a morphism. Let $\mathfrak{m}_R \subset R$ be the maximal ideal. For every n the pullback $\mathcal{H}$ of $\mathcal{H}$ to $\mathrm{Spec}(R/\mathfrak{m}_R^n)$ is finite and flat of rank d^2. By the Grothendieck existence theorem this closed subschemes are uniquely algebraizable to a closed subgroup scheme $H \subset \mathcal{H}_R$ (where $\mathcal{H}_R$ is the pullback of $\mathcal{H}$ to $\mathrm{Spec}(R)$), with H finite and flat of rank d^2. To prove that $H \subset \mathcal{H}_R$ is an isomorphism it then suffices to show that the map becomes an isomorphism after pullback by any morphism $\mathrm{Spec}(R') \to \mathrm{Spec}(R)$ with R' artinian local. This therefore also follows from the case of an artinian local ring. □

5.11.5. Let $\overline{\mathscr{A}}_{g,d}$ denote the rigidification of $\overline{\mathscr{T}}_{g,d}$ with respect to the group scheme $\mathcal{G} \subset \mathcal{I}$. Note that the map $\overline{\mathscr{T}}_{g,d} \to \overline{\mathscr{A}}_{g,d}$ is faithfully flat in general and smooth over $\mathbb{Z}[1/d]$. The inclusion $\mathscr{T}_{g,d} \subset \overline{\mathscr{T}}_{g,d}$ induces a dense open immersion $\mathscr{A}_{g,d} \subset \overline{\mathscr{A}}_{g,d}$. In addition, the log structure $M_{\overline{\mathscr{T}}_{g,d}}$ descends to a log structure $M_{\overline{\mathscr{A}}_{g,d}}$ on $\overline{\mathscr{A}}_{g,d}$ which is trivial precisely over $\mathscr{A}_{g,d}$.

Proposition 5.11.6. *The diagonal of $\overline{\mathscr{A}}_{g,d}$ is finite.*

Proof. We show that the diagonal is proper and quasi–finite.

To show that the diagonal is proper we check the valuative criterion. Let V be a discrete valuation ring with field of fractions K, let

$$\mathscr{S}_K = (G_K, P_K, M_K, M_{P_K}, L_K) \in \overline{\mathscr{T}}_{g,d}(K)$$

be an object, and let

$$\mathscr{S}_i = (G^i, P^i, M^i, M_{P^i}, L^i) \in \overline{\mathscr{T}}_{g,d}(V) \quad i = 1, 2$$

be two extensions of $\mathscr{S}_K$ to V. We have to show that after possibly replacing V by a finite extension and twisting the isomorphism $\mathscr{S}_{1,\eta} \to \mathscr{S}_{2,\eta}$ by an element of $\mathbb{G}_m(K)$ (acting on the isomorphism between the line bundles) there exists an isomorphism $\alpha : \mathscr{S}_1 \to \mathscr{S}_2$ over V defining the identity morphism on G_K. Let $\mathcal{G}_i/V$ $(i = 1, 2)$ denote the theta group of $\mathscr{S}_i$, and let H_i denote $\mathcal{G}_i/\mathbb{G}_m$.

The generic fibers of $\mathscr{G}_1$ and $\mathscr{G}_2$ are identified as are the generic fibers of H_1 and H_2. Denote by $\mathscr{G}_\eta$ and H_η these generic fibers.

Let $\underline{\mathrm{Aut}}(P^i)$ denote the group scheme of automorphisms of P^i and let Z_i denote the fiber product of the diagram

$$H_i$$
$$\downarrow$$
$$G \xrightarrow{\ \mathrm{action}\ } \underline{\mathrm{Aut}}(P^i).$$

Note that Z_i is a closed subgroup scheme of G.

Let $Y_i \subset X_i$ denote the groups used in the standard construction giving rise to $\mathscr{S}_i = (G^i, P^i, M^i, M_{P^i}, L^i)$.

Lemma 5.11.7 *The group schemes Z_i are flat over V.*

Proof. It suffices to verify that for every integer n the reduction of Z_i modulo $\mathfrak{m}_V^n$ is flat over $V/\mathfrak{m}_V^n$. This follows from the explicit description of the reductions $H_{i,n}$ in 5.4.2 which shows that the group scheme Z_i is isomorphic to the kernel of the map $H_{i,n} \to X_i/Y_i$. $\square$

It follows that Z_i is the closure of $H \subset G_\eta$ in G. We therefore obtain an isomorphism $\sigma_Z : Z_1 \to Z_2$ over V of group schemes. We extend this isomorphism to an isomorphism of group schemes $\sigma : H_1 \to H_2$ inducing the identity on H as follows. The closed subgroup schemes $Z_{i,n} \subset H_i$ are uniquely algebraizable to closed subgroup schemes $K_i \subset H_i$ (note that Z_i is not closed in H_i so K_i is strictly smaller than Z_i). The isomorphism σ_Z induces an isomorphism $\sigma_K : K_1 \to K_2$. The natural inclusions $D(X_i/Y_i) \subset K_i$ and the pairing on H_i define a homomorphism

$$H_i \to \mathrm{Hom}(D(X_i/Y_i), \mathbb{G}_m) \simeq X_i/Y_i.$$

Using the explicit description of H_i given in 5.4.2 one sees that this induces a canonical isomorphism $H_i/K_i \simeq X_i/Y_i$. The isomorphism $H_{1,\eta} \to H_{2,\eta}$ is compatible with the inclusions of the K_i and therefore also induces an isomorphism $X_1/Y_1 \to X_2/Y_2$.

From this we obtain a commutative diagram

$$
\begin{array}{ccccccccc}
0 & \longrightarrow & K_1 & \longrightarrow & H_1 & \longrightarrow & X_1/Y_1 & \longrightarrow & 0 \\
& & \sigma_K \downarrow & & & & \simeq \downarrow & & \Big| \\
0 & \longrightarrow & K_2 & \longrightarrow & H_2 & \longrightarrow & X_2/Y_2 & \longrightarrow & 0
\end{array}
\qquad (5.11.7.1)
$$

and an isomorphism $\sigma_\eta : H_{1,\eta} \to H_{2,\eta}$ filling in the generic fiber of this diagram.

Lemma 5.11.8 *There exists a unique isomorphism $\sigma : H_1 \to H_2$ filling in 5.11.7.1 restricting to σ_η on the generic fiber.*

Proof. Using the description of H_i in 5.4.2, one sees that the sequences

$$0 \to K_i \to H_i \to X_i/Y_i \to 0 \qquad (5.11.8.1)$$

are noncanonically split. Choosing splittings we obtain some isomorphism of group schemes $\sigma' : H_1 \to H_2$ restricting to σ_K on K_1. The restriction to the generic fiber may not equal σ_η, but the difference is measured by a homomorphism

$$\rho : X_1/Y_1 \to K_{2,\eta}. \qquad (5.11.8.2)$$

Since K_2 is proper over V this homomorphism extends to a homomorphism over V. Changing our choice of σ' but this map we obtain an extension of σ_η. The uniqueness is immediate since H_1 and H_2 are flat over V. $\square$

Note that the isomorphism $\sigma : H_1 \to H_2$ is compatible with the pairings $H_i \times H_i \to \mathbb{G}_m$ since by the flatness of H_i over V this can be checked on the generic fiber.

Lemma 5.11.9 *The isomorphism $\sigma : H_1 \to H_2$ lifts uniquely to an isomorphism $\tilde{\sigma} : \mathscr{G}_1 \to \mathscr{G}_2$ restricting to the given isomorphism $\mathscr{G}_{1,\eta} \to \mathscr{G}_{2,\eta}$ on the generic fiber.*

Proof. Note first that the isomorphism $G_1 \to G_2$, obtained from the fact that they are both equal to the connected component of the Neron model of the generic fiber, induces an isomorphism between the abelian schemes over V obtained from the abelian parts of the reductions of the G_i. Let A/V denote this abelian scheme. The line bundles $\mathscr{M}_i$ on A used in the standard construction can also be identified. Indeed, let W_η be the abelian part of the generic fiber G_η, and let $\mathscr{M}_\eta$ be the line bundle on G_η arising in the standard construction of the generic fiber. Then by [33, 3.2 and 3.3] the line bundle $\mathscr{M}_\eta$ extends uniquely to the Neron model of G_η. Since A is the abelian scheme obtained from the abelian part of the reductions of the Neron model of G_η. This gives a canonical description of the line bundles $\mathscr{M}_i$ (up to translation by an element of A). After possible modifying our choice of $\mathscr{M}_2$ by translation by an element of A we may assume that $\mathscr{M}_1 = \mathscr{M}_2$. Let $\mathscr{M}$ denote this line bundle on A. After changing the isomorphism on the generic fiber by an element of $\mathbb{G}_m(K)$ we can also assume that the isomorphism $\mathscr{M}_1 \to \mathscr{M}_2$ is compatible with rigidifications at 0.

Now by the proof of 5.4.2, we have an isomorphism of group schemes

$$\mathscr{G}_i \simeq \mathscr{G}_{(A,\mathscr{M})} \times (D(X_i/Y_i) \times X_i/Y_i) \qquad (5.11.9.1)$$

with group structure given by the formula

$$(g, h, x) \cdot (g', h', x') = (gg' + h'(x), h + h', x + x'). \qquad (5.11.9.2)$$

From this it follows that there exists an isomorphism $\mathscr{G}_1 \to \mathscr{G}_2$ inducing $\sigma : H_1 \to H_2$ and compatible with the pairings on the H_i. On the generic fiber the difference of this isomorphism and the inverse of the given isomorphism is an automorphism of $\mathscr{G}_1$ restricting to the identity on $\mathbb{G}_m$. Since $\mathbb{G}_m$ is central in $\mathscr{G}_1$ such an isomorphism is given by a homomorphism

$$H_{1,\eta} \to \mathbb{G}_{m,\eta}. \tag{5.11.9.3}$$

Such a homomorphism takes values in μ_N for N sufficiently big, and hence extends uniquely to a homomorphism $H_1 \to \mathbb{G}_m$. Changing our choice of isomorphism by this homomorphism we obtain the desired isomorphism $\tilde{\sigma}$: $\mathscr{G}_1 \to \mathscr{G}_2$. $\square$

Identifying $\mathscr{G}_1$ with $\mathscr{G}_2$ and H_1 with H_2 we write just $\mathscr{G}$ and H respectively for these group schemes. Let V_i denote the representation $\Gamma(P^i, L_i)$ of $\mathscr{G}$. We then have an isomorphism of $\mathscr{G}_\eta$–representations $\iota_K : V_1 \otimes K \simeq V_2 \otimes K$. After changing the isomorphism $\mathscr{S}_{1,\eta} \to \mathscr{S}_{2,\eta}$ by an element of $\mathbb{G}_m(K)$, we may assume that $\iota_K(V_1) \subset V_2$ and that the map $V_1 \to V_2/\pi V_2$ is not zero.

Lemma 5.11.10 *The map ι_K induces an isomorphism of $\mathscr{G}$–representations* $\iota : V_1 \to V_2$.

Proof. The representations $V_i/\pi V_i$ are irreducible $\mathscr{G} \otimes_V k$–representations by 5.4.27 and therefore the map $V_1 \otimes_V k \to V_2 \otimes_V k$ is either an isomorphism or zero. Since it is not zero by assumption it must be an isomorphism, and by Nakayama's lemma the map $V_1 \to V_2$ is also an isomorphism. $\square$

Inside the spaces $V_i \otimes k$ there exists a dense open subset of sections θ such that $(\mathscr{S}_i \otimes_V k, \theta)$ defines a point of $\mathscr{A}_{g,d}^{\text{Alex}}(k)$ (see 5.2.7). Thus we can find a section $\theta \in V_1$ such that $(\mathscr{S}_1, \theta)$ and $(\mathscr{S}_2, \iota(\theta))$ both define V–valued points of $\mathscr{A}_{g,d}^{\text{Alex}}$ which are isomorphic on the generic fiber after forgetting about the log structures. Since $\mathscr{A}_{g,d}^{\text{Alex}}$ is separated we obtain an isomorphism $(G_1, P_1, L_1) \to (G_2, P_2, L_2)$ extending the isomorphism on the generic fiber. This then extends uniqely to an isomorphism $\mathscr{S}_1 \to \mathscr{S}_2$ by 5.10.2. This completes the proof that the diagonal of $\overline{\mathscr{A}}_{g,d}$ is proper.

To see that the diagonal of $\overline{\mathscr{A}}_{g,d}$ is quasi–finite it suffices to show that for any algebraically closed field k and object $\mathscr{S} = (G, M_k, (P, M_P) \to (k, M_k), L) \in \overline{\mathscr{T}}_{g,d}(k)$, the automorphism group of $\mathscr{S}$ is an extension of $\mathbb{G}_m$ by a finite group. Any automorphism α of $\mathscr{S}$ induces an automorphism of the theta group $\mathscr{G}$ which is the identity on $\mathbb{G}_m$. Let $\text{Aut}_{\mathbb{G}_m}(\mathscr{G})$ denote the group of automorphisms of $\mathscr{G}$ which are the identity on $\mathbb{G}_m$ so we have a homomorphism

$$\text{Aut}(\mathscr{S}) \to \text{Aut}_{\mathbb{G}_m}(\mathscr{G}). \tag{5.11.10.1}$$

The group $\text{Aut}_{\mathbb{G}_m}(\mathscr{G})$ is a finite group since there is an extension

$$0 \to \text{Hom}(H, \mathbb{G}_m) \to \text{Aut}_{\mathbb{G}_m}(\mathscr{G}) \to \text{Aut}(H).$$

If K denotes the kernel of 5.11.10.1 then it suffices to show that K is a finite group. Let V denote the irreducible $\mathscr{G}$–representation $\Gamma(P, L)$. If $\alpha \in K$ is an automorphism then α defines an automorphism of V which since V is irreducible must be multiplication by a scalar. Therefore it suffices to show that the subgroup of K of automorphisms inducing the identity on V is finite. Choose any section $\theta \in V$ so that $(\mathscr{S}, \theta)$ defines an object of $\mathscr{A}_{g,d}^{\mathrm{Alex}}(k)$. Then any automorphism inducing the identity on V is in the image of

$$\mathrm{Aut}_{\overline{\mathscr{A}}_{g,d}(k)}(\mathscr{S}, \theta) \to \mathrm{Aut}_{\overline{\mathscr{T}}_{g,d}(k)}(\mathscr{S}).$$

Since $\mathrm{Aut}_{\overline{\mathscr{A}}_{g,d}(k)}(\mathscr{S}, \theta)$ is a finite group it follows that $\mathrm{Aut}_{\overline{\mathscr{T}}_{g,d}(k)}(\mathscr{S})$ is also an extension of $\mathbb{G}_m(k)$ by a finite group. This completes the proof of 5.11.6. $\square$

5.11.11. The stack $\overline{\mathscr{A}}_{g,d}$ inherits a natural log structure from $\overline{\mathscr{T}}_{g,d}$ since the action of $\mathbb{G}_m$ on the log structures of the base of objects of $\overline{\mathscr{T}}_{g,d}$ is trivial. Let $\mathscr{G} \to \overline{\mathscr{T}}_{g,d}$ denote the universal theta group, and let $\mathscr{H} \to \overline{\mathscr{T}}_{g,d}$ denote the finite flat group scheme $\mathscr{G}/\mathbb{G}_m$. For any object $(G, M_B, \mathscr{P}, M_{\mathscr{P}}, \mathscr{L}) \in \overline{\mathscr{T}}_{g,d}(B)$ (over some scheme B), the action of the group scheme $\mathscr{G}$ pulled back to B induces the trivial automorphism of $\mathscr{H}|_B$. It follows that the finite flat group scheme $\mathscr{H}$ with its pairing to $\mathbb{G}_m$ descends to a finite flat group scheme over $\overline{\mathscr{A}}_{g,d}$ with a pairing, which we again denote by $(\mathscr{H}, e)$.

Finally note that over $\mathbb{Z}[1/d]$ the log stack $(\overline{\mathscr{A}}_{g,d}, M_{\overline{\mathscr{A}}_{g,d}})$ is log smooth, since the map

$$\overline{\mathscr{T}}_{g,d} \to \overline{\mathscr{A}}_{g,d} \tag{5.11.11.1}$$

is smooth, so the log smoothness of $(\overline{\mathscr{A}}_{g,d}, M_{\overline{\mathscr{A}}_{g,d}})$ follows from the log smoothness of $(\overline{\mathscr{T}}_{g,d}, M_{\overline{\mathscr{T}}_{g,d}})$.

5.12 Example: Higher Dimensional Tate Curve

5.12.1. To elucidate the standard construction, let us work it out in more detail for $X = \mathbb{Z}^g$ and the quadratic form

$$B_0 := X_1^2 + X_2^2 + \cdots + X_g^2.$$

Lemma 5.12.2 *The paving S of $X_{\mathbb{R}} = \mathbb{R}^g$ defined by B_0 is given by the $\mathbb{Z}^g$-translates of the unit cube.*

Proof. For $\underline{a} \in \mathbb{Z}^g$, let $\square_{\underline{a}}$ denote the $\underline{a}$-translate of the unit cube. Let L_0 denote the linear form

$$L_0 := X_1 + X_2 + \cdots + X_g.$$

Note that L_0 agrees with B_0 on the vertices of $\square_0$. For $\underline{a} \in \mathbb{Z}^g$ set

$$L_{\underline{a}}(\underline{X}) := L_0(\underline{X} - \underline{a}) + \sum_{i=1}^{g} 2a_i X_i - \sum_{i=1}^{g} a_i^2.$$

Then $L_{\underline{a}}$ is a linear form whose values on the vertices of $\square_{\underline{a}}$ agree with the values of B_0.

Let

$$F : \mathbb{R}^g \to \mathbb{R}$$

be the piece-wise linear function whose value on $\underline{\alpha} \in \square_{\underline{a}}$ is equal to $L_{\underline{a}}(\underline{\alpha})$. To prove the lemma it suffices to show that F is a convex function.

For this let $F_i : \mathbb{R} \to \mathbb{R}$ be the function whose value on $a + \epsilon$ ($a \in \mathbb{Z}$ and $\epsilon \in [0, 1)$) is equal to

$$\epsilon + 2a(a + \epsilon) - a^2.$$

Then

$$F(\underline{\alpha}) = \sum_{i=1}^{g} F_i(\alpha_i).$$

To verify that for $t \in [0, 1]$ and $\underline{\alpha}, \underline{\beta} \in \mathbb{R}^g$ we have

$$F(t\underline{\alpha} + (1 - t)\underline{\beta}) \leq tF(\underline{\alpha}) + (1 - t)F(\underline{\beta})$$

it therefore suffices to show that each of the functions $F_i : \mathbb{R} \to \mathbb{R}$ are convex.

This is clear because for any integer $a \in \mathbb{Z}$ the function F_i is linear on the interval $[a, a + 1]$ and $F_i(a) = a^2$. $\square$

Remark 5.12.3. The same argument shows that for any rational numbers $a_1, \ldots, a_g \in \mathbb{Q}_{>0}$ the paving associated to the quadratic form

$$\sum_{i=1}^{g} a_i X_i^2$$

is given by the translates of the unit cube. In particular, it follows from this that the cone $U(B_0) \subset \Gamma^2 \mathbb{Q}^g$ defined in 5.8.13 has dimension at least g. We will see below that in fact it has exactly dimension g.

5.12.4. Next we describe the monoid H_S. As in the proof of 5.12.2, for $\underline{a} \in \mathbb{Z}^g$ let $\square_{\underline{a}}$ denote the $\underline{a}$-translate of the unit cube. A point $(d, \underline{v}) \in \mathbb{R} \times \mathbb{R}^g$ lies in $\mathrm{Cone}(1, \square_{\underline{a}})$ if and only if we have

$$0 \leq \frac{v_i}{d} - a_i \leq 1$$

for $i = 1, \ldots, g$.

For a subset $I \subset \{1, \ldots, g\}$ let κ_I denote the element of degree 1 in $\mathrm{Cone}(1, \square_0)$ with i-th coordinate equal to 1 if $i \in I$ and 0 otherwise. Similarly for $\underline{a} \in \mathbb{Z}^g$ let $t_{\underline{a}}^* \kappa_I$ denote the element of degree 1 in $\mathrm{Cone}(1, \square_{\underline{a}})$ with i-th coordinate equal to $a_i + 1$ if $i \in I$ and a_i otherwise.

Lemma 5.12.5 *The integral cone* $\mathrm{Cone}(1, \square_{\underline{a}})_\mathbb{Z} \subset \mathrm{Cone}(1, \square_{\underline{a}})$ *is generated by the* $t_{\underline{a}}^* \kappa_I$.

Proof. Let $(d, \underline{v}) \in \mathrm{Cone}(1, \square_{\underline{a}})_{\mathbb{Z}}$ be an element of degree d and set $\underline{w} = \underline{v} - d\underline{a}$ so that $(d, \underline{w}) \in \mathrm{Cone}(1, \square_0)_{\mathbb{Z}}$. If we can write

$$\underline{w} = \sum_I f_I \kappa_I$$

with $f_I \in \mathbb{N}$ and $\sum_I f_I = d$, then

$$\underline{v} = \sum_I f_I t_{\underline{a}}^* \kappa_I.$$

Therefore it suffices to consider the case $\underline{a} = 0$.

In this case we proceed by induction on d. The case $d = 1$ is immediate so assume $d > 1$. Let $\underline{z} \in \mathbb{Z}^g$ be the element with i-th coordinate equal to 1 if $v_i = d$ and 0 otherwise. Then $(1, \underline{z}) \in \mathrm{Cone}(1, \square_0)_{\mathbb{Z}}$ and

$$(d, \underline{v}) = (d - 1, \underline{v}') + (1, \underline{z}),$$

where $\underline{v}' := \underline{v} - \underline{z}$. Then $(d - 1, \underline{v}') \in \mathrm{Cone}(1, \square_0)$ so the result follows by induction. $\square$

Lemma 5.12.6 *The map*

$$s : \Gamma^2 \mathbb{Z}^g \to H_S^{gp}$$

defined in 5.8.8.2 is surjective.

Proof. By the construction in 5.8.8 the map s is the map

$$\Gamma^2 \mathbb{Z}^g \simeq H_1(\mathbb{Z}^g, P^{gp}) \to H_0(\mathbb{Z}^g, \tilde{H}_S^{gp})$$

obtained by taking group homology of the exact sequence of $\mathbb{Z}^g$-representations

$$0 \to \tilde{H}_s^{gp} \to (P \rtimes \tilde{H}_S)^{gp} \to P^{gp} \to 0.$$

Hence it suffices to show that the map

$$H_0(\mathbb{Z}^g, (P \rtimes \tilde{H}_s)^{gp}) \to H_0(\mathbb{Z}^g, P^{gp}) \simeq \mathbb{Z}$$

is an isomorphism, where the isomorphism $H_0(\mathbb{Z}^g, P^{gp}) \simeq \mathbb{Z}$ is induced by the degree map $P^{gp} \to \mathbb{Z}$. Equivalently we need to show that if $\rho : P \rtimes \tilde{H}_S \to M$ is a homomorphism to an abelian group which is $\mathbb{Z}^g$-invariant, then ρ factors through the degree map

$$P \rtimes \tilde{H}_S \to P \to \mathbb{N}.$$

This is clear for if $\underline{a} \in \mathbb{Z}^g$ then

$$\rho(t_{\underline{a}}^* \kappa_I) = \rho(\kappa_I) = \rho(\kappa_0).$$

$\square$

5.12.7. Note that for $i \neq j$ we have

$$s(e_i \otimes e_j) = \kappa_{\{i,j\}} * (1,0) - \kappa_i * \kappa_j = 0.$$

It follows that H_S^{gp} is generated by the images $s(e_i \otimes e_i)$ for $i = 1, \ldots, g$. On the other hand, the dimension of

$$\mathrm{Hom}(H_S^{gp}, \mathbb{Q})$$

is by 5.8.15 and 5.12.3 greater than or equal to g. It follows that H_S^{gp} has rank g generated by the elements $s(e_i \otimes e_i)$ $(i = 1, \ldots, g)$.

Note also that $s(e_i \otimes e_i)$ is contained in H_S so we obtain a commutative diagram

$$
\begin{array}{ccc}
\mathbb{N}^g & \xrightarrow{\ \chi\ } & H_S \\
\cap\downarrow & & \downarrow \\
\mathbb{Z}^g & \xrightarrow{\ \simeq\ } & H_S^{gp},
\end{array}
$$

where χ sends the i-th standard generator to $s(e_i \otimes e_i)$.

Lemma 5.12.8 *The map $\chi : \mathbb{N}^g \to H_S$ is an isomorphism.*

Proof. If χ is not an isomorphism then there exists a homomorphism $h : H_S \to \mathbb{Z}$ whose image is not contained in $\mathbb{N}$ but such that the composite $h\chi : \mathbb{N}^g \to \mathbb{Z}$ has image in $\mathbb{N}$. Let $a_i \in \mathbb{N}$ be the image of the i-th standard generator of $\mathbb{N}^g$. Then by the proof of 5.8.14 the map h is induced by the map defined by the positive semi-definite quadratic form

$$a_1 X_1^2 + \cdots + a_g X_g^2.$$

Therefore h also has image in $\mathbb{N}$. $\square$

5.12.9. We can also describe the monoid $P \rtimes H_S$. Write $P^{(1)} \rtimes H_S^{(1)}$ for the monoid obtained by taking $g = 1$ above. We have a canonical isomorphism $H_S^{(1)} \simeq \mathbb{N}$. Let $q \in H_S^{(1)}$ be a generator. Then $P^{(1)} \rtimes H_S$ is the quotient of the free monoid on generators q and x_n $(n \in \mathbb{Z})$ modulo the relations

$$x_{n+2} + x_n = q + 2x_{n+1}.$$

For $i = 1, \ldots, g$ let

$$p_i : P \rtimes H_S \to P^{(1)} \rtimes H_S^{(1)}$$

be the projection defined by the surjection

$$X \to \mathbb{Z}, \quad \sum_{i=1}^{g} a_i e_i \mapsto a_i.$$

If $q_1, \ldots, q_g \in H_S$ denote the generators defined by 5.12.8, then p_i restricts to the surjection

$$H_S \twoheadrightarrow H_S^{(1)}$$

sending q_j to 0 for $i \neq j$ and q_i to q. Let

$$\Pi'_{i=1,\ldots,g} P^{(1)} \rtimes H_S^{(1)}$$

denote the g-fold fiber product of $P^{(1)} \rtimes H_S^{(1)}$ over the degree map $\deg : P^{(1)} \rtimes H_S^{(1)} \to \mathbb{Z}$. We then obtain a map of graded monoids

$$p : P \rtimes H_S \to \Pi'_{i=1,\ldots,g} P^{(1)} \rtimes H_S^{(1)}$$

by taking the products of the p_i.

Lemma 5.12.10 *The map p is an isomorphism.*

Proof. The map p fits into a commutative diagram of monoids

$$
\begin{array}{ccccccccc}
0 & \longrightarrow & H_S & \longrightarrow & P \rtimes H_S & \longrightarrow & P & \longrightarrow & 0 \\
& & \downarrow{\scriptstyle p_H} & & \downarrow{\scriptstyle p} & & \downarrow{\scriptstyle \bar{p}} & & \\
0 & \longrightarrow & \prod_{i=1}^{g} H_S^{(1)} & \longrightarrow & \Pi'_{i=1,\ldots,g} P^{(1)} \rtimes H_S^{(1)} & \longrightarrow & \Pi'_{i=1,\ldots,g} P^{(1)} & \longrightarrow & 0.
\end{array}
$$

Since the map p_H is an isomorphism, it suffices to show that the map $\bar{p}$ is an isomorphism, which is immediate. $\square$

5.12.11. Fix an integer d, so we have the stacks $\overline{\mathscr{A}}_{g,d}$ defined over $\mathbb{Z}$. Choose any positive integers $d_1, \ldots, d_g$ such that $d_1 \cdots d_g = d$, and let $j : Y \subset \mathbb{Z}^g$ be the sublattice generated by the elements $d_i e_i$ (where $e_1, \ldots, e_g$ denotes the standard basis for $\mathbb{Z}^g$). Let f_i be the standard basis for Y so $j(f_i) = d_i e_i$.

We now associate a distinguished point $P_Y : \mathrm{Spec}(\mathbb{Z}) \to \overline{\mathscr{A}}_{g,d}$ and describe a formal neighborhood of P_Y in $\overline{\mathscr{A}}_{g,d}$. This generalizes the construction of the Tate curve in [15, VII].

5.12.12. Let $Y \wedge X$ denote the quotient of $Y \otimes X$ by the subgroup generated by elements of the form

$$y_1 \otimes j(y_2) - y_2 \otimes j(y_1), \quad y_1, y_2 \in Y.$$

Note that for any abelian group A we have a canonical identification

$$\mathrm{Hom}(Y \wedge X, A) \simeq \{\text{maps } \tau : Y \otimes X \to A \text{ such that } \tau|_{Y \otimes Y} \text{ is symmetric}\}.$$

The group $Y \wedge X$ may have torsion (for example take $Y = p \cdot \mathbb{Z}^g$). However, note that since $Y \wedge X$ is the pushout of the diagram

$$Y \otimes Y \longrightarrow Y \otimes X$$

$$\downarrow$$

$$S^2 Y,$$

there is a short exact sequence

$$0 \to S^2 Y \to Y \wedge X \to Y \otimes (X/Y) \to 0.$$

This implies that the torsion of $Y \wedge X$ is annihilated by d.

5.12.13. Let K (resp. I) denote the kernel (resp. image) of the composite map

$$Y \wedge X \longrightarrow S^2 X \xrightarrow{\ s\ } H_S^{gp} \simeq \mathbb{Z}^g.$$

Let w_i denote the element $s(e_i \otimes e_i)$ (where $e_i \otimes e_i \in S^2 X$). Then by construction the subgroup $I \subset \mathbb{Z}^g$ is the group generated by the elements $d_i \cdot w_i$. The surjection

$$Y \wedge X \longrightarrow I$$

has a section $I \to Y \wedge X$ given by sending $d_i \cdot w_i$ to the image of $f_i \otimes e_i \in Y \otimes X$. We therefore obtain a decomposition

$$Y \wedge X \simeq K \oplus I. \qquad (5.12.13.1)$$

Let

$$\pi : Y \wedge X \to K$$

denote the corresponding projection. Note that the the inclusion $S^2 Y \hookrightarrow Y \wedge X$ induces an inclusion

$$S^2 Y / (f_i \otimes f_i)_{i=1,\dots,g} \hookrightarrow K$$

whose cokernel is annihilated by d. For $1 \le i, j \le g$ let k_{ij} denote the image of $f_i \otimes f_j$ in K under this map (so $k_{ii} = 0$ for all i).

5.12.14. Let Δ denote the diagonalizable group scheme

$$\Delta := \mu_{d_1} \times \cdots \mu_{d_g}.$$

Then Δ is the kernel of the surjection

$$\mathbb{G}_m^g \to D(I)$$

defined by the inclusion $I \subset H_S^{gp} \simeq \mathbb{Z}^g$.

5.12.15. Let C_Y denote the completion of $\mathbb{Z}[H_S \oplus K]$ along the kernel of the homomorphism

$$\mathbb{Z}[H_S \oplus K] \to \mathbb{Z}$$

sending all nonzero elements of H_S to 0 and the elements of K to 1. Let

$$\tau : Y \wedge X \to C_Y^*$$

be the map induced by the composite

$$Y \wedge X \xrightarrow{\pi} K \hookrightarrow C_Y^* , \qquad (5.12.15.1)$$

and let

$$\psi : Y \to C_Y^*$$

be the map

$$\psi(\sum_{i=1}^{g} a_i f_i) := \prod_{i<j} z_{ij}^{a_i a_j},$$

where we write $z_{ij} \in C_Y^*$ for the unit corresponding to $k_{ij} \in K$. For two elements $\sum_{i=1}^{g} a_i f_i, \sum_{i=1}^{g} b_i f_i \in Y$ we then have

$$\psi(\sum_i (a_i + b_i) f_i) \psi(\sum_i a_i f_i)^{-1} \psi(\sum_i b_i f_i)^{-1} = \prod_{i<j} z_{ij}^{(a_i+b_i)(a_j+b_j)-a_i a_j - b_i b_j}$$

$$= \prod_{i<j} z_{ij}^{a_i b_j + a_j b_i}$$

$$= \tau(\sum_i a_i f_i, \sum_i b_i f_i).$$

We therefore have data as in 5.2.1, and can apply the standard construction to obtain an object

$$(G^{\text{Tate}}, M_{C_Y}, f : (\mathscr{P}^{\text{Tate}}, M_{\mathscr{P}^{\text{Tate}}}) \to (\text{Spec}(C_Y), M_{C_Y}), L_{\mathscr{P}^{\text{Tate}}}) \quad (5.12.15.2)$$

in $\overline{\mathscr{T}}_{g,d}(\text{Spec}(C_Y))$, where M_{C_Y} denotes the log structure associated to the natural map $H_S \to C_Y$. We call this object the *generalized Tate family*.

5.12.16. There is an action of Δ on $\text{Spec}(C_Y)$ induced by the inclusion $\Delta \subset D(H_S^{gp})$ and the natural action of $D(H_S^{gp})$ on $\text{Spec}(C_Y)$. The main result of this section is the following:

Theorem 5.12.17 *The map $t : \text{Spec}(C_Y) \to \overline{\mathscr{A}}_{g,d}$ defined by the generalized Tate family descends to a morphism*

$$T : [\text{Spec}(C_Y)/\Delta] \to \overline{\mathscr{A}}_{g,d}$$

which is formally étale.

Example 5.12.18. Let us describe the closed fiber of $\mathscr{P}^{\text{Tate}}$ when $g = 2$. For $(i, j) \in \mathbb{Z}^2$ let $X_{i+\frac{1}{2}, j+\frac{1}{2}}$ denote the scheme

$$X_{i+\frac{1}{2}, j+\frac{1}{2}} := \text{Proj}(k[x, y, z, w]/(yz - xw)).$$

This scheme contains four copies of $\mathbb{P}^1$ which we will index as follows:

$$P^0_{i+\frac{1}{2}} := V(z,w), \quad P^\infty_{i+\frac{1}{2}} := V(x,y),$$

$$P^0_{j+\frac{1}{2}} := V(y,w), \quad P^\infty_{j+\frac{1}{2}} := V(x,z).$$

The closed fiber $\widetilde{\mathscr{P}}_0$ of $\widetilde{\mathscr{P}}$ is obtained by gluing $X_{i+\frac{1}{2},j+\frac{1}{2}}$ to $X_{i-\frac{1}{2},j+\frac{1}{2}}$ along

$$P^0_{i+\frac{1}{2}} \simeq \mathrm{Proj}(k[x,y]) \xrightarrow{x \mapsto z, y \mapsto w} \mathrm{Proj}(k[z,w]) \simeq P^\infty_{(i-1)+\frac{1}{2}},$$

and $X_{i+\frac{1}{2},j+\frac{1}{2}}$ to $X_{i+\frac{1}{2},j-\frac{1}{2}}$ along

$$P^0_{j+\frac{1}{2}} = \mathrm{Proj}(k[x,z]) \xrightarrow{x \mapsto y, z \mapsto w} \mathrm{Proj}(k[y,w]) \simeq P^\infty_{(j-1)+\frac{1}{2}}.$$

The closed fiber of the scheme $\mathscr{P}^{\mathrm{Tate}}$ is then obtain by taking the quotient of $\widetilde{\mathscr{P}}_0$ by the action of $Y \subset \mathbb{Z}^g$ for which $(a,b) \in Y$ sends $X_{i+\frac{1}{2},j+\frac{1}{2}}$ to $X_{i+a+\frac{1}{2},j+b+\frac{1}{2}}$ by the identity map.

In the case $g = 2$ we can also describe the ring C_Y explicitly as follows. Let $m = \gcd(d_1,d_2)$ and write $d_1 = m\bar{d}_1$ and $d_2 = m\bar{d}_2$. Choose integers $\alpha,\beta \in \mathbb{Z}$ such that

$$1 = \alpha\bar{d}_1 + \beta\bar{d}_2.$$

Let $\kappa \in K$ be the image of

$$\bar{d}_2(f_1 \otimes e_2) - \bar{d}_1(f_2 \otimes e_1),$$

and let $\delta \in K$ denote the image of

$$\beta(f_2 \otimes e_1) + \alpha(f_1 \otimes e_2).$$

Then one verifies immediately that $m\kappa = 0$ and that the induced map

$$\mathbb{Z}/(m) \oplus \mathbb{Z} \to K, \quad (a,m) \mapsto a\kappa + b\delta$$

is an isomorphism. If $w \in C_Y$ (resp. $u \in C_Y$) denotes the element defined by δ (resp. κ) then we have

$$C_Y \simeq \mathbb{Z}[[q_1,q_1,(w-1),u^\pm]]/(u^m - 1).$$

The action of $(\zeta_1,\zeta_2) \in \Delta = \mu_{d_1} \times \mu_{d_2}$ is given by

$$q_i \mapsto \zeta_i q_i, \quad w \mapsto w, \quad u \mapsto u.$$

5.12.19. We now start the proof of 5.12.17 which occupies the remainder of this section.

Let us begin by explaining why the map $\mathrm{Spec}(C_Y) \to \overline{\mathscr{A}}_{g,d}$ descends to the stack quotient by Δ. Consider a finite flat ring homomorphism $\rho : C_Y \to A$ and a point $\zeta = (\zeta_1,\ldots,\zeta_g) \in \Delta(A)$. The key case is when A is the coordinate ring of $\text{Spec}(C_Y) \times \Delta$ or $\text{Spec}(C_Y) \times \Delta^2$ as we need to give descent data to $[\text{Spec}(C_Y)/\Delta]$ which means an isomorphism of the two pullbacks to $\text{Spec}(C_Y) \times \Delta$ which satisfy the cocycle condition on $\text{Spec}(C_Y) \times \Delta^2$. The map ρ is induced by maps

$$\beta : H_S \to A, \quad \gamma : K \to A^*.$$

View ζ as a map $\zeta : H_S^{gp} \to \mu_d(A)$, and choose an extension

$$\tilde{\zeta} : (P \rtimes H_S)^{gp} \to \mu_d(A).$$

Note that this is possible since P^{gp} is torsion free so the sequence

$$0 \to H_S^{gp} \to (P \rtimes H_S)^{gp} \to P^{gp} \to 0$$

is split. Let

$$\zeta * \rho : C_Y \to A \tag{5.12.19.1}$$

denote the map induced by the maps

$$\zeta * \beta : H_S \to A, \quad m \mapsto \zeta(m) \cdot \beta(m)$$

and the map $\gamma : K \to A^*$. Let M_A denote the log structure on $\text{Spec}(A)$ obtained by pulling back M_{C_Y} along ρ, and let M_A^ζ denote the log structure obtained by pulling back M_{C_Y} along $\zeta * \rho$. The commutative diagram

$$
\begin{array}{ccc}
A^* \oplus H_S & \xrightarrow{(u,m) \mapsto (u\zeta(m),m)} & A^* \oplus H_S \\
& & \\
\end{array}
$$

$$(u,m) \mapsto u\zeta(m)\beta(m) \searrow \quad \swarrow (u,m) \mapsto u\beta(m)$$

$$A,$$

defines a morphism of log schemes

$$f_\zeta : (\text{Spec}(A), M_A) \longrightarrow (\text{Spec}(A), M_A^\zeta) \tag{5.12.19.2}$$

which is the identity on the underlying schemes.

Let $\mathcal{R}$ (resp. $\mathcal{R}^\zeta$) denote the graded ring

$$\mathcal{R} := A \otimes_{\beta, \mathbb{Z}[H_S]} \mathbb{Z}[P \rtimes H_S] \quad (\text{resp. } \mathcal{R}^\zeta := A \otimes_{\zeta*\beta, \mathbb{Z}[H_S]} \mathbb{Z}[P \rtimes H_S]).$$

Then the base change of $\widetilde{\mathscr{P}} \to \text{Spec}(C_Y)$ along ρ (resp. $\zeta * \rho$) is isomorphic to $\text{Proj}(\mathcal{R})$ (resp. $\text{Proj}(\mathcal{R}^\zeta)$), and under this identification the line bundle $L_{\widetilde{\mathscr{P}}}$ on the base change is the tautological bundle. Here $(\widetilde{\mathscr{P}}, L_{\widetilde{\mathscr{P}}})$ is as in the standard construction.

The map $\tilde{\zeta}$ defines an isomorphism

$$\tilde{f}_\zeta : \mathcal{R}^\zeta \to \mathcal{R}.$$

induced by the map

$$P \rtimes H_S \to \mathcal{R}^{\zeta}, \quad m \mapsto \tilde{\zeta}(m) \cdot m.$$

If

$$c : P \rtimes H_S \to \mathcal{R}, \quad c^{\zeta} : P \rtimes H_S \to \mathcal{R}^{\zeta}$$

are the charts induced by the natural map $P \rtimes H_S \to \mathbb{Z}[P \rtimes H_S]$ then we have a commutative diagram

$$
\begin{array}{ccc}
\mathcal{R}^{\zeta *} \oplus P \rtimes H_S & \xrightarrow{(u,m) \mapsto (\tilde{\zeta}(m)u,m)} & \mathcal{R}^* \oplus P \rtimes H_S \\
{\scriptstyle c^{\zeta}} \downarrow & & \downarrow {\scriptstyle c} \\
\mathcal{R}^{\zeta} & \xrightarrow{\quad \tilde{f}_{\zeta} \quad} & \mathcal{R}.
\end{array}
$$

Let $\widetilde{\mathscr{P}}_{\rho}$ (resp. $\widetilde{\mathscr{P}}_{\zeta * \rho}$) denote the base change $\widetilde{\mathscr{P}} \times_{\mathrm{Spec}(C_Y), \rho} \mathrm{Spec}(A)$ (resp. $\widetilde{\mathscr{P}} \times_{\mathrm{Spec}(C_Y), \zeta * \rho} \mathrm{Spec}(A)$). We then obtain an isomorphism of log schemes with line bundles

$$(\widetilde{\mathscr{P}}_{\rho}, M_{\widetilde{\mathscr{P}}_{\rho}}, L_{\widetilde{\mathscr{P}}_{\rho}}) \to (\widetilde{\mathscr{P}}_{\zeta * \rho}, M_{\widetilde{\mathscr{P}}_{\zeta * \rho}}, L_{\widetilde{\mathscr{P}}_{\zeta * \rho}}),$$

which we again denote by $\tilde{f}_{\zeta}$. It follows from the construction that this map is compatible with the torus actions and the Y-action. We therefore obtain an isomorphism of the quotients by the Y-actions

$$g_{\zeta} : (5.12.15.2) \times_{\mathrm{Spec}(C_Y), \rho} \mathrm{Spec}(A) \to (5.12.15.2) \times_{\mathrm{Spec}(C_Y), \zeta * \rho} \mathrm{Spec}(A)$$

over the morphism of log schemes f_{ζ}.

This isomorphism in $\overline{\mathscr{T}}_{g,d}$ depends on the choice of the extension $\tilde{\zeta}$. However, the automorphism of $(5.12.15.2) \times_{\mathrm{Spec}(C_Y), \rho} \mathrm{Spec}(A)$ obtained by taking two different extensions of ζ is an element of the theta group of this object of $\overline{\mathscr{T}}_{g,d}(A)$. Therefore the induced isomorphism in $\overline{\mathscr{A}}_{g,d}$ is independent of the choice of the extension $\tilde{\zeta}$. This therefore defines the descent data for the morphism

$$\mathrm{Spec}(C_Y) \to \overline{\mathscr{A}}_{g,d}$$

to a morphism

$$T : [\mathrm{Spec}(C_Y)/\Delta] \to \overline{\mathscr{A}}_{g,d}.$$

5.12.20. This morphism T is faithful. Indeed if Ω a field and $x : \mathrm{Spec}(\Omega) \to \mathrm{Spec}(C_Y)$ then the stabilizer group $\Delta_x \subset \Delta$ of x acts faithfully on the log structure $x^* M_{C_Y}$ by the construction of the action. This implies that for any field valued point $x : \mathrm{Spec}(\Omega) \to [\mathrm{Spec}(C_Y)/\Delta]$ the map of automorphism group schemes

$$\underline{\mathrm{Aut}}_{[\mathrm{Spec}(C_Y)/\Delta]}(x) \to \underline{\mathrm{Aut}}_{\overline{\mathscr{A}}_{g,d}}(T(x))$$

is injective. Since the diagonal of $[\mathrm{Spec}(C_Y)/\Delta]$ over $\overline{\mathscr{A}}_{g,d}$ is finite (as both these stacks have finite diagonal) this implies that the diagonal of $[\mathrm{Spec}(C_Y)/\Delta]$ over $\overline{\mathscr{A}}_{g,d}$ is a monomorphism, and hence T is faithful.

5.12.21. To show that T is formally étale, we verify the infinitesimal lifting property. Let $B \longrightarrow B_0$ be a surjection of artinian local rings with square-zero kernel J and consider a commutative diagram of solid arrows

$$
\begin{array}{ccc}
\mathrm{Spec}(B_0) & \xrightarrow{\;b_0\;} & [\mathrm{Spec}(C_Y)/\Delta] \\
\big\uparrow & {\scriptstyle b}\;\nearrow & \big\downarrow{\scriptstyle T} \\
\mathrm{Spec}(B) & \xrightarrow{\;\bar b\;} & \mathscr{A}_{g,d},
\end{array}
\qquad (5.12.21.1)
$$

where the image of b_0 is contained in the maximal closed substack of $[\mathrm{Spec}(C_Y)/\Delta]$ (that is, the set-theoretic image of b_0 is the closed point of $[\mathrm{Spec}(C_Y)/\Delta]$). We wish to show that there exists a unique dotted arrow b filling in the diagram.

By descent theory we may replace B by a finite flat extension if needed. In particular, we may assume that b_0 lifts to a morphism

$$
\tilde b_0 : \mathrm{Spec}(B_0) \to \mathrm{Spec}(C_Y).
$$

This morphism $\tilde b_0$ defines data

$$
\beta_0 : H_S \to B_0, \quad \tau_0 : Y \wedge X \to B_0^*, \quad \psi_0 : Y \to B_0^*
$$

as in the standard construction 5.2.1. Note that since this data is obtained from a map to $\mathrm{Spec}(C_Y)$ the map τ_0 factors through the projection $Y \wedge X \to K$.

By 5.5.2, the map $\bar b$ is then obtained by applying the standard construction to a collection of data

$$
\beta : H_S \to B, \quad \tau : Y \wedge X \to B^*, \quad \psi : Y \to B^*
$$

lifting $(\beta_0, \tau_0, \psi_0)$. Let $(G_B, \mathscr{P}_B, M_{\mathscr{P}_B}, L_{\mathscr{P}_B})$ be the resulting lifting of

$$
\tilde b_0^*(5.12.15.2),
$$

so we have a specified isomorphism

$$
(G_B, \mathscr{P}_B, M_{\mathscr{P}_B}, L_{\mathscr{P}_B}) \times_{\mathrm{Spec}(B)} \mathrm{Spec}(B_0) \simeq \tilde b_0^*(5.12.15.2). \qquad (5.12.21.2)
$$

The short exact sequence of groups

$$
0 \to (Y \wedge X)/K \to H_S^{gp} \to \prod_{i=1}^{g}(\mathbb{Z}/(d_i)) \to 0
$$

induces a short exact sequence of diagonalizable group schemes

$$
0 \to \Delta \to D(H_S^{gp}) \to D((Y \wedge X)/K) \to 0.
$$

The composite map

$$(Y \wedge X)/K \xrightarrow{5.12.13.1} Y \wedge X \xrightarrow{\tau} B^*$$

defines a point

$$\Phi \in \mathrm{Ker}(D((Y \wedge X)/K)(B) \to D((Y \wedge X)/K)(B_0)).$$

After possibly replacing B by another finite flat extension, we can find an extension

$$\tilde{\Phi} \in D(H_S^{gp})(B)$$

of Φ. The reduction of reduction of $\tilde{\Phi}$ modulo J maps to zero in $D((Y \wedge X)/K)(B_0)$ and therefore gives an element $\tilde{\Phi}_0 \in \Delta(B_0)$.

Let

$$q : K \oplus (Y \wedge X/K) \simeq Y \wedge X \to (1 + J) \subset B^*$$

denote the composite map

$$Y \wedge X \longrightarrow S^2 X \xrightarrow{s} H_S^{gp} \xrightarrow{\tilde{\Phi}} 1 + J$$

and set

$$\beta' : H_S \to B, \quad m \mapsto \tilde{\Phi}(m)^{-1} \beta(m),$$

$$\tau' : Y \wedge X \to B^*, \quad (y \otimes x) \mapsto q(y \otimes x)^{-1} \tau(y \otimes x).$$

Observe that q restricts to the constant map $k \mapsto 1$ on K and to τ on $(Y \wedge X)/K$.

The map β' (resp. τ') is a lifting of $\tilde{\Phi}_0^* \beta_0$ (resp. τ_0), and τ' factors through the projection $\pi : Y \wedge X \to K$.

We therefore obtain a lifting

$$\tilde{b} :\to \mathrm{Spec}(C_Y)$$

of $\tilde{\Phi}_0^* \tilde{b}_0$.

5.12.22. Let

$$\psi' : Y \to B^*$$

be the lifting of ψ_0 defined by the composite

$$Y \xrightarrow{5.12.15.1} C_Y^* \xrightarrow{\tilde{b}^*} B^*.$$

Define

$$\mathcal{R} := B \otimes_{\beta, \mathbb{Z}[H_S]} \mathbb{Z}[P \rtimes H_S], \quad \mathcal{R}' := B \otimes_{\beta', \mathbb{Z}[H_S]} \mathbb{Z}[P \rtimes H_S]$$

so that $\mathcal{R}$ (resp. $\mathcal{R}'$) is a lifting of

$$\mathcal{R}_0 := B_0 \otimes_{\beta_0, \mathbb{Z}[H_S]} \mathbb{Z}[P \rtimes H_S] \quad (\text{resp. } \mathcal{R}'_0 := B_0 \otimes_{\tilde{\Phi}_0^* \beta_0, \mathbb{Z}[H_S]} \mathbb{Z}[P \rtimes H_S]).$$

Let $(\widetilde{\mathscr{P}}, M_{\widetilde{\mathscr{P}}}, L_{\widetilde{\mathscr{P}}})$ (resp. $(\widetilde{\mathscr{P}}', M_{\widetilde{\mathscr{P}}'}, L_{\widetilde{\mathscr{P}}'}), (\widetilde{\mathscr{P}}_0, M_{\widetilde{\mathscr{P}}_0}, L_{\widetilde{\mathscr{P}}_0}), (\widetilde{\mathscr{P}}'_0, M_{\widetilde{\mathscr{P}}'_0}, L_{\widetilde{\mathscr{P}}'_0}))$
denote $\text{Proj}(\mathcal{R})$ (resp. $\text{Proj}(\mathcal{R}'), \text{Proj}(\mathcal{R}_0), \text{Proj}(\mathcal{R}'_0)$) with the canonical line
bundle and log structure induced by the natural map from $P \rtimes H_S$.

If M_B (resp. M'_B) denotes the fine log structure on $\text{Spec}(B)$ defined by
the chart β (resp. β'), then $(\widetilde{\mathscr{P}}, M_{\widetilde{\mathscr{P}}})$ (resp. $(\widetilde{\mathscr{P}}', M_{\widetilde{\mathscr{P}}'})$) is a log scheme over
$(\text{Spec}(B), M_B)$ (resp. $(\text{Spec}(B), M'_B)$).

5.12.23. For $y = \sum_{i=1}^{g} a_i f_i \in Y$ define

$$\epsilon(y) := \prod_{i=1}^{g} \psi(a_i f_i).$$

Lemma 5.12.24 *For any $y \in Y$ we have*

$$\psi(y) = \psi'(y)\epsilon(y).$$

Proof. By construction if $y = \sum_i a_i f_i$ we have

$$\psi'(y) = \prod_{i<j} \tau(f_i, f_j)^{a_i a_j}.$$

The basis $\{f_i\}$ for Y defines an isomorphism

$$S^2 Y \simeq K_Y \oplus I_Y,$$

where K_Y is generated by the elements $f_i \otimes f_j$ $(i \neq j)$ and I_Y is generated by
the elements $f_i \otimes f_i$. Let

$$\pi_K : S^2 Y \to K_Y, \quad \pi_I : S^2 Y \to I_Y$$

be the two projections. Then it follows from the definitions that for any
$y_1, y_2 \in Y$ we have

$$\psi'(y_1 + y_2)\psi'(y_1)^{-1}\psi'(y_2)^{-1} = \tau(\pi_K(y_1 \otimes y_2))$$

and

$$\epsilon(y_1 + y_2)\epsilon(y_1)^{-1}\epsilon(y_2)^{-1} = \tau(\pi_I(y_1 \otimes y_2)).$$

It follows that ψ and the map $y \mapsto \psi'(y)\epsilon(y)$ are two maps

$$Y \to B^*$$

which define the same quadratic form τ. Therefore there exists a homomorphism $u : Y \to B^*$ such that for every $y \in Y$ we have

$$\psi(y) = u(y)\psi'(y)\epsilon(y).$$

To verify that u is the trivial homomorphism, it suffices to show that $u(f_i) = 1$
for $i = 1, \ldots, g$. This is immediate as by definition we have $\psi'(f_i) = 1$ and
$\epsilon(f_i) = \psi(f_i)$. $\square$

5.12.25. Let $q_1, \ldots, q_g \in H_S$ be the generators defined by the isomorphism 5.12.8. After possibly replacing B by a finite flat extension, we can choose elements $\lambda_i \in B^*$ $(i = 1, \ldots, g)$ such that $\lambda_i^2 = \tilde{\Phi}(q_i)$. After replacing B by another finite flat extension we can also choose elements $\zeta_i \in B^*$ such that

$$\psi(f_i) = \lambda_i^{d_i^2} \zeta_i^{d_i}.$$

Define

$$\tilde{\psi}_i : \mathbb{Z} \to B^*, \quad n \mapsto \lambda_i^{n^2} \zeta_i^n.$$

Note that

$$\begin{aligned}
\tilde{\psi}_i(n)\tilde{\psi}_i(m) &= \lambda_i^{n^2+m^2} \zeta_i^{n+m} \\
&= \lambda_i^{-2nm} \lambda_i^{(n+m)^2} \zeta_i^{n+m} \\
&= \tilde{\Phi}(q_i)^{-nm} \tilde{\psi}_i(n+m),
\end{aligned}$$

so

$$\tilde{\psi}_i(n+m)\tilde{\psi}_i(n)^{-1}\tilde{\psi}_i(m)^{-1} = \tilde{\Phi}(q_i)^{nm}. \tag{5.12.25.1}$$

5.12.26. Now recall (see 5.12.9) that we can describe the monoid $P^{(1)} \rtimes H_S^{(1)}$ as the quotient of the free monoid with generators $q \in H_S^{(1)}$ and x_n $(n \in \mathbb{Z})$ modulo the relations

$$x_{n+2} + x_n = q + 2x_{n+1}.$$

Define a map

$$h_i : P^{(1)} \rtimes H_S^{(1)} \to B^*$$

by sending x_n to $\tilde{\psi}_i(n)$ and q to $\tilde{\Phi}(q_i)$. Observe that

$$\begin{aligned}
\tilde{\psi}_i(n+2)\tilde{\psi}_i(n)\tilde{\psi}_i(n+1)^{-2} &= \tilde{\Phi}(q_i)^{-n(n+2)} \tilde{\Phi}(q_i)^{(n+1)^2} \\
&= \tilde{\Phi}(q_i).
\end{aligned}$$

It follows that h_i is well-defined.

Let

$$h : P \rtimes H_S \simeq \Pi'_{i=1,\ldots,g} P^{(1)} \rtimes H_S^{(1)} \to B^*$$

be the map obtained by taking the product of the h_i. Then the restriction of h to H_S is the map $\tilde{\Phi}$.

Lemma 5.12.27 *For any $d > 0$ and $n \in \mathbb{Z}$ we have*

$$h_i(d, n + dd_i) = h_i(d, n)\psi(f_i)^d \tilde{\Phi}(q_i)^{d_i \cdot n}. \tag{5.12.27.1}$$

Here we view $(d, n + dd_i)$ and (d, n) as elements of $P^{(1)} \rtimes H_S^{(1)}$ using the natural set map $P^{(1)} \to P^{(1)} \rtimes H_S^{(1)}$.

Proof. Let $u : P^{(1)} \rtimes H_S^{(1)} \to \mathbb{Z}$ be the composite of the projection

$$P^{(1)} \rtimes H_S^{(1)} \to P^{(1)}$$

and the map

$$P^{(1)} \to \mathbb{Z}, \quad (d, n) \mapsto n.$$

Let

$$G : P^{(1)} \rtimes H_S^{(1)} \to B^*$$

be the morphism $h_i \circ t_{f_i}$, where $t_{f_i} : P^{(1)} \rtimes H_S^{(1)} \to P^{(1)} \rtimes H_S^{(1)}$ denotes the translation action of f_i, and let

$$F : P^{(1)} \rtimes H_S^{(1)} \to B^*$$

denote the map

$$m \mapsto h_i(m) \psi(f_i)^{\deg(m)} (\tilde{\Phi}(q_i)^{d_i})^{u(m)}.$$

It suffices to show that $F = G$. Since both F and G are morphisms of monoids, it suffices to show that the agree on a set of generators. Since they clearly agree on $H_S^{(1)}$ it suffices to show that $F(x_n) = G(x_n)$. Equivalently we need that

$$\tilde{\psi}_i(n + d_i) = \tilde{\psi}_i(n) \tilde{\psi}_i(d_i) \tilde{\Phi}(q_i)^{d_i n},$$

which is 5.12.25.1. □

5.12.28. Define an isomorphism

$$\rho : \mathcal{R}' = B \otimes_{\beta', \mathbb{Z}[H_S]} \mathbb{Z}[P \rtimes H_S] \to B \otimes_{\beta, \mathbb{Z}[H_S]} \mathbb{Z}[P \rtimes H_S] = \mathcal{R} \quad (5.12.28.1)$$

by sending e_m ($m \in P \rtimes H_S$) to $h(m) \cdot e_m$, where we write e_m for the images in $\mathcal{R}$ and $\mathcal{R}'$ of $m \in P \rtimes H_S$. Since we have a commutative diagram

$$
\begin{array}{ccc}
P \rtimes H_S & \xrightarrow{m \mapsto (h(m), m)} & B^* \oplus P \rtimes H_S \\
\downarrow & & \downarrow \\
\mathcal{R}' & \xrightarrow{\rho} & \mathcal{R},
\end{array}
$$

this map induces an isomorphism of log schemes with lines bundles and torus actions

$$\rho : (\widetilde{\mathscr{P}}, M_{\widetilde{\mathscr{P}}}, L_{\widetilde{\mathscr{P}}}) \to (\widetilde{\mathscr{P}'}, M_{\widetilde{\mathscr{P}'}}, L_{\widetilde{\mathscr{P}'}})$$

over the isomorphism

$$\sigma : (\operatorname{Spec}(B), M_B) \to (\operatorname{Spec}(B), M_B')$$

which is the identity on $\operatorname{Spec}(B)$ and with isomorphism of log structures induced by the map

$$H_S \to B^* \oplus H_S, \quad m \mapsto (\tilde{\Phi}(m), m).$$

5.12.29. We claim that ρ is compatible with the Y-actions. By the construction of the Y-action in 5.2.4, this amounts to showing that for $y \in Y$ and $(d, x) \in P$ we have

$$\psi(y)^d \tau(y, x) h(d, x) = \psi'(y)^d \tau'(y, x) h(d, x + dy). \tag{5.12.29.1}$$

Furthermore, since Y is generated by the f_i it suffices to prove this formula in the case when $y = f_i$ for some i.

Now by definition we have $\psi'(f_i) = 1$ so when $y = f_i$ the formula 5.12.29.1 can be rewritten as

$$\psi(f_i)^d \tau(f_i, x) h(d, x) = \tau'(f_i, x) h(d, x + df_i). \tag{5.12.29.2}$$

Write

$$x = \sum_{j=1}^{g} a_j e_j.$$

Then

$$\tau(f_i, x) = \prod_{j=1}^{g} \tau(f_i, e_j)^{a_j}$$

and

$$\tau'(f_i, x) = \prod_{j \neq i} \tau(f_i, e_j)^{a_j},$$

so verifying 5.12.29.2 it is equivalent to showing that

$$\psi(f_i)^d \tau(f_i, e_i)^{a_i} h(d, x) = h(d, x + df_i). \tag{5.12.29.3}$$

Now by definition we also have

$$h(d, x) = \prod_{j=1}^{g} h_j(d, a_j)$$

and

$$h(d, x + df_i) = h_i(d, a_i + dd_i) \cdot \prod_{j \neq i} h_j(d, a_j).$$

Using this and the equality $\tau(f_i, e_i) = \tilde{\Phi}(q_i)^{d_i}$ we further reduce the verification of 5.12.29.3 to verifying the formula

$$\psi(f_i)^d \tilde{\Phi}(q_i)^{d_i a_i} h_i(d, a_i) = h_i(d, a_i + dd_i)$$

which is 5.12.27.

We therefore obtain an isomorphism

$$\tilde{b}^*(5.12.15.2) \simeq (G_B, \mathscr{P}_B, M_{\mathscr{P}_B}, L_{\mathscr{P}_B}).$$

The reduction of this isomorphism is an isomorphism

$$\tilde{\Phi}_0^* \tilde{b}_0(5.12.15.2) \to \tilde{b}_0^*(5.12.15.2)$$

for which the induced isomorphism of log structures

$$\tilde{\Phi}_0^* \tilde{b}_0^* M_{C_Y} \to \tilde{b}_0^* M_{C_Y}$$

is defined as in 5.12.19.2. It follows that $\tilde{b}$ defines a dotted arrow b filling in the diagram 5.12.21.1.

5.12.30. For the uniqueness, we need to show that if $\tilde{b}$ and $\tilde{b}'$ are two morphisms filling in the following diagram

$$\begin{array}{ccc}
 & & \mathrm{Spec}(C_Y) \\
 & \tilde{b} \nearrow \quad \tilde{b}' \nearrow & \uparrow t \\
\mathrm{Spec}(B) & \xrightarrow{\quad \bar{b} \quad} & \overline{\mathscr{A}}_{g,d},
\end{array}$$

and if $\zeta_0 \in \Delta_0(B_0)$ is an element such that $\zeta_0 * \tilde{b}_0 = \tilde{b}'_0$ (where $\tilde{b}_0$ and $\tilde{b}'_0$ are the reductions of $\tilde{b}$ and $\tilde{b}'$ respectively) and such that the induced morphism in $\overline{\mathscr{A}}_{g,d}$

$$\bar{b}|_{B_0} \xrightarrow{\simeq} t(\tilde{b}_0) \xrightarrow{t(\zeta_0)} t(\tilde{b}'_0) \xrightarrow{\simeq} \bar{b}|_{B_0}$$

is the identity, then there exists a lifting $\zeta \in \Delta(B)$ of ζ_0 and an isomorphism $\zeta * \tilde{b} \simeq \tilde{b}'$ mapping to the given isomorphism in $\overline{\mathscr{A}}_{g,d}(B)$. Note that by the faithfulness 5.12.20 the element ζ is unique.

Let M_B (resp. M'_B) denote the log structure $\tilde{b}^* M_{C_Y}$ (resp. $\tilde{b}'^* M_{C_Y}$). Note that the chart $H_S \to M_{C_Y}$ over C_Y defines charts

$$\gamma : H_S \to M_B, \quad \gamma' : H_S \to M'_B.$$

The isomorphism $t \circ \tilde{b} \simeq t \circ \tilde{b}'$ in $\overline{\mathscr{A}}_{g,d}(B)$ defines an isomorphism of log structures

$$\sigma : M_B \to M'_B.$$

The induced map

$$H_S \simeq \overline{M}_B \to \overline{M}'_B \simeq H_S$$

is the identity as this can be verified over B_0. It follows that there exists a unique homomorphism

$$\tilde{\Phi} : H_S^{gp} \to \mathrm{Ker}(B^* \to B_0^*)$$

such that the diagram

$$\begin{array}{ccc}
H_S & \xrightarrow{m \mapsto (\tilde{\Phi}(m), m)} & B^* \oplus H_S \\
 & \gamma \searrow \quad \swarrow \gamma' & \\
 & B &
\end{array}$$

commutes, and σ is obtained by passing to the associated log structures. Furthermore $\tilde{\Phi}$ reduces to the given element $\zeta_0 \in \Delta(B_0)$.

We claim that $\tilde{\Phi}$ sends the image of $Y \wedge X$ to $1 \in B^*$ and hence is an element of $\Delta(B)$, and that $\tilde{\Phi}$ is the desired element ζ.

5.12.31. Giving the map $\tilde{b}$ (resp. $\tilde{b}'$) is equivalent to giving the chart γ (resp. γ') and a map

$$\tau : Y \wedge X \to B^* \quad (\text{resp. } \tau' : Y \wedge X \to B^*)$$

lifting τ_0 such that τ (resp. τ') factors through the projection $\pi_K : Y \wedge X \to K$.

The map τ (resp. τ') defines a map

$$\psi : Y \to B^*, \quad (\text{resp. } \psi' : Y \to B^*)$$

by the formula 5.12.15.1.

5.12.32. After possibly replacing B by a finite flat extension, there exists by the same argument as in section 4.4 a homomorphism

$$h : (P \rtimes H_S)^{gp} \to B^*$$

extending $\tilde{\Phi}$ and such that the isomorphism

$$t \circ \tilde{b} \simeq \bar{b} \simeq t \circ \tilde{b}'$$

is induced by the same construction as in the construction of the isomorphism 5.12.28.1 by the homorphism h. The compatibility with the Y-action in this case amounts to the statement that for every $y \in Y$ and $x \in X$ we have

$$\psi'(y)\tau'(y,x)h(1,x) = h(1, x+y)\psi(y)\tau(y,x). \tag{5.12.32.1}$$

Note that after possibly composing h with a homomorphism of the form

$$(P \rtimes H_S)^{gp} \longrightarrow P^{gp} \xrightarrow{\text{deg}} \mathbb{Z} \longrightarrow B^*$$

we may assume that $h(1,0) = 1$.

Taking $y = f_i$ and noting that $\psi'(f_i) = \psi(f_i) = 1$ the formula 5.12.32.1 becomes

$$\tau'(f_i, x)h(1,x) = \tau(f_i, x)h(1, x + f_i).$$

Taking $x = 0$ this gives

$$1 = h(1,0) = h(1, f_i).$$

Taking $x = e_i$ and using the fact that

$$\tau(f_i, e_i) = 1 = \tau'(f_i, e_i)$$

the formula 5.12.32.1 gives that

$$h(1, e_i) = h(1, e_i + f_i).$$

Now by construction, the image of $f_i \otimes e_i$ under the composite map

$$Y \wedge X \longrightarrow H_S^{gp} \xrightarrow{\tilde{\Phi}} B^*$$

is equal to

$$h(1, e_i + f_i)h(1, e_i)^{-1}h(1, f_i)^{-1}.$$

We conclude that $\tilde{\Phi}$ restricts to the identity on the image of $Y \wedge X$.

5.12.33. The statement that $\tilde{\Phi} \in \Delta(B)$ is the desired element ζ is then equivalent to the assertion that $\tau = \tau'$. This can be seen as follows. Since $\tilde{\Phi}$ is trivial on the image of $Y \wedge X$, it follows that for any i, j we have

$$h(1, e_i + f_j) = h(1, e_i)h(1, f_j).$$

Applying the formula 5.12.32.1 once again (and using the fact that $h(1, f_j) = 1$) we conclude that for all i, j we have

$$\tau'(f_j, e_i) = \tau(f_j, e_i),$$

and therefore $\tau' = \tau$.

This completes the proof of 5.12.17. □

5.13 The Case $g = 1$

In the case $g = 1$ the stack $\overline{\mathscr{A}}_{1,d}$ can be described in a simple way using the classically defined stack $\overline{\mathscr{M}}_{1,1}$.

5.13.1. For this description, let us start by recalling the d-th root construction (see for example [12]).

Let X be a scheme and let (L, α) be a line bundle on X with a map $\alpha : L \to \mathscr{O}_X$ of line bundles. For any integer $d \geq 1$ we can associate an algebraic stack

$$X^{(d)} \to X$$

over X as follows. The stack $X^{(d)}$ is the stack over X which to any scheme $f : T \to X$ associates the groupoid of triples (M, β, ι), where M is a line bundle on T, $\iota : M^d \simeq f^*L$ is an isomorphism of line bundles, and $\beta : M \to \mathscr{O}_T$ is a map of $\mathscr{O}_T$-modules such that the diagram

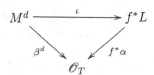

commutes.

Obsever that if $L = \mathscr{O}_X$ and α corresponds to a section $s \in \mathscr{O}_X$, then the stack $X^{(d)}$ is isomorphic to the stack-theoretic quotient

$$X^{(d)} \simeq [\mathrm{Spec}_X(\mathscr{O}_X[t]/(t^d - s))/\mu_d],$$

where $\zeta \in \mu_d$ acts by $t \mapsto \zeta t$ (for a more detailed explanation of this isomorphism see [12, §2.3]).

Since locally on X the line bundle L is trivial, one obtains from this local description the following facts:

(i) $X^{(d)}$ is an algebraic stack over X with finite diagonal, and the projection $X^{(d)} \to X$ identifies X with the coarse moduli space of $X^{(d)}$.
(ii) The map $X^{(d)} \to X$ is an isomorphism over the open set of X where the map α is an isomorphism.
(iii) If d is invertible in $\mathscr{O}_X$ then $X^{(d)}$ is a Deligne-Mumford stack.

Note that in the above we could also take X to be an algebraic stack, in which case $X^{(d)}$ is another algebraic stack over X.

5.13.2. The case we will be interested in is when $X = \overline{\mathscr{M}}_{1,1}$, $L = I_\infty$ is the ideal sheaf of the divisor $\overline{\mathscr{M}}_{1,1} - \mathscr{M}_{1,1}$ (with the reduced structure), and α is the inclusion $I_\infty \subset \mathscr{O}_{\overline{\mathscr{M}}_{1,1}}$.

Theorem 5.13.3 *There is a canonical isomorphism* $\overline{\mathscr{A}}_{1,d} \simeq \overline{\mathscr{M}}_{1,1}^{(d)}$ *of stacks over* $\mathbb{Z}$.

5.13.4. Let us start the proof by defining a map

$$c : \overline{\mathscr{A}}_{1,d} \to \overline{\mathscr{M}}_{1,1}.$$

Let

$$\mathscr{E}^u \to \overline{\mathscr{T}}_{1,d}, \quad \overline{\mathscr{P}}^u \to \overline{\mathscr{T}}_{1,d}$$

be the universal semi-abelian group scheme acting on the space $\overline{\mathscr{P}}^u$. Let $\mathscr{P}^u \to \overline{\mathscr{T}}_{1,d}$ be the maximal open substack of $\overline{\mathscr{P}}^u$ where the projection to $\overline{\mathscr{T}}_{1,d}$ is smooth. The stack $\mathscr{P}^u$ is the stack which to any scheme B associates the groupoid of data

$$\Delta : (M_B, G, (\mathscr{P}, M_{\mathscr{P}}) \to (B, M_B), \mathscr{L}, e : B \to \mathscr{P}),$$

where

$$(M_B, G, (\mathscr{P}, M_{\mathscr{P}}) \to (B, M_B), \mathscr{L}) \in \overline{\mathscr{T}}_{1,d}(B)$$

and e is a section with image in the smooth locus of $\mathscr{P} \to B$.

For such a collection of data Δ, we have a line bundle $\mathscr{O}_{\mathscr{P}}(e)$ on $\mathscr{P}$. Define

$$Q_\Delta := \underline{\mathrm{Proj}}_B(\oplus_{n \geq 0} f_* \mathscr{O}_{\mathscr{P}}(ne)),$$

where $f : \mathscr{P} \to B$ is the structure morphism. By the proof of [15, IV, Proposition 1.2], the scheme Q_Δ together with the image e_{Q_Δ} of the section e is a flat and proper scheme over B with a section in the smooth locus, all of whose geometric fibers are irreducible nodal curves of genus 1. In other words, (Q_Δ, e_{Q_Δ}) is an object of $\overline{\mathscr{M}}_{1,1}(B)$. Moreover, the formation of this object of $\overline{\mathscr{M}}_{1,1}(B)$ is compatible with arbitrary base change $B' \to B$. We therefore obtain a morphism of stacks

$$\tilde{c} : \mathscr{P}^u \to \overline{\mathscr{M}}_{1,1}.$$

5.13.5. Let B be a scheme and let

$$\chi := (M_B, G, (\mathscr{P}, M_\mathscr{P}) \to (B, M_B), \mathscr{L}) \in \overline{\mathscr{T}}_{1,d}(B)$$

be an object. Let $\mathscr{G}_\chi$ be the theta group of χ, and let H_χ denote the quotient $\mathscr{G}_\chi/\mathbb{G}_m$. We then have an action of $G \times H_\chi$ on the scheme $\mathscr{P}$.

Proposition 5.13.6. *Let $e, e' : B \to \mathscr{P}$ be two sections into the smooth locus. Then there exists a unique automorphism $\alpha : \mathscr{P} \to \mathscr{P}$ of the B-scheme $\mathscr{P}$ taking e to e' and such that α commutes with the action of $G \times H_\chi$ on $\mathscr{P}$.*

Proof. Let $\underline{\mathrm{Pic}}_{\mathscr{P}/B}$ be the relative Picard scheme of $\mathscr{P}$ over B, and let $\underline{\mathrm{Pic}}^0_{\mathscr{P}/B} \subset \underline{\mathrm{Pic}}_{\mathscr{P}/B}$ be the subscheme classifying degree 0 line bundles (see for example [15, I.3.2]). The action of $G \times H$ on $\mathscr{P}$ induces an action of $G \times H_\chi$ on $\underline{\mathrm{Pic}}^0_{\mathscr{P}/B}$.

Lemma 5.13.7 *This action of $G \times H_\chi$ on $\underline{\mathrm{Pic}}^0_{\mathscr{P}/B}$ is trivial.*

Proof. Let $g \in G \times H_\chi(B)$ be a section. To prove that the induced map

$$t_g : \underline{\mathrm{Pic}}^0_{\mathscr{P}/B} \to \underline{\mathrm{Pic}}^0_{\mathscr{P}/B}$$

is trivial, it suffices by [15, II.1.14] to consider the case when B is the spectrum of an algebraically closed field. In this situation we have two cases to consider.

If G is a smooth elliptic curve and $\mathscr{P}$ is a torsor then the action is trivial by 2.1.5.

If G is equal to $\mathbb{G}_m$ and $\mathscr{P}$ is the standard d-gon then the result follows from [15, II.1.7 (b)]. □

From this lemma and [15, II.3.2] there exists a unique structure of a generalized elliptic curve on $\mathscr{P}$ with identity section e and such that $G \times H_\chi$ acts by translation on $\mathscr{P}$. The section $e' \in \mathscr{P}(B)$ therefore defines a translation map

$$t_{e'} : \mathscr{P} \to \mathscr{P}$$

taking e to e' and commuting with the $G \times H_\chi$-action. This proves the existence of the isomorphism α.

For the uniqueness note that if $\sigma : \mathscr{P} \to \mathscr{P}$ is an automorphism taking e to e and commuting with the $G \times H_\chi$ action, then we have a commutative diagram

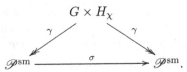

Here $\mathscr{P}^{sm}$ denotes the smooth locus of $\mathscr{P} \to B$, and γ denotes the map $g \mapsto g \cdot e$. The map γ is surjective as this can be verified fiber by fiber, where it is immediate. Using [15, II.1.7 (b) and II.1.14] it follows that α induces the trivial endomorphism of $\underline{\mathrm{Pic}}^0_{\mathscr{P}/S}$, and hence by [15, II.3.2] α is translation by an element of $\mathscr{P}^{sm}$. Since it fixes e it follows that $\alpha = \mathrm{id}$. This completes the proof of 5.13.6. $\square$

5.13.8. It follows from 5.13.6 and the fact that the map $\mathscr{P}^u \to \overline{\mathscr{T}}_{1,d}$ is flat and surjective, that the map $\tilde{c}$ descends to a morphism

$$c' : \overline{\mathscr{T}}_{1,d} \to \overline{\mathscr{M}}_{1,1}.$$

If

$$\Delta = (M_B, G, (\mathscr{P}, M_{\mathscr{P}}) \to (B, M_B), \mathscr{L}, e : B \to \mathscr{P}) \in \mathscr{P}^u(B)$$

is an object with associated object $(Q_\Delta, e_{Q_\Delta}) \in \overline{\mathscr{M}}_{1,1}(B)$, then the automorphism $c'(\theta)$ of (Q_Δ, e_{Q_Δ}) defined by an automorphism θ of

$$\chi := (M_B, G, (\mathscr{P}, M_{\mathscr{P}}) \to (B, M_B), \mathscr{L}) \in \overline{\mathscr{T}}_{1,d}(B)$$

can be described as follows. Let

$$\alpha_\theta : \mathscr{P} \to \mathscr{P}$$

be the unique isomorphism sending $\theta(e)$ to e and which commutes with the $G \times H_\chi$-action. Then $\alpha_\theta \circ \theta : \mathscr{P} \to \mathscr{P}$ sends e to e and therefore induces an automorphism of (Q_Δ, e_{Q_Δ}) which is $c'(\theta)$. From this description it follows in particular that for any object $\chi \in \overline{\mathscr{T}}_{1,d}(B)$ the induced map of group schemes

$$\underline{\mathrm{Aut}}_{\overline{\mathscr{T}}_{1,d}}(\chi) \to \underline{\mathrm{Aut}}_{\overline{\mathscr{M}}_{1,1}}(c'(\chi))$$

sends the theta group of χ to the identity. We conclude that the map c' is obtained from a unique morphism

$$c : \overline{\mathscr{A}}_{1,d} \to \overline{\mathscr{M}}_{1,1}.$$

5.13.9. In the case of the Tate curve the contraction map c can be described as follows.

First let us fix some notation. Taking $g = 1$ and $Y = d\mathbb{Z}$ in section 5.12 we obtain an object $\chi_{\text{Tate}} \in \overline{\mathscr{T}}_{g,d}(\text{Spec}(\mathbb{Z}[[q]]))$, where we use the isomorphism $H_S^{(1)} \simeq \mathbb{N}$ provided by 5.12.8 to identity the ring C_Y with $\mathbb{Z}[[q]]$. Let G_d be the semi-abelian scheme over $\mathbb{Z}[[q]]$ defined by this object and let

$$\text{Tate}_d \rightarrow \text{Spec}(\mathbb{Z}[[q]])$$

be the underlying scheme with action of G_d. The reduction modulo q of G_d is $\mathbb{G}_m$.

5.13.10. By [15, VII.1.14] $c(\chi_{\text{Tate}}) \in \overline{\mathscr{M}}_{1,1}(\mathbb{Z}[[q]])$ is canonically isomorphic to

$$\text{Tate}_1 \times_{\text{Spec}(\mathbb{Z}[[q]]), q \mapsto q^d} \text{Spec}(\mathbb{Z}[[q]]).$$

We therefore have a commutative diagram

$$
\begin{array}{ccc}
\text{Spec}(\mathbb{Z}[[q]]) & \xrightarrow{q \mapsto q^d} & \text{Spec}(\mathbb{Z}[[q]]) \\
\downarrow{\scriptstyle \text{Tate}_d} & & \downarrow{\scriptstyle \text{Tate}_1} \\
\overline{\mathscr{A}}_{1,d} & \xrightarrow{\quad c \quad} & \overline{\mathscr{M}}_{1,1}.
\end{array}
$$

It follows that if $\mathscr{D}_{1,d} \subset \overline{\mathscr{A}}_{1,d}$ is the scheme-theoretic image of

$$\text{Spec}(\mathbb{Z}) \xrightarrow{q = 0} \text{Spec}(\mathbb{Z}[[q]]) \longrightarrow \overline{\mathscr{A}}_{1,d},$$

then the ideal sheaf $I_{1,d}$ of $\mathscr{D}_{1,d}$ is an invertible sheaf on $\overline{\mathscr{A}}_{1,d}$ which comes equipped with an isomorphism

$$I_{1,d}^d \simeq c^* I_{1,1}.$$

This ideal sheaf therefore defines a lifting

$$\tilde{c} : \overline{\mathscr{A}}_{1,d} \rightarrow \overline{\mathscr{M}}_{1,1}^{(d)}$$

of c. Note that by construction the composite

$$[\text{Spec}(\mathbb{Z}[[q]])/\mu_d] \xrightarrow{\quad T \quad} \overline{\mathscr{A}}_{1,d} \xrightarrow{\quad \tilde{c} \quad} \overline{\mathscr{M}}_{1,1}^{(d)}$$

is formally étale. Therefore to prove that $\tilde{c}$ is an isomorphism it suffices to show that $\tilde{c}$ is representable. For then $\tilde{c}$ is a finite étale morphism (as $\tilde{c}$ is proper, quasi-finite, and representable) which restricts to an isomorphism over $\mathscr{M}_{1,1} \subset \overline{\mathscr{M}}_{1,1}^{(d)}$, and this implies that $\tilde{c}$ is an isomorphism.

5.13.11. To conclude the proof of 5.13.3, we therefore need to show that $\tilde{c}$ is representable.

Let T be a scheme, and let

$$f : T \to \overline{\mathscr{A}}_{1,d}$$

be a morphism corresponding to a collection of data

$$(G, M_T, (\mathscr{P}, M_{\mathscr{P}}) \to (T, M_T), \mathscr{L}) \tag{5.13.11.1}$$

over T. For every geometric point $\bar{t} \to T$ the stalk $\overline{M}_{T,\bar{t}}$ is either 0 or isomorphic to $\mathbb{N}$. Since the monoid $\mathbb{N}$ admits no automorphisms, one sees from this that there exists a unique surjection $\mathbb{N} \longrightarrow \overline{M}_T$. We therefore obtain a diagram

$$0 \longrightarrow \mathscr{O}_T^* \longrightarrow M_T^{gp} \longrightarrow \overline{M}_T^{gp} \longrightarrow 0$$
$$\uparrow$$
$$\mathbb{Z}.$$

Pulling back the top exact sequence we get an extension of $\mathbb{Z}$ by $\mathscr{O}_T^*$ which corresponds to an $\mathscr{O}_T^*$-torsor. It follows from the construction that the corresponding line bundle is $f^* I_{1,d}$.

From this it follows that if α is an automorphism of 5.13.11.1 for which $c(\alpha) = \text{id}$, then α induces the trivial automorphism of G and M_T. This implies that α is an element of the theta group and hence defines the trivial automorphism in $\overline{\mathscr{A}}_{1,d}(T)$. This concludes the proof of 5.13.3. $\square$

Remark 5.13.12. Theorem 5.13.3 implies that the universal d-torsion group $\mathscr{E}[d] \to \mathscr{M}_{1,1}$ extends to a finite flat group scheme of rank d^2 over $\overline{\mathscr{M}}_{1,1}^{(d)}$. This is expected, in light of [15, VII.1.15].

6

Level Structure

Using the results of the previous sections we now proceed to construct compact moduli stacks for abelian varieties with polarizations and level structure. We present two approaches. The first is based on the theory of logarithmic étale cohomology. The second approach is more in the spirit of Deligne-Rapoport's construction of compact moduli stacks for elliptic curves with level structure, and is based on our discussion of theta groups in chapter 5. Though we only discuss in this chapter level structures in characteristics prime to the level, we intend in future writings to discuss how to extend this second approach to level structure to compactify moduli spaces also at primes dividing the level. Finally we discuss a modular compactification of certain moduli spaces for polarized abelian varieties with theta level structure as considered in [36].

6.1 First Approach Using Kummer étale Topology

We start by reviewing the definition of the Kummer étale topology of a log scheme and the basic results about this topology (without proofs). For a more complete treatment we recommend the excellent survey of Kummer étale cohomology [21].

6.1.1. A morphism $h : Q \to P$ of fs monoids (recall from 2.3.33 that 'fs' stands for 'fine and saturated') is called *Kummer* if h is injective and for all $a \in P$ there exists $n \in \mathbb{N}$, $n \geq 1$, such that $na \in h(Q)$. A morphism of fine saturated log schemes

$$f : (Y, M_Y) \to (X, M_X)$$

is called *Kummer* if for every geometric point $\bar{y} \to Y$ the induced morphism of monoids

$$\overline{M}_{X, f(\bar{y})} \to \overline{M}_{Y, \bar{y}}$$

is a Kummer morphism. The morphism f is called *Kummer étale* if it is both log étale and Kummer.

M.C. Olsson, *Compactifying Moduli Spaces for Abelian Varieties*. Lecture Notes in Mathematics 1958.

Example 6.1.2. Let P be a fs monoid and $N \geq 1$ an integer. Then the map of fine log schemes over $\mathbb{Z}[1/N]$ (see 2.3.9 for the notation)

$$\operatorname{Spec}(P \to \mathbb{Z}[1/N][P]) \to \operatorname{Spec}(P \to \mathbb{Z}[1/N][P])$$

induced by multiplication by N on P is Kummer étale.

6.1.3. Let (X, M_X) be a fs log scheme. The *Kummer étale site* of (X, M_X), denoted $\operatorname{KEt}(X, M_X)$, is defined as follows:

Objects: Kummer étale morphisms of fs log schemes $(U, M_U) \to (X, M_X)$.

Morphisms: (X, M_X)-morphisms. Any such morphism is again Kummer étale [48, 1.5].

Topology: The topology generated by the covering families $\{(U_i, M_{U_i}) \to (U, M_U)\}$ such that the map

$$\coprod_i U_i \to U$$

is surjective.

We denote the topos associated to $\operatorname{KEt}(X, M_X)$ by

$$(X, M_X)_{\mathrm{ket}}.$$

6.1.4. Let $f : (Y, M_Y) \to (X, M_X)$ be a morphism of fs log schemes. Then there is a functor

$$\operatorname{KEt}(X, M_X) \to \operatorname{KEt}(Y, M_Y), \quad (U, M_U) \mapsto (U, M_U) \times_{(X, M_X)} (Y, M_Y),$$

where the fiber product is taken in the category of fs log schemes. This functor is continuous and induces a morphism of topoi

$$f_{\mathrm{ket}} : (Y, M_Y)_{\mathrm{ket}} \to (X, M_X)_{\mathrm{ket}},$$

If no confusion seems likely to arise we also write just f for f_{ket}.

6.1.5. A morphism of fine monoids $h : Q \to P$ is *exact* if the diagram

$$
\begin{array}{ccc}
Q & \xrightarrow{\ h\ } & P \\
\downarrow & & \downarrow \\
Q^{\mathrm{gp}} & \xrightarrow{\ h^{\mathrm{gp}}\ } & P^{\mathrm{gp}}
\end{array}
$$

is cartesian. A morphism of fine log schemes $f : (Y, M_Y) \to (X, M_X)$ is called *exact* if for every geometric point $\bar{y} \to Y$ the induced map of monoids

$$\overline{M}_{X, f(\bar{y})} \to \overline{M}_{Y, \bar{y}}$$

is exact.

Example 6.1.6. If $Q \to P$ is an exact morphism of fine monoids then the induced map of log schemes

$$\mathrm{Spec}(P \to \mathbb{Z}[P]) \to \mathrm{Spec}(Q \to \mathbb{Z}[Q])$$

is also exact.

Example 6.1.7. Let $H_S \to P \rtimes H_S$ be as in 4.1.5. Then this is an exact morphism. Indeed from the exact sequence

$$0 \to H_S^{\mathrm{gp}} \to (P \rtimes H_S)^{\mathrm{gp}} \to P^{\mathrm{gp}} \to 0$$

it follows that

$$H_S^{\mathrm{gp}} \times_{(P \rtimes H_S)^{\mathrm{gp}}} (P \rtimes H_S) = \{(p, h) \in P \rtimes H_S | p \mapsto 0 \text{ in } P^{\mathrm{gp}}\} = H_S.$$

From this it follows that for any scheme B and any object

$$(G, M_B, f : (\mathscr{P}, M_{\mathscr{P}}) \to (B, M_B), L_{\mathscr{P}}) \in \overline{\mathscr{T}}_{g,d}(B),$$

the morphism $f : (\mathscr{P}, M_{\mathscr{P}}) \to (B, M_B)$ is exact. Indeed it suffices to consider the case when B is the spectrum of an algebraically closed field, in which case the object is given by the standard construction. To verify that f is exact it suffices to show that the map

$$\tilde{f} : (\widetilde{\mathscr{P}}, M_{\widetilde{\mathscr{P}}}) \to (B, M_B)$$

is exact (where $(\mathscr{P}, M_{\mathscr{P}})$ is a quotient of $(\widetilde{\mathscr{P}}, M_{\widetilde{\mathscr{P}}})$). Recall that $\widetilde{\mathscr{P}}$ is fibered over an abelian variety A and for any point $a \in A$ there exists a neighborhood U of a such that

$$U \times_A \widetilde{\mathscr{P}} \simeq U \times_{\mathrm{Spec}(\mathbb{Z}[H_S])} \mathrm{Proj}(\mathbb{Z}[P \rtimes H_S]).$$

We conclude that to show that f is exact it suffices to show that the cone

$$\mathrm{Spec}(P \rtimes H_S \to \mathbb{Z}[P \rtimes H_S]) \to \mathrm{Spec}(H_S \to \mathbb{Z}[H_S])$$

is exact, which follows from the exactness of $H_S \to P \rtimes H_S$.

The main result that we will need in this section about the Kummer étale topology is the following theorem of Nakayama.

Theorem 6.1.8 ([37, 5.1] and [38, 4.3]) *Let $f : (X, M_X) \to (S, M_S)$ be a proper exact log smooth morphism of fs log schemes. Assume that S is noetherian, and that M_S is trivial at every generic point of S. Then for any integer n invertible on S and $q \geq 0$, the Kummer étale sheaf $R^q f_{\mathrm{ket}*}\mathbb{Z}/n\mathbb{Z}$ on $(S, M_S)_{\mathrm{ket}}$ is locally constant constructible, and its formation commutes with arbitrary base change $(T, M_T) \to (S, M_S)$ in the category of fs log schemes.*

6.1.9. It is also convenient to consider the *big Kummer étale topos*

$$(fs/(X, M_X))_{\text{ket}}$$

associated to a fs log scheme (X, M_X). This is the topos associated to the category $(fs/(X, M_X))$ of fs log schemes over (X, M_X) with the topology generated by surjective families

$$\{(U_i, M_{U_i}) \to (U, M_U)\}$$

of Kummer étale morphisms.

Note that this definition makes sense also for (X, M_X) an algebraic stack with an fs log structure.

For any morphism of fs log schemes $f : (Y, M_Y) \to (X, M_X)$ there is an induced morphism of topoi

$$f_{\text{KET}} : (fs/(Y, M_Y))_{\text{ket}} \to (fs/(X, M_X))_{\text{ket}}.$$

Nakayama's theorem 6.1.8 implies the following.

Corollary 6.1.10 *Let $f : (X, M_X) \to (S, M_S)$ be a morphism as in 6.1.8 and let n be an integer invertible in S. Then the for any $q \geq 0$ the pushforward*

$$R^q f_{\text{KET}*} \mathbb{Z}/(n)$$

for the big Kummer étale topology is a locally constant constructible sheaf on $(fs/(S, M_S))_{\text{ket}}$.

By a standard reduction to the case of schemes this result holds also for schematic proper exact and log smooth morphisms of log algebraic stacks $(X, M_X) \to (S, M_S)$.

6.1.11. Fix integers $g, n \geq 1$. Throughout the rest of this subsection 6.1 we work over $\mathbb{Z}[1/n]$ (so for example $\mathscr{K}_g$ should be interpreted as the fiber product $\mathscr{K}_g \times_{\text{Spec}(\mathbb{Z})} \text{Spec}(\mathbb{Z}[1/n])$). Let $M_{\mathscr{K}_g}$ be the canonical log structure on $\mathscr{K}_g$. We then have a log smooth proper exact morphism

$$f : (\mathscr{P}, M_{\mathscr{P}}) \to (\mathscr{K}_g, M_{\mathscr{K}_g}) \tag{6.1.11.1}$$

given by the universal polarized log scheme with semiabelian group action. Therefore for every $q \geq 0$ we get a locally constant constructible sheaf

$$\Lambda_n^q := R^q f_{\text{KET}} * \mathbb{Z}/n\mathbb{Z} \tag{6.1.11.2}$$

on $(fs/(\mathscr{K}_g, M_{\mathscr{K}_g}))_{\text{ket}}$.

6.1.12. Let $j : \mathscr{A}_g \hookrightarrow \mathscr{K}_g$ be the dense open inclusion of $\mathscr{A}_g$ in $\mathscr{K}_g$. The open substack $\mathscr{A}_g$ is the maximal open substack of $\mathscr{K}_g$ where the log structure $M_{\mathscr{K}_g}$

is trivial. The inclusion j induces a morphism of topoi (which we denote by the same letter)

$$j : \mathscr{A}_{g,\mathrm{ET}} \to (\mathscr{K}_g, M_{\mathscr{K}_g})_{\mathrm{Ket}}, \qquad (6.1.12.1)$$

where $\mathscr{A}_{g,\mathrm{ET}}$ denotes the big étale topos of $\mathscr{A}_g$.

Let $h : A \to \mathscr{A}_g$ be the universal principally polarized abelian variety, and let $f : P \to \mathscr{A}_g$ be the A–torsor corresponding to the interpretation of the moduli stack $\mathscr{A}_g$ as the moduli stack classifying data (A, P, L, θ), where A is an abelian scheme of dimension g, P is an A-torsor, L is a relatively ample invertible sheaf on P defining an isomorphism $\lambda_L : A \to A^t$, and θ is a section of L which is nonvanishing in every fiber. The restriction $j^* \Lambda_n^q$ of Λ_n^q to $\mathscr{A}_g$ is equal to $R^q f_* \mathbb{Z}/n\mathbb{Z}$ (pushforward with respect to the big étale topoi).

Etale locally on $\mathscr{A}_g$ we can choose a trivialization $\iota : A \to P$ which induces an isomorphism

$$\iota^* : R^q h_* \mathbb{Z}/n\mathbb{Z} \to R^q f_* \mathbb{Z}/n\mathbb{Z}. \qquad (6.1.12.2)$$

Lemma 6.1.13 *The isomorphism 6.1.12.2 is independent of the choice of ι, and therefore these locally defined isomorphisms define a global isomorphism*

$$R^q h_* \mathbb{Z}/n\mathbb{Z} \to R^q f_* \mathbb{Z}/n\mathbb{Z}. \qquad (6.1.13.1)$$

Proof. The translation action of A on P induces an action of A on the locally constant sheaf $R^q f_* \mathbb{Z}/n\mathbb{Z}$. Since A has geometrically connected fibers this action is trivial which implies the lemma. $\square$

6.1.14. Set

$$T_n := \mathscr{H}om(\Lambda_n^1, \mathbb{Z}/n\mathbb{Z}). \qquad (6.1.14.1)$$

This is again a locally constant sheaf of $\mathbb{Z}/n$-modules on $(fs/(\mathscr{K}_g, M_{\mathscr{K}_g}))_{\mathrm{ket}}$. The restriction $j^* T_n$ of T_n to $\mathscr{A}_g$ comes equipped with the skew-symmetric Weil pairing

$$e_n : \bigwedge^2 j^* T_n \to \mu_n. \qquad (6.1.14.2)$$

Lemma 6.1.15 *The Weil pairing $e_n : \bigwedge^2 j^* T_n \to \mu_n$ extends uniquely to a skew-symmetric pairing (which we again denote by e_n and call the Weil pairing) of locally constant constructible sheaves on $(fs/(\mathscr{K}_g, M_{\mathscr{K}_g}))_{\mathrm{ket}}$*

$$e_n : \bigwedge^2 T_n \to \mu_n. \qquad (6.1.15.1)$$

Proof. Let $U \to \mathscr{K}_g$ be a smooth surjection with U a scheme, and let U_i denote the $(i + 1)$-fold fiber product of U with itself over $\mathscr{K}_g$ (so $U_0 = U$). Let M_{U_i} be the pullback of $M_{\mathscr{K}_g}$ to the small étale site $U_{i,\mathrm{et}}$. We then have projections

$$\mathrm{pr}_1, \mathrm{pr}_2 : (U_1, M_{U_1}) \to (U_0, M_{U_0}), \quad \mathrm{pr}_{12}, \mathrm{pr}_{23}, \mathrm{pr}_{13} : (U_2, M_{U_2}) \to (U_1, M_{U_1}).$$

By standard descent theory, the category of locally constant constructible sheaves of $\mathbb{Z}/(n)$-modules on $(fs/(\mathscr{K}_g, M_{\mathscr{K}_g}))_{\mathrm{ket}}$ is equivalent to the category of pairs (F, ι), where F is a locally constant constructible sheaf of $\mathbb{Z}/(n)$-modules on $(U_0, M_{U_0})_{\mathrm{ket}}$ (the small Kummer étale site) and ι is an isomorphism on $(U_1, M_{U_1})_{\mathrm{ket}}$

$$\iota : \mathrm{pr}_2^* F \to \mathrm{pr}_1^* F$$

such that the diagram

$$
\begin{array}{ccccc}
\mathrm{pr}_{23}^* \mathrm{pr}_2^* F & \xrightarrow{\mathrm{pr}_{23}^* \iota} & \mathrm{pr}_{23}^* \mathrm{pr}_1^* F & \xrightarrow{\simeq} & \mathrm{pr}_{12}^* \mathrm{pr}_2^* F \\
\downarrow{\scriptstyle \simeq} & & & & \downarrow{\scriptstyle \mathrm{pr}_{12}^* \iota} \\
\mathrm{pr}_{13}^* \mathrm{pr}_2^* F & \xrightarrow{\mathrm{pr}_{13}^* \iota} & \mathrm{pr}_{13}^* \mathrm{pr}_1^* F & \xrightarrow{\simeq} & \mathrm{pr}_{12}^* \mathrm{pr}_1^* F
\end{array}
$$

commutes. The lemma therefore follows from Fujiwara and Kato's purity theorem [21, 7.4] which implies that for every i the functor

(locally constant constructible sheaves of $\mathbb{Z}/(n)$-modules on $(U_i, M_{U_i})_{\mathrm{ket}}$)

(locally constant constructible sheaves of $\mathbb{Z}/(n)$-modules on $(U_i \times_{\mathscr{K}_g} \mathscr{A}_g)_{\mathrm{et}}$)

is fully faithful. $\square$

6.1.16. Using the pair (T_n, e_n) one can compactify moduli spaces for principally polarized abelian varieties with various level structure. We illustrate this here by compactifying the moduli space for principally polarized abelian varieties of dimension g with a full symplectic level structure.

Let

$$s : (\mathbb{Z}/n \times \mu_n) \times (\mathbb{Z}/n \times \mu_n) \to \mu_n \tag{6.1.16.1}$$

be the symplectic form sending

$$(r, \zeta) \times (r', \zeta') \mapsto \zeta'(r) \cdot \zeta(r')^{-1}, \tag{6.1.16.2}$$

where we view μ_n as $\mathrm{Hom}(\mathbb{Z}/n, \mu_n)$. Let L_n denote $(\mathbb{Z}/n \times \mu_n)^g$ and let $s_g : \bigwedge^2 L_n \to \mu_n$ be the symplectic form induced by s.

6.1.17. Consider the sheaf

$$I_{g,n} := \underline{\mathrm{Isom}}((L_n, s_g), (T_n, e_n)) \tag{6.1.17.1}$$

of isomorphisms between L_n (viewed as a constant sheaf on $(fs/(\mathscr{K}_g, M_{\mathscr{K}_g}))_{\mathrm{ket}}$) and T_n respecting the pairings. This is again a locally constant sheaf on the big Kummer étale site of $(\mathscr{K}_g, M_{\mathscr{K}_g})$ and by [21, 3.13] is therefore representable by a log stack Kummer étale over $(\mathscr{K}_g, M_{\mathscr{K}_g})$. We write

$$(I_{g,n}, M_{I_{g,n}}) \to (\mathscr{K}_g, M_{\mathscr{K}_g}) \tag{6.1.17.2}$$

for this log stack.

Note that since $(\mathcal{K}_g, M_{\mathcal{K}_g})$ is log smooth over $\mathbb{Z}[1/n]$, the log stack $(I_{g,n}, M_{I_{g,n}})$ is also log smooth over $\mathbb{Z}[1/n]$. In particular $I_{g,n}$ is normal. Furthermore, the restriction $\overset{\circ}{I}_{g,n} := I_{g,n} \times_{\mathcal{K}_{g,n}} \mathcal{A}_g$ is isomorphic to the stack over $\mathbb{Z}[1/n]$ associating to any scheme T the groupoid of data (A, λ, ι) as follows:

(i) $(h : A \to T, \lambda)$ is a principally polarized abelian scheme over T;
(ii) $\iota : (L_n, s_g) \to (A[n], e_n)$ is a symplectic isomorphism.

We summarize this discussion in the following theorem:

Theorem 6.1.18 *The normalization $I_{g,n}$ of $\mathcal{K}_g$ in $\overset{\circ}{I}_{g,n}$ is a proper Artin stack over $\mathbb{Z}[1/n]$ with at worst toric singularities. The complement $I_{g,n} - \overset{\circ}{I}_{g,n}$ defines a fine saturated log structure $M_{I_{g,n}}$ on $I_{g,n}$ and the induced morphism*

$$(I_{g,n}, M_{I_{g,n}}) \to (\mathcal{K}_g, M_{\mathcal{K}_g}) \tag{6.1.18.1}$$

is Kummer étale.

6.2 Second Approach using the Theta Group

Let $\overline{\mathcal{A}}_{g,d}[1/d]$ denote the stack

$$\overline{\mathcal{A}}_{g,d} \times_{\mathrm{Spec}(\mathbb{Z})} \mathrm{Spec}(\mathbb{Z}[1/d]). \tag{6.2.0.2}$$

6.2.1. Let $\mathcal{H}$ denote the finite flat group scheme over $\overline{\mathcal{A}}_{g,d}$ discussed in 5.11.11. For any connected scheme S and morphism $f : S \to \overline{\mathcal{A}}_{g,d}[1/d]$, the pullback $f^*\mathcal{H}$ is a finite flat group scheme over S of rank d^2. Etale locally on S the group scheme $f^*\mathcal{H}$ is constant, and therefore has a set of elementary divisors $(d_1, d_1, d_2, d_2, \ldots, d_g, d_g)$, which necessarily occur in pairs because $f^*\mathcal{H}$ is étale locally equipped with a perfect pairing to $\mathbb{G}_m$. We call the sequence of integers $(d_1, \ldots, d_g)$ the *type* of $f^*\mathcal{H}$. Evidently the type of $f^*\mathcal{H}$ is constant for connected S. For a type δ, define $\overline{\mathcal{A}}_{g,\delta}[1/d] \subset \overline{\mathcal{A}}_{g,d}[1/d]$ to be the substack of morphisms $f : S \to \overline{\mathcal{A}}_{g,d}[1/d]$ such that the type of $f^*\mathcal{H}$ is equal to δ.

The stack $\overline{\mathcal{A}}_{g,d}[1/d]$ has a decomposition as a disjoint union

$$\overline{\mathcal{A}}_{g,d}[1/d] = \coprod_\delta \overline{\mathcal{A}}_{g,\delta}[1/d], \tag{6.2.1.1}$$

where the disjoint union is over *types* $\delta = (d_1, \ldots, d_g)$, where the d_i are positive integers, $d_{i+1}|d_i$ for all i, and $d_1 \cdots d_g = d$.

We also define

$$\overline{\mathcal{T}}_{g,\delta} := \overline{\mathcal{T}}_{g,d} \times_{\overline{\mathcal{A}}_{g,d}} \overline{\mathcal{A}}_{g,\delta}[1/d], \tag{6.2.1.2}$$

and

$$\mathcal{T}_{g,\delta} := \overline{\mathcal{T}}_{g,\delta} \times_{\overline{\mathcal{T}}_{g,d}} \mathcal{T}_{g,\delta}. \tag{6.2.1.3}$$

6.2.2. An important type is obtained by taking $d = n^g$ for some integer g and setting $\delta_0 = (n, \ldots, n)$ (n repeated g times).

If S is a $\mathbb{Z}[1/n]$-scheme and $(A, P, L) \in \mathscr{A}_{g,1}(S)$, then $(A, P, L^{\otimes n})$ is an object of $\overline{\mathscr{T}}_{g,\delta_0}[1/n](S)$ since the kernel of the map

$$\lambda_{L^{\otimes n}} : A \to A^t$$

is equal to the n-torsion subgroup $A[n]$. We therefore obtain a morphism of stacks

$$\tilde{j} : \mathscr{T}_{g,1}[1/n] \to \overline{\mathscr{T}}_{g,\delta_0}[1/n],$$

where $\overline{\mathscr{T}}_{g,\delta_0}[1/n]$ denotes the fiber product

$$\overline{\mathscr{T}}_{g,\delta_0}[1/n] := \overline{\mathscr{T}}_{g,d} \times_{\overline{\mathscr{A}}_{g,d}} \mathscr{A}_{g,\delta_0}[1/n].$$

For any object $(A, P, L) \in \overline{\mathscr{T}}_{g,1}[1/n](S)$ over some scheme S, there is a canonical map of theta groups

$$\mathbb{G}_m = \mathscr{G}_{(A,P,L)} \to \mathscr{G}_{(A,P,L^{\otimes n})}$$

given by the map

$$\times n : \mathbb{G}_m \to \mathbb{G}_m.$$

We therefore obtain an induced map on rigidifications

$$j : \mathscr{A}_{g,1}[1/n] \to \overline{\mathscr{A}}_{g,\delta_0}[1/n]. \tag{6.2.2.1}$$

Proposition 6.2.3. *The map 6.2.2.1 is a dense open immersion, and identifies $\mathscr{A}_{g,1}[1/n]$ with the open substack of $\overline{\mathscr{A}}_{g,\delta_0}[1/n]$ where the log structure $M_{\overline{\mathscr{A}}_{g,\delta_0}[1/n]}$ is trivial.*

Proof. Let us first show that j is a fully faithful functor.

Let S be a scheme and let $(A, P, L) \in \mathscr{T}_{g,1}(S)$ be an object. If α is an automorphism of (A, P, L) given by

$$\alpha_A : A \to A, \quad \alpha_P : P \to P, \quad \iota : \alpha_P^* L \to L$$

such that α maps to the identity in

$$\underline{\mathrm{Aut}}_{\overline{\mathscr{T}}_{g,\delta_0}}(A, P, L^{\otimes n}),$$

then $\tilde{j}(\alpha)$ is in $\mathscr{G}_{(A,P,L^{\otimes n})}$. It follows that $\alpha_A = \mathrm{id}$ and that α_P is translation by a point of $A[n]$. Since L is a principal polarization $\alpha_P^* L$ is isomorphic to L if and only if this point is equal to e and therefore $\alpha_P = \mathrm{id}$ also. We conclude that $\alpha \in \mathscr{G}_{(A,P,L)} = \mathbb{G}_m$ and therefore maps to the identity in

$$\underline{\mathrm{Aut}}_{\overline{\mathscr{T}}_{g,1}}(A, P, L).$$

It follows that for any two objects $(A_i, P_i, L_i) \in \mathscr{A}_{g,1}(S)$ $(i = 1, 2)$ the map

$$\underline{\mathrm{Isom}}_{\overline{\mathscr{A}}_{g,1}}((A_1, P_1, L_1), (A_2, P_2, L_2)) \to \underline{\mathrm{Isom}}_{\overline{\mathscr{A}}_{g,\delta_0}}((A_1, P_1, L_1^{\otimes n}), (A_2, P_2, L_2^{\otimes n})) \tag{6.2.3.1}$$

is a monomorphism.

If

$$\sigma_A : A_1 \to A_2, \quad \sigma_P : P_1 \to P_2, \quad \iota : \sigma_P^* L_2^{\otimes n} \to L_1^{\otimes n}$$

is an isomorphism in $\overline{\mathscr{T}}_{g,\delta_0}(S)$, then $\sigma_P^* L_2 \otimes L_1^{-1}$ is an n-torsion line bundle on P_1. Write

$$\sigma_P^* L_2 \otimes L_1^{-1} \simeq t_a^* L_1 \otimes L_1^{-1} \tag{6.2.3.2}$$

with $a \in A_1[n]$. Then

$$t_{-a}^* \sigma_P^* L_2 \simeq L_1 \tag{6.2.3.3}$$

which shows that after changing (σ_P, ι) by an element of $\mathscr{G}_{(A_1, P_1, L_1^{\otimes n})}$ lifting $-a$, we can assume that $\sigma_P^* L_2$ is isomorphic to L_1. Choose some isomorphism $\iota' : \sigma_P^* L_2 \to L_1$. Then $\iota'^{\otimes n}$ differs from ι by some element of $\mathbb{G}_m(S)$. After making a flat base change $S' \to S$ we can choose an n-th root of this unit and hence can find an isomorphism $\iota' : \sigma_P^* L_2 \to L_1$ inducing ι. It follows that the map 6.2.3.1 is an isomorphism.

To prove 6.2.3, it now suffices to show that any object $(A, P, R) \in \mathscr{T}_{g,\delta_0}(S)$ over some scheme S with A and P smooth, is fppf locally on S in the image of $\mathscr{T}_{g,1}(S)$. For this note that the homomorphism

$$\lambda_R : A \to A^t$$

factors as

$$\begin{array}{c} A \\ \times n \downarrow \quad \searrow^{\lambda_R} \\ A \xrightarrow{\ \rho\ } A^t, \end{array}$$

where ρ is an isomorphism. By [35, p. 231 Theorem 3] this map ρ is fppf-locally on S equal to λ_L for some line bundle L on A. Since A is divisible we can after making another flat base change on S choose L such that $L^{\otimes n}$ is isomorphic to R. The resulting object $(A, P, L) \in \mathscr{A}_{g,1}(S)$ then maps to (A, P, R). $\square$

6.2.4. Since the stack $\overline{\mathscr{A}}_{g,\delta_0}[1/n]$ provides a compactification of $\mathscr{A}_{g,1}[1/n]$ together with an extension of the n-torsion subgroup scheme (with its Weil pairing) of the universal abelian scheme over $\mathscr{A}_{g,1}[1/n]$ to a finite flat group scheme $\mathscr{H}$ over $\overline{\mathscr{A}}_{g,\delta_0}[1/n]$, we can use the pair $(\overline{\mathscr{A}}_{g,\delta_0}, \mathscr{H})$ to construct compactifications of various moduli problems for principally polarized abelian varieties with n-level structure. We illustrate this with the example of $\Gamma_0(n)$-structure.

Definition 6.2.5. Let S be a scheme, and $H \to S$ a finite flat abelian group scheme of rank n^{2g} with a perfect skew-symmetric pairing

$$\langle \cdot, \cdot \rangle : H \times_S H \to \mathbb{G}_{m,S}.$$

A $\Gamma_0(n)$-*level structure on* H is a subgroup scheme $B \subset H$ flat over H of rank n^g which is isotropic with respect to the pairing.

6.2.6. Define $\overline{\mathscr{A}}_{g,\Gamma_0(n)}$ to be the stack over $\mathbb{Z}[1/n]$ associating to any $\mathbb{Z}[1/n]$-scheme S the groupoid of pairs (f, B), where $f : S \to \overline{\mathscr{A}}_{g,\delta_0}[1/n]$ is a morphism and $B \subset f^* \mathscr{H}$ is a $\Gamma_0(n)$-level structure on $f^* \mathscr{H}$. Note that if f factors through $\mathscr{A}_{g,1}[1/n]$ so that f corresponds to a triple (A, P, L) with A an abelian scheme, P an A-torsor, and L a line bundle on P, then the $\Gamma_0(n)$-structure B is equal to a $\Gamma_0(n)$-structure on $A[n]$.

Let $M_{\overline{\mathscr{A}}_{g,\Gamma_0(n)}}$ denote the pullback of the log structure on $\overline{\mathscr{A}}_{g,\delta_0}$ to $\overline{\mathscr{A}}_{g,\Gamma_0(n)}$.

Theorem 6.2.7 *The projection map*

$$\overline{\mathscr{A}}_{g,\Gamma_0(n)} \to \overline{\mathscr{A}}_{g,\delta_0}[1/n] \tag{6.2.7.1}$$

is finite étale. In particular, the log stack $(\overline{\mathscr{A}}_{g,\Gamma_0(n)}, M_{\overline{\mathscr{A}}_{g,\Gamma_0(n)}})$ *is log smooth and proper over* $\mathbb{Z}[1/n]$.

Proof. It suffices to show that if S is a scheme, and $H \to S$ is a finite flat group scheme with a pairing as in 6.2.5, then the functor F on S-schemes associating to any $S' \to S$ the set of $\Gamma_0(n)$-level structures on $H \times_S S'$, is representable by a finite étale S-scheme. This can be verified after replacing S by an étale cover, so it suffices to consider the case when H is a constant group scheme. In this case the result is immediate. $\square$

Example 6.2.8. In the case $g = 1$ one can recover the Deligne-Rapoport compactification $\overline{\mathscr{M}}_{1,1,\Gamma_0(n)}$ of the stack $\mathscr{M}_{1,1,\Gamma_0(n)}$ classifying elliptic curves with $\Gamma_0(n)$-structure (in the notation of [15, IV.4.3] $\overline{\mathscr{M}}_{1,1,\Gamma_0(n)}$ is the stack $\mathscr{M}_{\Gamma_0(n)}[1/n]$). Observe that the map

$$\overline{\mathscr{A}}_{1,\Gamma_0(n)} \to \overline{\mathscr{A}}_{1,1}[1/n] = \overline{\mathscr{M}}_{1,1}[1/n]$$

is not representable, but has a relative coarse moduli space (see for example [1, §3])

$$\overline{A}_{1,\Gamma_0(n)} \to \overline{\mathscr{M}}_{1,1}[1/n].$$

The stack $\overline{A}_{1,\Gamma_0(n)}$ is normal (since $\overline{\mathscr{A}}_{1,\Gamma_0(n)}$ is normal being log smooth over $\mathbb{Z}[1/n]$) and the projection $\overline{A}_{1,\Gamma_0(n)} \to \overline{\mathscr{M}}_{1,1}[1/n]$ is finite. Furthermore, the restriction

$$\overline{\mathscr{A}}_{1,\Gamma_0(n)} = \overline{A}_{1,\Gamma_0(n)} \times_{\overline{\mathscr{M}}_{1,1}} \mathscr{M}_{1,1}$$

is isomorphic to $\mathscr{M}_{1,1,\Gamma_0(n)}$. It follows that $\overline{A}_{1,\Gamma_0(n)}$ is isomorphic to the normalization of $\overline{\mathscr{M}}_{1,1}[1/n]$ in $\mathscr{M}_{1,1,\Gamma_0(n)}$ and therefore equal to $\overline{\mathscr{M}}_{1,1,\Gamma_0(n)}$.

6.3 Resolving Singularities of Theta Functions

6.3.1. In this section we consider moduli spaces of abelian varieties with totally symmetric theta level structure (see 6.3.9 below for precise definitions) which were introduced by Mumford in [36, Part II]. For $g \geq 1$ and a sequence of positive integers $\delta = (d_1, \ldots, d_g)$ with each d_i divisible by 8, Mumford constructed a moduli space $\mathscr{M}_{g,\delta}^{\mathrm{tot}}$ classifying abelian varieties of dimension g with a polarization and a totally symmetric theta level structure of type δ. He also showed [36, Part II, Theorem 3] that the spaces $\mathscr{M}_{g,\delta}^{\mathrm{tot}}$ are quasi-projective varieties and in fact have a canonical embedding $\mathscr{M}_{g,\delta}^{\mathrm{tot}} \hookrightarrow \mathbb{P}V_\delta$ into a certain projective space. The scheme-theoretic closure $\overline{\mathscr{M}}_{g,\delta}^{\mathrm{tot}}$ of $\mathscr{M}_{g,\delta}$ in $\mathbb{P}V_\delta$ is in general highly singular (and not even normal in general). Its normalization is called the *Satake compactification* of $\mathscr{M}_{g,\delta}^{\mathrm{tot}}$ (this Satake compactification can also be constructed using theta functions and analytic methods). In this section we construct a modular log smooth compactification of $\mathscr{M}_{g,\delta}^{\mathrm{tot}}$ which maps to $\overline{\mathscr{M}}_{g,\delta}^{\mathrm{tot}}$ and therefore provides a resolution of singularities of $\overline{\mathscr{M}}_{g,\delta}^{\mathrm{tot}}$.

Let us begin by reviewing some of Mumford's theory of the theta group (see [36] for further details).

6.3.2. Let A/S be an abelian scheme over a base scheme S, and let L be an invertible sheaf on A. The line bundle L is called *symmetric* if there exists an isomorphism $\iota^*L \to L$, where $\iota : A \to A$ is the map $a \mapsto -a$. Note that if L is symmetric, then in fact there exists a canonical isomorphism $a : \iota^*L \to L$. Namely, there exists a unique isomorphism a such that the induced map

$$e^*L = e^*\iota^*L \xrightarrow{\ e^*a\ } e^*L \qquad (6.3.2.1)$$

is the identity. From this and descent theory one deduces the following:

Lemma 6.3.3 *Let S be a scheme, A/S an abelian scheme, and L a line bundle on A. Assume that fppf-locally on S the line bundle L is symmetric. Then L is symmetric.*

6.3.4. Let $\mathscr{G}(L)$ denote the theta group associated to L [35, §23]. In our earlier notation, $\mathscr{G}(L)$ is the group $\mathscr{G}_{(A,L)}$, where A is viewed as a (trivial) A-torsor. If L is symmetric, then the theta group scheme $\mathscr{G}(L)$ comes equipped with an involution $D : \mathscr{G}(L) \to \mathscr{G}(L)$. Namely, choose any isomorphism $\psi : \iota^*L \to L$, and define D by sending a point

$$(a \in A, \lambda : t_a^*L \to L) \in \mathscr{G}(L) \qquad (6.3.4.1)$$

to the point $-a \in A$ with the isomorphism

$$t_{-a}^*L \xrightarrow{\ \psi^{-1}\ } t_{-a}^*\iota^*L \longrightarrow \iota^*t_a^*L \xrightarrow{\ \lambda\ } \iota^*L \xrightarrow{\ \psi\ } L. \qquad (6.3.4.2)$$

one checks immediately that D does not depend on the choice of ψ, and that D is a group automorphism inducing the identity on $\mathbb{G}_m$.

6.3.5. Fix a sequence of positive integers $\delta = (d_1, \ldots, d_g)$, and define a group scheme $\mathscr{G}(\delta)$ over $\mathrm{Spec}(\mathbb{Z})$ as follows. Let $K(\delta)$ denote the group

$$K(\delta) := \oplus_{i=1}^{r}(\mathbb{Z}/d_i\mathbb{Z}), \qquad (6.3.5.1)$$

and let $\widehat{K(\delta)}$ denote the Cartier dual of $K(\delta)$. Set

$$H(\delta) := K(\delta) \times \widehat{K(\delta)}, \qquad (6.3.5.2)$$

and define $\mathscr{G}(\delta)$ to be the group scheme whose underlying scheme is

$$\mathbb{G}_m \times K(\delta) \times \widehat{K(\delta)} = \mathbb{G}_m \times H(\delta), \qquad (6.3.5.3)$$

and whose group structure is given by

$$(\alpha, x, l) \cdot (\alpha', x', l') = (\alpha\alpha' l'(x), x + x', l + l'). \qquad (6.3.5.4)$$

Remark 6.3.6. Define an antisymmetric pairing

$$\langle \cdot, \cdot \rangle : H(\delta) \times H(\delta) \to \mathbb{G}_m, \quad \langle (x, l), (x', l') \rangle := l'(x) \cdot l(x')^{-1}. \qquad (6.3.6.1)$$

Then it follows from the definition of the group law on $\mathscr{G}(\delta)$ that in $\mathscr{G}(\delta)$ we have

$$(\alpha, x, l)(\alpha', x', l')(\alpha, x, l)^{-1}(\alpha', x', l')^{-1} = \langle (x, l), (x', l') \rangle. \qquad (6.3.6.2)$$

Lemma 6.3.7 *Let* $\underline{\mathrm{Aut}}_{\mathbb{G}_m}(\mathscr{G}(\delta))$ *denote the functor over* $\mathbb{Z}[1/2d]$ *which to any* $\mathbb{Z}[1/2d]$-*scheme* S *associates the group of automorphisms of the group scheme* $\mathscr{G}(\delta)_S$ *which restrict to the identity on* $\mathbb{G}_m$, *and let* $\underline{\mathrm{Aut}}(H(\delta), \langle \cdot, \cdot \rangle)$ *denote the group scheme of automorphisms of* $H(\delta)$ *compatible with the skew-symmetric pairing. Then there is an exact sequence of functors of groups*

$$1 \longrightarrow \underline{\mathrm{Hom}}(H(\delta), \mathbb{G}_m) \longrightarrow \underline{\mathrm{Aut}}_{\mathbb{G}_m}(\mathscr{G}(\delta)) \overset{\pi}{\longrightarrow} \underline{\mathrm{Aut}}(H(\delta), \langle \cdot, \cdot \rangle) \longrightarrow 1, \qquad (6.3.7.1)$$

where π *sends an automorphism* α *to the automorphism* $\bar{\alpha} : H(\delta) \to H(\delta)$ *obtained by passing to the quotient by* $\mathbb{G}_m$. *In particular,* $\underline{\mathrm{Aut}}_{\mathbb{G}_m}(\mathscr{G}(\delta))$ *is representable by a finite étale group scheme over* $\mathbb{Z}[1/2d]$.

Proof. Since $\mathbb{G}_m$ is central in $\mathscr{G}(\delta)$, any automorphism α of $\mathscr{G}(\delta)_S$ inducing the identity on $H(\delta)_S$ is given by

$$\alpha(g) = g \cdot w(\bar{g}), \qquad (6.3.7.2)$$

where $\bar{g}$ is the image of g in $H(\delta)_S$ and $w : H(\delta) \to \mathbb{G}_m$ is a homomorphism. Conversely any homomorphism w determines an automorphism of $\mathscr{G}(\delta)$ by the same formula. This gives the identification

$$\underline{\mathrm{Hom}}(H(\delta), \mathbb{G}_m) \simeq \mathrm{Ker}(\pi). \qquad (6.3.7.3)$$

The surjectivity of π follows from the same argument used in [36, I, Corollary of Th. 1] which shows that in order to lift a symplectic automorphism $\bar{\alpha}$: $H(\delta) \to H(\delta)$ to $\mathcal{G}(\delta)$ it suffices to choose a lifting $\tilde{s} : K(\delta) \to \mathcal{G}(\delta)$ of the homomorphism $K(\delta) \to H(\delta)$ obtained by restricting $\bar{\alpha}$, where $K(\delta)$ is as in 6.3.5. $\square$

6.3.8. There is a standard representation V_δ of $\mathcal{G}(\delta)$ over $\mathbb{Z}[1/2d]$ defined as follows. Let V_δ denote the $\mathbb{Z}[1/2d]$-module of functions $K(\delta) \to \mathbb{Z}[1/2d]$. Then for any scheme S over $\mathbb{Z}[1/2d]$ and quasi-coherent sheaf F on S the sheaf $V_\delta \otimes_{\mathbb{Z}[1/2d]} F$ is the sheaf of functions $K(\delta) \to F$. If $(\alpha, x, l) \in \mathcal{G}(\delta)(S)$ is a section with $\alpha \in \mathcal{O}_S^*$, $x = (x_1, \ldots, x_g) \in K(\delta)$, and $l = (l_1, \ldots, l_g) \in \widehat{K(\delta)}$, then we define an action of (α, x, l) on $V_\delta \otimes \mathcal{O}_S$ by sending a function f : $K(\delta) \to \mathcal{O}_S$ to the function

$$K(\delta) \to \mathcal{O}_S, \quad (y_1, \ldots, y_g) \mapsto \alpha \cdot (\prod_{i=1}^{g} l_i(y_i)) f(x_1 + y_1, \ldots, x_g + y_g). \quad (6.3.8.1)$$

Note that $\mathbb{G}_m \subset \mathcal{G}(\delta)$ acts by the standard character. Therefore in the case when S is the spectrum of a field the representation $V_\delta \otimes \mathcal{O}_S$ is the unique representation of $\mathcal{G}(\delta)$ mentioned in 5.4.27.

6.3.9. The group scheme $\mathcal{G}(\delta)$ also has an involution $\tau : \mathcal{G}(\delta) \to \mathcal{G}(\delta)$ given by

$$(a, h) \mapsto (a, -h), \quad a \in \mathbb{G}_m, h \in H(\delta). \quad (6.3.9.1)$$

If L is an ample symmetric line bundle on an abelian scheme A, then a *symmetric theta-level structure on* (A, L) *of type* δ is an isomorphism $\sigma : \mathcal{G}(\delta)_S \to \mathcal{G}(L)$ such that the diagram

$$\begin{array}{ccc} \mathcal{G}(\delta)_S & \xrightarrow{\sigma} & \mathcal{G}(L) \\ \downarrow{\scriptstyle \tau} & & \downarrow{\scriptstyle D} \\ \mathcal{G}(\delta)_S & \xrightarrow{\sigma} & \mathcal{G}(L) \end{array} \qquad (6.3.9.2)$$

commutes.

Let $g \geq 1$ be an integer, and set $d = d_1 \cdots d_g$. Let $\mathcal{M}_{g,\delta}$ denote the stack over the category of $\mathbb{Z}[1/2d]$-schemes which associates to any such scheme S the groupoid of data

$$(A, L, \sigma, \epsilon : e^*L \simeq \mathcal{O}_S), \qquad (6.3.9.3)$$

where

(i) A/S is an abelian scheme of relative dimension g.
(ii) L is a symmetric relatively ample invertible sheaf on A.
(iii) $\sigma : \mathcal{G}(\delta) \times S \to \mathcal{G}(L)$ is a symmetric theta level structure.
(iv) $\epsilon : e^*L \to \mathcal{O}_S$ is an isomorphism of $\mathcal{O}_S$-modules.

Lemma 6.3.10 *Assume that each d_i is divisible by 4. Fix the following data:*

1. *An abelian scheme $f : A \to S$ of relative dimension g.*
2. *A relatively ample invertible sheaf L on A.*
3. *A symmetric theta level structure $\sigma : \mathscr{G}(\delta)_S \to \mathscr{G}(L)$.*

Then L is relatively very ample, f_*L is locally free of rank d and its formation commutes with arbitrary base change, and $R^i f_* L = 0$ for all $i > 0$.

Proof. By standard cohomology and base change results it suffices to consider the case when $S = \operatorname{Spec}(k)$ is the spectrum of an algebraically closed field.

The vanishing of $H^i(A, L)$ for $i > 0$ in this case follows from [35, Corollary on p. 159 and the Vanishing Theorem on p. 150], and the dimension of $H^0(A, L)$ is given by [35, Riemann-Roch theorem, p. 150].

The assumption that each d_i is divisible by 4 implies that $H(L)$ contains the 4-torsion $A[4]$ of A. Therefore by [35, Theorem 3, p. 231] we can write $L = M^{\otimes 4}$ for some other line bundle M on A. Since L is ample the line bundle M is also ample, and therefore by [35, Theorem on p. 163] L is very ample.
$\square$

Remark 6.3.11. If each d_i is divisible by 4 then the stack $\mathscr{M}_{g,\delta}$ is in fact equivalent to a functor. For this it suffices to show that if S is a scheme then any automorphism of a collection of data 6.3.9.3 is the identity. This can be seen as follows. It suffices to consider the case when S is the spectrum of a strictly henselian local ring R. The space of global section $\Gamma(A, L)$ is then a representaton of $\mathscr{G}(L)$ over R, and hence via σ we can view this as a representation of $\mathscr{G}(\delta)$ over R. If

$$\chi : A \to A, \quad \chi^b : \chi^*L \to L \tag{6.3.11.1}$$

is an automorphism of the data 6.3.9.3 then (χ, χ^b) induces an automorphism α of the $\mathscr{G}(\delta)$-representation $\Gamma(A, L)$ such that the diagram

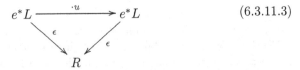

$$\tag{6.3.11.2}$$

commutes, where j is the canonical closed immersion (recall that L is relatively very ample). On the other hand, by [36, Part II, Proposition 2] any automorphism of the $\mathscr{G}(\delta)$-representation $\Gamma(A, L)$ is multiplication by a scalar, and therefore the map $\mathbb{P}\alpha$ is the identity. We conclude that $\chi : A \to A$ is the identity map. This implies that $\chi^b : L \to L$ is multiplication by a scalar $u \in R^*$ such that the diagram

$$
\begin{array}{ccc}
e^*L & \xrightarrow{\cdot u} & e^*L \\
 & \searrow{\epsilon} \quad \swarrow{\epsilon} & \\
 & R &
\end{array}
\tag{6.3.11.3}
$$

commutes. It follows that $u = 1$.

6.3.12. One can also consider a variant moduli problem classifying (A, L, σ, ϵ) as above, where in addition the line bundle L is assumed *totally symmetric*. This means that there exists an isomorphism $\psi : \iota^* L \to L$ whose pullback to the 2-torsion points $A[2]$ (on which ι is the identity) is equal to the identity map. Let us write $\mathcal{M}_{g,\delta}^{\mathrm{tot}}$ for the resulting functor. There is a natural inclusion $\mathcal{M}_{g,\delta}^{\mathrm{tot}} \subset \mathcal{M}_{g,\delta}$.

As explained in [36, I. p. 310] there is for every $n \geq 1$ a canonical map

$$\eta_n : \mathcal{G}(L^n) \to \mathcal{G}(L) \tag{6.3.12.1}$$

such that the diagram

$$
\begin{array}{ccccccccc}
1 & \longrightarrow & \mathbb{G}_m & \longrightarrow & \mathcal{G}(L^n) & \longrightarrow & H(L^n) & \longrightarrow & 1 \\
& & \downarrow{\scriptstyle \times n} & & \downarrow{\scriptstyle \eta_n} & & \downarrow{\scriptstyle n} & & \\
1 & \longrightarrow & \mathbb{G}_m & \longrightarrow & \mathcal{G}(L) & \longrightarrow & H(L) & \longrightarrow & 1
\end{array}
\tag{6.3.12.2}
$$

commutes. By [36, I, Corollary 2] a symmetric line bundle L on an abelian scheme is totally symmetric if and only if

$$\mathrm{Ker}(\eta_2) = \{z \in \mathcal{G}(L^2) | z^2 = 1\}. \tag{6.3.12.3}$$

Since $\mathbb{G}_m$ is divisible and central in $\mathcal{G}(L^2)$, the scheme K whose functor of points is given by

$$\{z \in \mathcal{G}(L^2) | z^2 = 1\} \tag{6.3.12.4}$$

is a μ_2-torsor over the kernel of multiplication of by 2 on $H(L^2)$. In particular, it is an étale scheme over the base scheme (since $2d$ is assumed invertible) of rank 2^{2g+1}.

Proposition 6.3.13. *The substack $\mathcal{M}_{g,\delta}^{\mathrm{tot}} \subset \mathcal{M}_{g,\delta}$ is representable by open and closed immersions.*

Proof. Let S be a $\mathbb{Z}[1/2d]$-scheme and (A, L, σ, ι) an object of $\mathcal{M}_{g,\delta}(S)$. We need to show that the condition that L is totally symmetric is representable by an open subset of S.

From the commutative diagram 6.3.12.2 with $n = 2$ and the snake lemma, one sees that there is an exact sequence

$$1 \to \mu_2 \to \mathrm{Ker}(\eta_2) \to H(L^2)[2] \to 1, \tag{6.3.13.1}$$

so in particular the rank of the group scheme $\mathrm{Ker}(\eta_2)$ is equal to 2^{2g+1}, which is also equal to the rank of K. The condition that L is totally symmetric is therefore equivalent to the assertion that the map

$$K \to \mathcal{G}(L^2)/\mathrm{Ker}(\eta_2) \simeq \mathcal{G}(L) \tag{6.3.13.2}$$

sends K to the identity section (since both K and $\mathrm{Ker}(\eta_2)$ are finite étale schemes over S of the same degree). Note also that the composite 6.3.13.2 in fact has image in $\mu_2 \subset \mathbb{G}_m \subset \mathscr{G}(L)$ by the commutativity of 6.3.12.2. It follows that the condition that the map 6.3.13.2 is trivial is an open and closed condition, since it is a map of finite étale schemes.   □

Theorem 6.3.14 ([36, Part II, Theorem 3]) *If each d_i is divisible by 8, then $\mathscr{M}_{g,\delta}^{\mathrm{tot}}$ is representable by a quasi-projective scheme.*

6.3.15. For the convenience of the reader, let us recall Mumford's description of $\mathscr{M}_{g,\delta}^{\mathrm{tot}}$. Let S be a scheme and let

$$(f : A \to S, L, \sigma, \epsilon) \tag{6.3.15.1}$$

be an object of $\mathscr{M}_{g,\delta}^{\mathrm{tot}}(S)$. By 6.3.10 the line bundle L is relatively very ample on A and f_*L is a locally free sheaf on S of rank d whose formation is compatible with arbitrary base change. The vector bundle f_*L is a representation of $\mathscr{G}(L)$ over S, and hence via σ can be viewed as a representation of $\mathscr{G}(\delta)_S$. By [36, Part II, Proposition 2] there exists a unique line bundle K on S and an isomorphism $K \otimes_{\mathbb{Z}[1/2d]} V_\delta \simeq f_*L$ of $\mathscr{G}(\delta)_S$-representation. Furthermore this isomorphism is unique up to multiplication by an element of $\mathscr{O}_S^*$. We therefore obtain a canonical embedding

$$i : A \hookrightarrow \mathbb{P}(V_\delta)_S. \tag{6.3.15.2}$$

Composing this embedding with the identity section $S \to A$ of A we obtain a morphism

$$S \to \mathbb{P}(V_\delta). \tag{6.3.15.3}$$

This defines a morphism of functors

$$\mathscr{M}_{g,\delta}^{\mathrm{tot}} \to \mathbb{P}(V_\delta) \tag{6.3.15.4}$$

and it is shown in [36, Part II, Theorem on p. 83] that this is an immersion. In fact, Mumford describes in loc. cit. explicit equations for the scheme-theoretic closure $\overline{\mathscr{M}}_{g,\delta}^{\mathrm{tot}} \subset \mathbb{P}(V_\delta)$ ("Riemann's theta relations").

In the remainder of this section we explain how to construct a canonical modular compactification of $\mathscr{M}_{g,\delta}$. By the above discussion this then also gives a compactification of $\mathscr{M}_{g,\delta}^{\mathrm{tot}}$.

6.3.16. The functor $\mathscr{M}_{g,\delta}$ can be reinterpreted as follows. Let S be a scheme and A/S an abelian scheme. For an A-torsor P, let P^ι denote the A-torsor with same underlying scheme as P but with action of A given by the composite homomorphism

$$A \xrightarrow{\ \iota\ } A \xrightarrow{\ \mathrm{action}\ } \underline{\mathrm{Aut}}(P). \tag{6.3.16.1}$$

If L is an ample invertible sheaf on P, then there is a canonical isomorphism of theta groups

$$\text{can} : \mathscr{G}_{(P,L)} \simeq \mathscr{G}_{(P^\iota,L)}. \tag{6.3.16.2}$$

This isomorphism sends a pair

$$\sigma_P : P \to P, \quad \sigma_L : \sigma_P^* L \to L \tag{6.3.16.3}$$

to the same pair (σ_P, σ_L) viewed as an automorphism of (P^ι, L).

Proposition 6.3.17. *Let S be a scheme, A/S an abelian scheme, P/S an A-torsor, and let L be an ample invertible sheaf on P. Then fppf-locally on S there exists an isomorphism*

$$\rho = (\rho_P, \rho_L) : (P, L) \to (P^\iota, L) \tag{6.3.17.1}$$

over $\iota : A \to A$. Moreover, the isomorphism ρ is unique up to composition with an element of

$$\mathscr{G}_{(P,L)} = \underline{\mathrm{Aut}}_A(P, L). \tag{6.3.17.2}$$

Proof. The uniqueness statement is clear, for by the very definition of $\mathscr{G}_{(P,L)}$ if ρ_1 and ρ_2 are two isomorphisms then $\rho_1^{-1} \circ \rho_2$ is an automorphism of (P, L) commuting with the A-action.

For the existence we may work fppf locally on S, and may therefore assume that P is trivial. Fix a point $e \in P$ defining an isomorphism $A \to P$ (this enables us to view L as a sheaf on A).

Lemma 6.3.18 *After making an étale base change on S, there exists an invertible sheaf $M \in A^t(S)$ such that $\iota^* L \simeq L \otimes M$.*

Proof. This is standard. We recall the argument for the convenience of the reader.

It suffices to show that for any scheme-valued point $\alpha \in A$ the invertible sheaf

$$\mathscr{R} := t_\alpha^*(\iota^* L \otimes L^{-1}) \otimes \iota^* L^{-1} \otimes L \tag{6.3.18.1}$$

is trivial. For this note that

$$t_\alpha^* \iota^* L \otimes \iota^* L^{-1} \simeq \iota^*(t_{-\alpha}^* L \otimes L^{-1})$$
$$\simeq t_{-\alpha}^* L^{-1} \otimes L,$$

where the second isomorphism is because ι acts as multiplication by -1 on A^t. Therefore $\mathscr{R}$ is isomorphic to

$$t_\alpha^* L^{-1} \otimes t_{-\alpha}^* L^{-1} \otimes L^2 \tag{6.3.18.2}$$

which is trivial by the theorem of the square. $\square$

After replacing S by an fppf cover, there exists therefore an element $\alpha \in A(S)$ such that $M \simeq t_\alpha^* L \otimes L^{-1}$ in which case $\iota^* L \simeq t_\alpha^* L$. Therefore we can take $\rho_P : P \to P$ to be the isomorphism $\iota \circ t_\alpha$ and ρ_L any isomorphism $\iota^* t_\alpha^* L \to L$. $\square$

Proposition 6.3.19. *Let S be a scheme, A/S an abelian scheme, P/S an A-torsor, and $\sigma : \mathscr{G}(\delta)_S \to \mathscr{G}_{(P,L)}$ an isomorphism restricting to the identity on $\mathbb{G}_m$. Then fppf locally on S there exists an isomorphism $\rho = (\rho_P, \rho_L) : (P, L) \to (P^\iota, L)$ over $\iota : A \to A$ such that the diagram*

$$
\begin{array}{ccc}
\mathscr{G}(\delta)_S & \xrightarrow{\ \sigma\ } & \mathscr{G}_{(P,L)} \\
{\scriptstyle \tau}\downarrow & & \downarrow{\scriptstyle \rho} \\
\mathscr{G}(\delta)_S & \xrightarrow{\ \sigma\ } & \mathscr{G}_{(P^\iota,L)}
\end{array}
\tag{6.3.19.1}
$$

commutes, where the right vertical arrow is the isomorphism induced by ρ. Moreover ρ is unique up to composition with the action of an element of $\mathbb{G}_m$ (where $u \in \mathbb{G}_m$ acts as the identity on P and multiplication by u on L).

Proof. We may work fppf-locally on S and therefore by 6.3.17 may assume that there exists an isomorphism $\rho = (\rho_P, \rho_L) : (P, L) \to (P^\iota, L)$ over $\iota : A \to A$. The diagram 6.3.19.1 may not commute, but we need to show that after possible composing ρ with an element of $\mathscr{G}_{(P,L)}$ then we can make the diagram commute. Note that by 6.3.19.1 any two choices of ρ differ by composition by an element of $\mathscr{G}_{(P,L)}$ and therefore the uniqueness part of 6.3.19 follows from the fact that $\mathbb{G}_m$ is equal to the center of $\mathscr{G}_{(P,L)}$.

If $\alpha \in \mathscr{G}_{(P,L)}$ is a section and $\rho' = \rho \circ \alpha$, then the induced map

$$
\rho' : \mathscr{G}_{(P,L)} \to \mathscr{G}_{(P^\iota,L)}
\tag{6.3.19.2}
$$

is equal to conjugation by α following by the map induced by ρ. Note also that in any case the map $\mathscr{G}_{(P,L)} \to \mathscr{G}_{(P,L)}$ induces by construction of ρ in the proof of 6.3.17 the identity on $\mathbb{G}_m$ and multiplication by -1 on $H(P, L)$. Proposition 6.3.19 therefore follows from the following lemma applied to $(\sigma \circ \tau)^{-1} \circ (\rho \circ \sigma)$. $\square$

Lemma 6.3.20 *If $\lambda : \mathscr{G}(\delta)_S \to \mathscr{G}(\delta)_S$ is an automorphism inducing the identity on $\mathbb{G}_m$ and $H(\delta)_S$, then fppf-locally on S there exists an element $\beta \in \mathscr{G}(\delta)(S)$ such that λ is equal to conjugation by β.*

Proof. Since $\mathbb{G}_m$ is central in $\mathscr{G}(\delta)_S$, there exists a unique homomorphism $w : H(\delta) \to \mathbb{G}_m$ such that for any $g \in \mathscr{G}(\delta)_S$ with image $\bar{g} \in H(\delta)_S$ we have

$$
\lambda(g) = w(\bar{g}) \cdot g.
\tag{6.3.20.1}
$$

Since the pairing $\langle \cdot, \cdot \rangle$ on $H(\delta)$ defined in 6.3.6 is perfect, there exists an element $w^* \in H(\delta)_S$ such that

$$
\langle w^*, - \rangle : H(\delta)_S \to \mathbb{G}_m
\tag{6.3.20.2}
$$

is equal to w. After replacing S by a covering, we can lift w^* to an element $\tilde{w}^* \in \mathscr{G}(\delta)_S$. Then for any $g \in \mathscr{G}(\delta)_S$ we have by definition of the pairing

$$\tilde{w}^* g \tilde{w}^{*-1} g^{-1} = w(\bar{g}), \qquad (6.3.20.3)$$

or equivalently

$$\tilde{w}^* g \tilde{w}^{*-1} = w(\bar{g}) \cdot g = \lambda(g). \qquad (6.3.20.4)$$

$\square$

Remark 6.3.21. Note that the maps $\rho_P : P \to P$ obtained locally on S from 6.3.19 glue since the ambiguity in ρ only involves the map ρ_L. It follows that given data (A, P, L, σ) as in 6.3.19 there exists a canonical globally defined isomorphism $\rho_P : P \to P^\iota$ over the map $\iota : A \to A$ such that the diagram 6.3.19.1 commutes.

6.3.22. Let $\Sigma_{g,\delta}$ denote the stack over $\mathbb{Z}[1/2d]$ which associates to any $\mathbb{Z}[1/2d]$-scheme S the groupoid of data:

$$(A, P, L, \sigma) \qquad (6.3.22.1)$$

where:

(i) A is an abelian scheme over S;
(ii) P is an A-torsor;
(iii) L is an ample invertible sheaf on P;
(iv) $\sigma : \mathscr{G}(\delta)_S \to \mathscr{G}_{(P,L)}$ is an isomorphism of group schemes inducing the identity on $\mathbb{G}_m$.

Proposition 6.3.23. *The stack $\Sigma_{g,\delta}$ is algebraic, and the projection morphism*

$$\Sigma_{g,\delta} \to \mathscr{T}_{g,\delta}[1/2d] \qquad (6.3.23.1)$$

obtained by "forgetting" σ is finite and étale (where $\mathscr{T}_{g,\delta}$ is defined as in 6.2.1).

Proof. Given an object $(A, P, L) \in \mathscr{T}_{g,\delta}(S)$, the fiber product I of the diagram

$$
\begin{array}{ccc}
 & & \Sigma_{g,\delta} \\
 & & \downarrow \\
S & \xrightarrow{(A,P,L)} & \mathscr{T}_{g,\delta}
\end{array}
\qquad (6.3.23.2)
$$

is the functor associating to any S-scheme T the set of isomorphisms

$$\sigma : \mathscr{G}(\delta)_T \to \mathscr{G}_{(P_T, L_T)} \qquad (6.3.23.3)$$

inducing the identity on $\mathbb{G}_m$. Now fppf locally on S there exists one such isomorphism, and therefore I is a torsor under the finite étale S-group scheme

$$\underline{\mathrm{Aut}}_{\mathbb{G}_m}(\mathscr{G}(\delta)). \qquad (6.3.23.4)$$

In particular, I is represented by a finite étale S-group scheme. $\square$

6.3.24. Let $(A_u, P_u, L_u, \sigma_u)$ be the universal data over $\Sigma_{g,\delta}$. Then P_u is again an algebraic stack proper and smooth over $\Sigma_{g,\delta}$. Moreover, by 6.3.19 there is a canonical involution

$$\rho_u : P_u \to P_u \tag{6.3.24.1}$$

over $\Sigma_{g,\delta}$. We will be interested in the stack P_u^ρ of fixed points of this action. By definition P_u^ρ is the fiber product of the diagram

$$
\begin{array}{c}
P_u \\
\downarrow {\scriptstyle \mathrm{id} \times \rho} \\
P_u \xrightarrow{\ \Delta\ } P_u \times_{\Sigma_{g,\delta}} P_u.
\end{array}
\tag{6.3.24.2}
$$

Note that since $P_u \to \Sigma_{g,\delta}$ is proper (and in particular separated) and representable, the diagonal map

$$\Delta : P_u \to P_u \times_{\Sigma_{g,\delta}} P_u \tag{6.3.24.3}$$

is a closed immersion. Therefore $P_u^\rho \subset P_u$ is a closed substack.

6.3.25. If S is a scheme then the groupoid $P_u(S)$ is the groupoid of data

$$(A, P, L, \sigma, e), \tag{6.3.25.1}$$

where $(A, P, L, \sigma) \in \Sigma_{g,\delta}(S)$ and $e : A \to P$ is a trivialization.

Let $\mathscr{N}_{g,\delta}$ denote the stack over $\mathbb{Z}[1/2d]$ associating to any scheme S the groupoid of data (A, L, σ), where:

(i) A/S is an abelian scheme;
(ii) L is an ample invertible sheaf on A;
(iii) $\sigma : \mathscr{G}(\delta)_S \to \mathscr{G}(L)$ is an isomorphism of group schemes inducing the identity on $\mathbb{G}_m$.

There is a natural functor

$$F : P_u \to \mathscr{N}_{g,\delta}, \quad (A, P, L, \sigma, e) \mapsto (A, e^*L, \sigma). \tag{6.3.25.2}$$

Lemma 6.3.26 *The functor F is an isomorphism of stacks.*

Proof. This is immediate. An inverse to F is given by the functor

$$\mathscr{N}_{g,\delta} \to P_u, \quad (A, L, \sigma) \mapsto (A, A, L, \sigma, \mathrm{id}). \tag{6.3.26.1}$$

$\square$

6.3.27. Let $\mathscr{N}_{g,\delta}^{\mathrm{sym}} \subset \mathscr{N}_{g,\delta}$ denote the substack classifying triples (A, L, σ), where L is a symmetric ample sheaf on A and σ is symmetric theta level structure.

Proposition 6.3.28. *The functor F induces an isomorphism*

$$P_u^\rho \to \mathcal{N}_{g,\delta}^{\mathrm{sym}}. \tag{6.3.28.1}$$

Proof. To say that a point $(A, P, L, \sigma, e) \in P_u$ is invariant under ρ means that the diagram

$$
\begin{array}{ccc}
A & \xrightarrow{\ e\ } & P \\
\iota \downarrow & & \downarrow \rho_P \\
A & \xrightarrow{\ e\ } & P
\end{array}
\tag{6.3.28.2}
$$

commutes, and hence $\iota^* e^* L$ is locally on the base isomorphic to $e^* L$. Therefore $e^* L$ is symmetric and σ is a symmetric theta level structure. Conversely, if (A, L, σ) is an object of $\mathcal{N}_{g,\delta}^{\mathrm{sym}}$ then the object $(A, A, L, \sigma, e) \in P_u$ is in P_u^ρ.
$\square$

6.3.29. For any object (A, P, L, σ, e) of P_u^ρ and $z \in \mathbb{G}_m$ we get an automorphism of (A, P, L, σ, e) simply by multiplication by z on L. Let $\Theta_{g,\delta}$ denote the rigidification of P_u^ρ with respect to $\mathbb{G}_m$.

Proposition 6.3.30. *The stack $\mathcal{M}_{g,\delta}$ is isomorphic to $\Theta_{g,\delta}$.*

Proof. This follows from the same argument used in 2.4.11. $\square$

Before beginning our discussion of the compactification of $\Theta_{g,\delta}$ let us also note the following:

Proposition 6.3.31. *The stack $\mathcal{N}_{g,\delta}^{\mathrm{sym}}$, and hence also $\Theta_{g,\delta}$, is smooth over $\mathbb{Z}[1/2d]$.*

Proof. It suffices to show that $\mathcal{N}_{g,\delta}^{\mathrm{sym}}$ is formally smooth over $\mathbb{Z}[1/2d]$. Consider a diagram of affine schemes

$$\mathrm{Spec}(R_0) \hookrightarrow \mathrm{Spec}(R) \hookrightarrow \mathrm{Spec}(R'), \tag{6.3.31.1}$$

where R_0 is a field, R and R' are artinian local rings and $R' \to R$ is surjective with kernel J annihilated by the maximal ideal of R'. Let (A, L, σ) be an object of $\mathcal{N}_{g,\delta}^{\mathrm{sym}}(R)$. We have to exhibit a lifting of (A, L, σ) to R'.

For this choose first any lifting (A', L') of (A, L) to R'. Then the line bundle L' need not be symmetric on A', but we can write

$$\iota^* L' \simeq L' \otimes M' \tag{6.3.31.2}$$

for some M' reducing to $\mathcal{O}_A$. The liftings of $\mathcal{O}_A$ to A' are classified by the group $H^1(A_0, \mathcal{O}_{A_0} \otimes J)$. Since 2 is assumed invertible in R' multiplication by 2 on $H^1(A_0, \mathcal{O}_{A_0})$ is an isomorphism and hence there exists a lifting E' of $\mathcal{O}_A$ to A' such that $E'^2 \simeq M'$. Then

$$\iota^*(L' \otimes E') \simeq \iota^* L' \otimes E'^{-1} \simeq L' \otimes M \otimes E'^{-1} \simeq L' \otimes E'. \tag{6.3.31.3}$$

Replacing L' by $L' \otimes E'$ we therefore see that there exists a symmetric lifting of L to A'. Fix one such lifting L'.

It remains to see that we can find an isomorphism $\sigma' : \mathscr{G}(\delta)_{R'} \to \mathscr{G}(L')$ reducing to σ and such that the diagram

$$
\begin{array}{ccc}
\mathscr{G}(\delta)_{R'} & \xrightarrow{\ \sigma'\ } & \mathscr{G}(L') \\
{\scriptstyle \tau}\downarrow & & \downarrow{\scriptstyle D} \\
\mathscr{G}(\delta)_{R'} & \xrightarrow{\ \sigma'\ } & \mathscr{G}(L')
\end{array}
\tag{6.3.31.4}
$$

commutes.

Étale locally on $\mathrm{Spec}(R')$ we can choose an isomorphism $\mathscr{G}(\delta)_{R'} \to \mathscr{G}(L')$. The reduction of this isomorphism may not agree with σ, but since $\underline{\mathrm{Aut}}_{\mathbb{G}_m}(\mathscr{G}(\delta))$ is étale over $\mathrm{Spec}(\mathbb{Z}[1/2d])$ we can change our choice of isomorphism to find a lifting σ' of σ. The diagram 6.3.31.4 then automatically commutes since it commutes over R and $\underline{\mathrm{Aut}}_{\mathbb{G}_m}(\mathscr{G}(\delta))$ is an étale group scheme. This completes the proof of 6.3.31. $\square$

6.3.32. We now compactify $\Theta_{g,\delta}$ by generalizing the preceding constructions to objects of $\overline{\mathscr{T}}_{g,\delta}$.

Let S be a scheme and let (G, M_S, P, M_P, L) be an object of $\overline{\mathscr{T}}_{g,\delta}(S)$. Let $(P^\iota, M_{P_\iota}, L)$ denote the log scheme over (S, M_S) with the same underlying polarized log scheme (P, M_P, L) but G-action the composite

$$
G \xrightarrow{\ g \mapsto -g\ } G \xrightarrow{\ \text{action}\ } \mathrm{Aut}(P, M_P).
\tag{6.3.32.1}
$$

Note that as before 6.3.16 there is a canonical isomorphism of theta groups

$$
\mathscr{G}_{(P,M_P,L)} \simeq \mathscr{G}_{(P^\iota, M_{P^\iota}, L)}.
\tag{6.3.32.2}
$$

Proposition 6.3.33. *After replacing S by an fppf cover, there exists an isomorphism*

$$
\rho = (\rho_{(P,M_P)}, \rho_L) : ((P, M_P), L) \to ((P^\iota, M_{P^\iota}), L)
\tag{6.3.33.1}
$$

of log schemes over (S, M_S) with ample sheaves such that the diagram

$$
\begin{array}{ccc}
G \times_{(S,M_S)} (P, M_P) & \xrightarrow{\ \text{action}\ } & (P, M_P) \\
{\scriptstyle \iota \times \rho_{(P,M_P)}}\downarrow & & \downarrow{\scriptstyle \rho_{(P,M_P)}} \\
G \times_{(S,M_S)} (P^\iota, M_{P^\iota}) & \xrightarrow{\ \text{action}\ } & (P^\iota, M_{P^\iota})
\end{array}
\tag{6.3.33.2}
$$

commutes. Moreover, ρ is unique up to composition with an element of $\mathscr{G}_{(P,M_P,L)}$.

The proof will be in several steps (6.3.34–6.3.43).

6.3.34. The uniqueness statement in the proposition is immediate from the definition of the theta group.

By a standard application of Artin's approximation theorem, to prove the existence of ρ it suffices to consider the case when S is the spectrum of a complete noetherian local ring R with separably closed residue field. By 5.7.2 we are even reduced to the case when R is artinian local with separably closed residue field. In this case (G, M_S, P, M_P, L) is obtained from the standard construction from data as in 5.2.1 and some map $\beta : H_S^{\mathrm{sat}} \to R$ sending all nonzero elements to the nilradical of R. Note that this implies in particular that β factors through $\overline{H}_S^{\mathrm{sat}}$, where we define

$$\overline{H}_S^{\mathrm{sat}} := H_S^{\mathrm{sat}}/(torsion). \qquad (6.3.34.1)$$

Let T denote the toric part of G.

6.3.35. By 6.3.18 there exists some $\alpha \in A$ such that $\iota_A^* \mathscr{M}$ is isomorphic to $\mathscr{M} \otimes L_\alpha$, where $L_\alpha := t_\alpha^* \mathscr{M} \otimes \mathscr{M}^{-1}$. After making an fppf base change on S we can find an element $\alpha' \in A$ with $2\alpha' = \alpha$. Then

$$\iota_A^*(\mathscr{M} \otimes L_{\alpha'}) \simeq \iota_A^* \mathscr{M} \otimes L_{-\alpha'} \simeq (\mathscr{M} \otimes L_\alpha) \otimes L_{\alpha'}^{-1} \simeq \mathscr{M} \otimes L_{\alpha'}. \qquad (6.3.35.1)$$

Therefore by replacing $\mathscr{M}$ by $\mathscr{M} \otimes L_{\alpha'}$ we may assume that $\iota^* \mathscr{M}$ is isomorphic to $\mathscr{M}$. Fix one such isomorphism

$$\iota_{\mathscr{M}} : \iota_A^* \mathscr{M} \to \mathscr{M}. \qquad (6.3.35.2)$$

Let $\mathcal{R}$ denote the sheaf of algebras over A defined in 5.2.4. We construct an algebra isomorphism

$$p : \iota_A^* \mathcal{R} \to \mathcal{R} \qquad (6.3.35.3)$$

as follows. For every $x \in X$ note that $\iota_A^* L_x$ is rigidified at 0, and therefore there exists a unique isomorphism of rigidified line bundles

$$p_x : \iota_A^* L_x \to L_{-x}. \qquad (6.3.35.4)$$

For $x, z \in X$ the diagram

$$
\begin{array}{ccc}
\iota_A^*(L_x \otimes L_z) & \xrightarrow{p_x \otimes p_z} & L_{-x} \otimes L_{-z} \\
{\scriptstyle can} \downarrow & & \downarrow {\scriptstyle can} \\
\iota_A^*(L_{x+z}) & \xrightarrow{\quad p_{x+z} \quad} & L_{-x-z}
\end{array}
\qquad (6.3.35.5)
$$

commutes as it is a diagram of rigidified line bundles.

We define p to be the map

$$\iota_A^* \mathcal{R} \simeq \oplus_{(d,x) \in P} \iota_A^*(\mathscr{M}^d \otimes L_x) \xrightarrow{\oplus \iota_{\mathscr{M}}^d \otimes p_x} \oplus_{(d,x) \in P} \mathscr{M}^d \otimes L_x \simeq \mathcal{R}. \qquad (6.3.35.6)$$

The commutativity of the diagram 6.3.35.5 and the following 6.3.36 ensures that this in fact is an isomorphism of graded algebras. We therefore obtain an isomorphism of polarized schemes

$$\tilde{\rho} : (\widetilde{P}, \mathscr{O}_{\widetilde{P}}(1)) \to (\widetilde{P}, \mathscr{O}_{\widetilde{P}}(1)) \times_{A, \iota} A. \tag{6.3.35.7}$$

Lemma 6.3.36 *(i) The involution $\iota : X_{\mathbb{R}} \to X_{\mathbb{R}}$ sending x to $-x$ preserves the paving S (i.e. for every $\omega \in S$, the image $\iota(\omega)$ is a cell of S). In particular, the map $(d, x) \mapsto (d, -x)$ induces a map $\iota : H_S \to H_S$.*

(ii) The induced map $\bar{\iota} : \overline{H}_S^{\text{sat}} \to \overline{H}_S^{\text{sat}}$ is the identity. Equivalently, for any $x, z \in X$ and $d, d' > 0$ the two elements

$$(d, x) * (d', z), \ (d, -x) * (d', -z) \in H_S \tag{6.3.36.1}$$

become equal in $\overline{H}_S^{\text{sat}}$.

Proof. For (i), let $B : S^2 X \to \mathbb{Q}$ be a quadratic positive semidefinite form defining S, with associated quadratic function $a : X \to \mathbb{Q}$ sending x to $B(x, x)/2$. Let $g : X_{\mathbb{R}} \to \mathbb{R}$ be the function defined by the lower envelope of the convex hull of the set

$$G_a := \{(x, a(x)) | x \in X\} \subset X_{\mathbb{R}} \times \mathbb{R} \tag{6.3.36.2}$$

so that S is the domains of linearity of g. Then $\iota(S)$ is the domains of linearity of the function $x \mapsto g(-x)$. This function is in fact equal to g because $a(-x) = a(x)$ for all x which implies that G_a is invariant under the transformation $x \mapsto -x$. This proves (i).

For (ii) note that by 4.1.8 we have an inclusion $\overline{H}_S^{\text{sat}} \to H_S^{gp} \otimes \mathbb{Q}$. It therefore suffices to show that the map $\iota : H_S^{gp} \otimes \mathbb{Q} \to H_S^{gp} \otimes \mathbb{Q}$ is the identity. It follows from the definitions, that via the isomorphism $s : S^2 X_{\mathbb{Q}} \simeq H_S^{gp} \otimes \mathbb{Q}$ defined in 5.8.15 the map ι corresponds to the map

$$S^2 X_{\mathbb{Q}} \to S^2 X_{\mathbb{Q}}, \ x \otimes y \mapsto (-x) \otimes (-y) \tag{6.3.36.3}$$

which is the identity map. □

6.3.37. Let $P \rtimes \overline{H}_S^{\text{sat}}$ denote the pushout of the diagram

$$\begin{array}{ccc} H_S & \longrightarrow & \overline{H}_S^{\text{sat}} \\ \downarrow & & \\ P \rtimes H_S. & & \end{array} \tag{6.3.37.1}$$

Then it follows that there is an involution $\iota : P \rtimes \overline{H}_S^{\text{sat}} \to P \rtimes \overline{H}_S^{\text{sat}}$ induced by $-1 : X \to X$ which is equal to the identity on $\overline{H}_S^{\text{sat}}$.

Let $U \subset A$ be an open subset over which the sheaves $\mathcal{M}$ and L_x are trivial, and let U^ι denote $\iota(U)$. Choose trivializations of $\mathcal{M}|_U$ and the $L_x|_U$'s over U compatible with the isomorphisms

$$L_x \otimes L_z \to L_{x+z}. \tag{6.3.37.2}$$

These trivializations then also induce (via the isomorphism $\iota^* \mathcal{M} \to \mathcal{M}$ and the p_x's) trivializations of $\mathcal{M}|_{\iota(U)}$ and $L_x|_{\iota(U)}$. With these choices the restriction $\widetilde{P}_U$ of $\widetilde{P}$ to U is given by

$$\underline{\mathrm{Proj}}_U(\mathscr{O}_U \otimes_{\mathbb{Z}[\overline{H}_S^{\mathrm{sat}}]} \mathbb{Z}[P \rtimes \overline{H}_S^{\mathrm{sat}}]), \tag{6.3.37.3}$$

and similarly $\widetilde{P}_{\iota(U)}$ is given by

$$\underline{\mathrm{Proj}}_{\iota(U)}(\mathscr{O}_{\iota(U)} \otimes_{\mathbb{Z}[\overline{H}_S^{\mathrm{sat}}]} \mathbb{Z}[P \rtimes \overline{H}_S^{\mathrm{sat}}]), \tag{6.3.37.4}$$

and the map $p : \iota_A^* \mathcal{R} \to R$ is induced by the map of monoids $\iota : P \rtimes \overline{H}_S^{\mathrm{sat}} \to P \rtimes \overline{H}_S^{\mathrm{sat}}$. This extends the morphism p to an isomorphism of log schemes

$$\tilde{\rho} : (\widetilde{P}, M_{\widetilde{P}}, \mathscr{O}_{\widetilde{P}}(1)) \to (\widetilde{P}, M_{\widetilde{P}}, \mathscr{O}_{\widetilde{P}}(1)) \times_{A, \iota} A. \tag{6.3.37.5}$$

It follows from the construction that these isomorphisms defined locally on A glue to give a global isomorphism.

To complete the proof of 6.3.33 it remains to see that this isomorphism $\tilde{\rho}$ is compatible with the translation action of Y and the G-action.

6.3.38. To prove the compatibility with the G-actions, recall that G can be viewed as the functor classifying data $g = (a, \{\lambda_x\})$, where $a \in A$ is a point and $\lambda_x : t_a^* L_x \to L_x$ are isomorphisms compatible with the isomorphisms $L_x \otimes L_z \to L_{x+z}$. For such data g, the inverse point $\iota(g)$ is the point $(-a, \{\lambda_x^\perp\})$, where $\lambda_x^\perp : t_{-a}^* L_x \to L_x$ is the inverse of the isomorphism

$$L_x \xrightarrow{\simeq} t_{-a}^* t_a^* L_x \xrightarrow{t_{-a}^* \lambda_x} t_{-a}^* L_x. \tag{6.3.38.1}$$

To prove the compatibility with the G-actions it suffices to show that the diagram

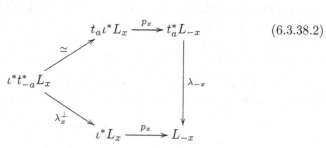

$$\tag{6.3.38.2}$$

commutes (in fact it suffices to show that this diagram commutes up to multiplication by a scalar independent of x but we prove the stronger result).

If $a = \{e\}$, then the diagram commutes by the definition of $\lambda_x^\perp$. In fact if $a = \{e\}$, then λ_x differs from the identity morphism by a scalar and the map $\lambda_x^\perp$ differs from the identity morphism by the inverse of that scalar. Since the diagram

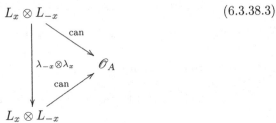

$$L_x \otimes L_{-x} \qquad (6.3.38.3)$$

commutes it follows that $\lambda_x^\perp = \lambda_{-x}$.

By taking the difference of the two composites going around the diagram 6.3.38.2 we obtain a morphism of schemes

$$F : G \to \mathbb{G}_m. \qquad (6.3.38.4)$$

This map is defined by a global section $u \in H^0(G, \mathcal{O}_G^*)$. To complete the proof of the compatibility with the G-action it suffices to show that $u = 1$. By the commutativity of 6.3.38.2 when $g \in T$ we know that the image of u in $H^0(T, \mathcal{O}_T^*)$ is equal to 1. The compatibility with the G-action therefore follows from the following lemma.

Lemma 6.3.39 *The map*

$$H^0(G, \mathcal{O}_G) = \oplus_{x \in X} H^0(A, L_x) \to \oplus_{x \in X} \Gamma(S, e^* L_x) = \Gamma(T, \mathcal{O}_T) \quad (6.3.39.1)$$

is injective.

Proof. By a standard devissage using the flatness of G and T over S one reduces to the case when S is the spectrum of an algebraically closed field. In this case the result is clear, for $H^0(A, L_x)$ is nonzero if and only if $L_x \simeq \mathcal{O}_A$ in which case $H^0(A, L_x)$ has dimension 1. □

6.3.40. Finally we check the compatibility with the Y-actions. Fix an element $y \in Y$. First note that the maps

$$t_{-y}^* p_x : t_{-y}^* \iota^* L_x \to t_{-y}^* L_{-x} \qquad (6.3.40.1)$$

can be described as follows. Giving such a map is equivalent to giving an isomorphism between their fibers over $0 \in A$, or equivalently an isomorphism

$$L_x(y) \to L_{-x}(-y). \qquad (6.3.40.2)$$

We have already encountered such a map. Namely, let $\mathcal{B} \to A \times A^t$ denote the Poincare biextension, and recall that in 2.2.6 a map

$$\sigma : \mathscr{B} \to (-1)^* \mathscr{B} \tag{6.3.40.3}$$

was constructed. This map σ defines for every $(y, x) \in A \times A^t$ an isomorphism

$$\sigma_{y,x} : L_x(y) = \mathscr{B}_{(y,x)} \to \mathscr{B}_{(-y,-x)} = L_{-x}(-y). \tag{6.3.40.4}$$

If we fix $x \in X$ we therefore obtain two sections of the $\mathbb{G}_m$-torsor $\mathscr{B}_{(-,x)}$ over A. The first is given by the maps $\sigma_{y,x}$ and the second is given by the maps 6.3.40.1 (note that these maps are defined for any $y \in A$). These two sections define a morphism of schemes

$$A \to \mathbb{G}_m \tag{6.3.40.5}$$

which takes the identity to 1 (this follows from the definition of σ). It follows that in fact 6.3.40.1 is equal to the map defined by $\sigma_{y,x}$.

Lemma 6.3.41 *For any $y \in Y$ and $x \in X$, the diagram*

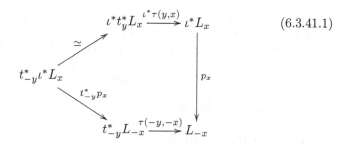

$$\tag{6.3.41.1}$$

commutes.

Proof. Let E^{triv} denote the trivial biextension of $Y \times X$ by $\mathbb{G}_m$. By the above discussion the commutativity of 6.3.41.1 is equivalent to the commutativity of the diagram of biextensions

$$
\begin{array}{ccc}
E^{\mathrm{triv}} & \xrightarrow{\sigma_{E^{\mathrm{triv}}}} & E^{\mathrm{triv}} \\
{\scriptstyle \tau}\downarrow & & \downarrow{\scriptstyle \tau} \\
\mathscr{B} & \xrightarrow{\sigma_{\mathscr{B}}} & \mathscr{B},
\end{array}
\tag{6.3.41.2}
$$

and the observation that $\sigma_{E^{\mathrm{triv}}}$ on $\mathbb{G}_m \times Y \times X$ is just the map $(\mathrm{id}, -1, -1)$. $\square$

Remark 6.3.42. The restriction to $A \times A$ of the isomorphism of biextensions $\sigma : \mathscr{B} \to (-1)^* \mathscr{B}$ over $A \times A^t$ can be described explicitly as follows. For any point $(y, x) \in A \times A$ the fiber $\mathscr{B}_{y, \lambda_{\mathscr{M}}(x)}$ is equal to

$$\mathscr{M}(x + y) \otimes \mathscr{M}(x)^{-1} \otimes \mathscr{M}(y)^{-1}, \tag{6.3.42.1}$$

and $((-1)^* \mathscr{B})_{(y, \lambda_{\mathscr{M}}(x))}$ is equal to

$$\mathcal{M}(-x-y) \otimes \mathcal{M}(-x)^{-1} \otimes \mathcal{M}(-y)^{-1}. \tag{6.3.42.2}$$

The fixed isomorphism $\iota^*\mathcal{M} \to \mathcal{M}$ defines for every a an isomorphism between $\mathcal{M}(a)$ and $\mathcal{M}(-a)$, and hence we get for every $(y,x) \in A \times A$ an isomorphism between 6.3.42.1 and 6.3.42.2. In this way we obtain an isomorphism $\rho : \mathcal{B} \to (-1)^*\mathcal{B}$ over $A \times A$ (a priori not respecting the biextension structure, but compatible with the $\mathbb{G}_m$-action). The difference between ρ and σ is a map $A \times A \to \mathbb{G}_m$, which as before must be constant. Since ρ and σ agree over $(e,e) \in A \times A$ it follows that $\rho = \sigma$.

6.3.43. Let ψ' be the trivialization

$$Y \to c^{t*}\mathcal{M}^{-1} \tag{6.3.43.1}$$

sending $y \in Y$ to the element $\psi(-y)$ of

$$\mathcal{M}^{-1}(-y) \simeq (\iota^*\mathcal{M}^{-1})(y) \xrightarrow{\iota_\mathcal{M}} \mathcal{M}^{-1}(y). \tag{6.3.43.2}$$

It follows from the discussion in 6.3.42 that the induced trivialization of $\mathcal{B}|_{Y \times X}$ is then equal to $\sigma^{-1} \circ \tau \circ \sigma_{E^{\mathrm{triv}}}$ which by the above is equal to τ. We conclude that ψ and ψ' differ by a homomorphism $h : Y \to \mathbb{G}_m$.

After making a flat base change on S, we can extend h to a homomorphism $\tilde{h} : X \to \mathbb{G}_m$. Let ρ' denote the isomorphism

$$(\widetilde{P}, M_{\widetilde{P}}, \mathscr{O}_{\widetilde{P}}(1)) \to (\widetilde{P}, M_{\widetilde{P}}, \mathscr{O}_{\widetilde{P}}(1)) \times_{A,\iota} A \tag{6.3.43.3}$$

obtained by composing ρ with $\tilde{h} \in T(S)$, and let

$$\rho'_x : \iota^*L_x \to L_{-x} \tag{6.3.43.4}$$

denote the isomorphism $\tilde{h}(x) \cdot \rho_x$ so that ρ' is obtained by summing the maps

$$\iota_{\mathcal{M}}^{\otimes d} \otimes \rho'_x : \iota^*(\mathcal{M}^d \otimes L_x) \to \mathcal{M}^d \otimes L_{-x}. \tag{6.3.43.5}$$

By the definition of $\tilde{h}$, the diagram

$$\begin{array}{ccc}
\iota^*(t_y^*\mathcal{M}) & \xrightarrow{\psi(y)} & \iota^*(\mathcal{M} \otimes L_y) \\
\nearrow\scriptstyle{\simeq} & & \Big\downarrow{\scriptstyle \iota_\mathcal{M} \otimes \rho'_y} \\
t_{-y}^*\iota^*\mathcal{M} & & \\
\searrow\scriptstyle{\iota_\mathcal{M}} & & \\
& t_{-y}^*\mathcal{M} \xrightarrow{\psi(-y)} & \mathcal{M} \otimes L_{-y}
\end{array} \tag{6.3.43.6}$$

commutes.

Replacing ρ by ρ' we then find that for every $(d,x) \in P$ and $y \in Y$ the diagram

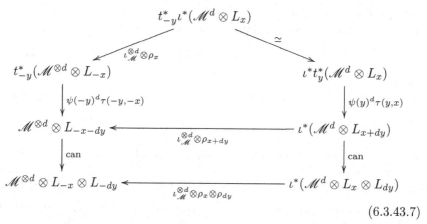

$$(6.3.43.7)$$

commutes. Indeed the square commutes by the commutativity of 6.3.35.5, and the outside pentagon commutes as it is obtained by tensoring d-copies of 6.3.43.6 with a copy of 6.3.41.1. This implies that the diagram of log schemes with line bundles

$$
\begin{array}{ccc}
Y \times (\widetilde{P}, M_{\widetilde{P}}, \mathscr{O}_{\widetilde{P}}(1)) & \xrightarrow{\text{action}} & (\widetilde{P}, M_{\widetilde{P}}, \mathscr{O}_{\widetilde{P}}(1)) \\
{\scriptstyle (-1)\times\rho} \big\downarrow & & \big\downarrow {\scriptstyle \rho} \\
Y \times (\widetilde{P}^{\iota}, M_{\widetilde{P}^{\iota}}, \mathscr{O}_{\widetilde{P}}(1)) & \xrightarrow{\text{action}} & (\widetilde{P}^{\iota}, M_{\widetilde{P}^{\iota}}, \mathscr{O}_{\widetilde{P}}(1))
\end{array}
\qquad (6.3.43.8)
$$

commutes. Therefore ρ descends to an isomorphism

$$(P, M_P, L) \to (P^{\iota}, M_{P^{\iota}}, L) \qquad (6.3.43.9)$$

thereby completing the proof of 6.3.33. □

By the same argument proving 6.3.19 we also obtain the following:

Proposition 6.3.44. *Let S be a scheme, (G, M_S, P, M_P, L) an object of $\overline{\mathscr{T}}_{g,\delta}(S)$, and $\sigma : \mathscr{G}(\delta)_S \to \mathscr{G}_{(P,M_P,L)}$ an isomorphism restricting to the identity on $\mathbb{G}_m$. Then fppf locally on S there exists an isomorphism $\rho = (\rho_{(P,M_P)}, \rho_L) : ((P, M_P), L) \to ((P^{\iota}, M_{P^{\iota}}), L)$ over $\iota : G \to G$ such that the diagram*

$$
\begin{array}{ccc}
\mathscr{G}(\delta)_S & \xrightarrow{\sigma} & \mathscr{G}_{(P,M_P,L)} \\
{\scriptstyle \tau} \big\downarrow & & \big\downarrow {\scriptstyle \rho} \\
\mathscr{G}(\delta)_S & \xrightarrow{\sigma} & \mathscr{G}_{(P^{\iota},M_{P^{\iota}},L)}
\end{array}
\qquad (6.3.44.1)
$$

commutes, where the right vertical arrow is the isomorphism induced by ρ. Moreover ρ is unique up to composition with the action of an element of $\mathbb{G}_m$ (where $u \in \mathbb{G}_m$ acts as the identity on (P, M_P) and multiplication by u on L).

6.3.45. Let $\overline{\Sigma}_{g,\delta}$ denote the stack over $\mathbb{Z}[1/2d]$ which to any $\mathbb{Z}[1/2d]$-scheme S associates the groupoid of data

$$(G, M_S, P, M_P, L, \sigma), \qquad (6.3.45.1)$$

where

$$(G, M_S, P, M_P, L) \in \overline{\mathscr{T}}_{g,\delta}(S) \qquad (6.3.45.2)$$

and

$$\sigma : \mathscr{G}(\delta)_S \to \mathscr{G}_{(P,M_P,L)} \qquad (6.3.45.3)$$

is an isomorphism of group schemes inducing the identity on $\mathbb{G}_m$. As in 6.3.23 the forgetful map

$$\overline{\Sigma}_{g,\delta} \to \overline{\mathscr{T}}_{g,\delta}[1/2d] \qquad (6.3.45.4)$$

is finite and étale (in particular $\overline{\Sigma}_{g,\delta}$ is algebraic). Let $M_{\overline{\Sigma}_{g,\delta}}$ denote the pullback to $\overline{\Sigma}_{g,\delta}$ of the log structure on $\overline{\mathscr{T}}_{g,\delta}[1/2d]$ so that $(\overline{\Sigma}_{g,\delta}, M_{\overline{\Sigma}_{g,\delta}})$ is log smooth over $\mathbb{Z}[1/2d]$.

6.3.46. Let

$$(G_u, M_{\overline{\Sigma}_{g,\delta}}, \overline{P}_u, M_{\overline{P}_u}, L_u, \sigma_u) \qquad (6.3.46.1)$$

denote the universal data over $\overline{\Sigma}_{g,\delta}$. By 6.3.44 there is a canonical isomorphism

$$\rho : (\overline{P}_u, M_{\overline{P}_u}) \to (\overline{P}_u, M_{\overline{P}_u}) \qquad (6.3.46.2)$$

over $(\overline{\Sigma}, M_{\overline{\Sigma}})$. Let $(\overline{P}_u, M_{\overline{P}_u})^\rho$ denote the fiber product of the diagram of fine log stacks

$$(\overline{P}_u, M_{\overline{P}_u})$$
$$\downarrow {\scriptstyle \mathrm{id} \times \rho} \qquad\qquad (6.3.46.3)$$
$$(\overline{P}_u, M_{\overline{P}_u}) \xrightarrow{\;\Delta\;} (\overline{P}_u, M_{\overline{P}_u}) \times_{(\overline{\Sigma}_{g,\delta}, M_{\overline{\Sigma}_{g,\delta}})} (\overline{P}_u, M_{\overline{P}_u}).$$

Recall (see 2.3.34) that the forgetful functor from integral log schemes to schemes does not preserve fiber products. To understand the log stack $(\overline{P}_u, M_{\overline{P}_u})^\rho$ we now compare it to the scheme-theoretic fixed points.

Lemma 6.3.47 *Let M be an integral monoid and $\rho : M \to M$ an automorphism with $\rho^2 = \mathrm{id}$.*
(i) Let $E \subset M \times M$ be the set of elements of (m, m') such that there exists an element $k \in M$ with $m + k = m' + \rho(k)$. Then E is a congruence relation on M (i.e. E is a submonoid of $M \times M$ and defines an equivalence relation on the underlying set of M).
(ii) Let M_ρ be the colimit in the category of integral monoids of the diagram

$$M \xrightarrow{\;\mathrm{id}\;} M$$
$$\rho \downarrow \qquad\qquad (6.3.47.1)$$
$$M.$$

Then M_ρ is the quotient of M by the congruence relation E.

Proof. The subset $E \subset M \times M$ is a submonoid: If

$$m + k = m' + \rho(k) \quad \text{and} \quad n + l = n' + \rho(l) \tag{6.3.47.2}$$

then

$$m + n + (k + l) = m' + n' + \rho(k) + \rho(l) = m' + n' + \rho(k + l) \tag{6.3.47.3}$$

so $(m + n, m' + n') \in E$.

The subset $E \subset M \times M$ is also an equivalence relation:

(i) $(m, m) \in E$ (take $k = 0$);
(ii) If $m + k = m' + \rho(k)$ then if $k' := \rho(k)$ we have $\rho(k') = k$ (since $\rho^2 = \text{id}$) which implies that $m' + k' = m + \rho(k')$. Therefore if $(m, m') \in E$ we have $(m', m) \in E$);
(iii) If $m + k = m' + \rho(k)$ and $m' + k' = m'' + \rho(k')$ then

$$m + k + k' = m' + \rho(k) + k' = m'' + \rho(k) + \rho(k') = m'' + \rho(k + k'). \tag{6.3.47.4}$$

Let M' denote the quotient of M by E. Then there is a canonical map $M' \to M_\rho$ so it suffices to show that M' is integral. If $m, n, n' \in M$ are elements with images $\bar{m}, \bar{n}, \bar{n}' \in M'$ then if

$$\bar{m} + \bar{n} = \bar{m} + \bar{n}' \tag{6.3.47.5}$$

there exists an element $k \in M$ with

$$m + n + k = m + n' + \rho(k). \tag{6.3.47.6}$$

Since M is integral this implies that $n + k = n' + \rho(k)$ so $\bar{n} = \bar{n}'$. $\quad\square$

6.3.48. Let $\overline{P}_u^\rho$ be the fiber product of the diagram

$$
\begin{array}{ccc}
 & & \overline{P}_u \\
 & & \downarrow {\scriptstyle \text{id} \times \rho} \\
\overline{P}_u & \xrightarrow{\ \Delta\ } & \overline{P}_u \times_{\overline{\Sigma}_{g,\delta}} \overline{P}_u.
\end{array}
\tag{6.3.48.1}
$$

Note that $\overline{P}_u^\rho$ may differ from the underlying stack of $(\overline{P}_u, M_{\overline{P}_u})^\rho$.

Since $\overline{P}_u$ is separated and representable over $\overline{\Sigma}_{g,\delta}$ the map Δ is a closed immersion whence the functor

$$j : \overline{P}_u^\rho \to \overline{P}_u \tag{6.3.48.2}$$

is also a closed immersion. Since the diagram

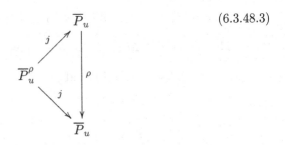

(6.3.48.3)

commutes, the map $\rho^b : \rho^* M_{\overline{P}_u} \to M_{\overline{P}_u}$ induces an automorphism, which we again denote by ρ, of the pullback $j^* M_{\overline{P}_u}$. Note that by construction $\rho^2 = \mathrm{id}$.

Lemma 6.3.49 *Let (U, M_U) be a fine saturated log scheme and let $\rho : M_U \to M_U$ be an automorphism of the log structure M_U such that $\rho^2 = \mathrm{id}$. Then étale locally on U there exists a triple (N, β, ι), where*

(i) N is a fine saturated monoid;
(ii) $\beta : N \to M_U$ is a chart;
(iii) $\iota : N \to N$ is an automorphism such that the diagram

$$
\begin{array}{ccc}
N & \xrightarrow{\;\beta\;} & M_U \\
{\scriptstyle\iota}\downarrow & & \downarrow{\scriptstyle\rho} \\
N & \xrightarrow{\;\beta\;} & M_U
\end{array}
\qquad (6.3.49.1)
$$

commutes.

Proof. Let $\bar{x} \to U$ be a geometric point, and let Q denote the monoid $\overline{M}_{U,\bar{x}}$, and let $\rho_Q : Q \to Q$ be the automorphism defined by ρ. Since M_U is saturated, there exists a section $s : Q \to M_{U,\bar{x}}$. Let Q' denote the monoid $Q \oplus Q$ and define $s' : Q' \to M_{U,\bar{x}}$ to be the map

$$
(q_1, q_2) \mapsto s(\rho_Q(q_1)) + \rho(s(q_2)). \qquad (6.3.49.2)
$$

Let $\iota : Q' \to Q'$ be the automorphism

$$
(q_1, q_2) \mapsto (\rho_Q(q_2), \rho_Q(q_1)). \qquad (6.3.49.3)
$$

Then the diagram

$$
\begin{array}{ccc}
Q' & \xrightarrow{\;s'\;} & M_{U,\bar{x}} \\
{\scriptstyle\iota}\downarrow & & \downarrow{\scriptstyle\rho} \\
Q' & \xrightarrow{\;s'\;} & M_{U,\bar{x}}
\end{array}
\qquad (6.3.49.4)
$$

commutes. Set

$$
N := Q'^{gp} \times_{M_{U,\bar{x}}^{gp}} M_{U,\bar{x}}, \qquad (6.3.49.5)
$$

let $\iota : N \to N$ be the automorphism induced by ι on Q'^{gp}, and let $\beta : N \to M_{U,\bar{x}}$ be the projection. Then by [24, 2.10] the map β extends to a chart in some étale neighborhood of $\bar{x}$ and the diagram 6.3.49.1 will also commute in some étale neighborhood since it commutes at $\bar{x}$.   □

6.3.50. There is a natural projection

$$z : (\overline{P}_u, M_{\overline{P}_u})^\rho \to (\overline{P}_u^\rho, j^* M_{\overline{P}_u}). \tag{6.3.50.1}$$

Let $c : U \to \overline{P}_u^\rho$ be a smooth morphism from a scheme such that there exists a triple (N, β, ι) as in 6.3.49 for $c^* j^* M_{\overline{P}_u}$. Then the fiber product of the diagram

$$
\begin{array}{c}
(U, c^* j^* M_{\overline{P}_u}) \\
\downarrow c \\
(\overline{P}_u, M_{\overline{P}_u})^\rho \xrightarrow{\ z\ } (\overline{P}_u^\rho, j^* M_{\overline{P}_u})
\end{array}
\tag{6.3.50.2}
$$

is equal to the fiber product of the diagram

$$
\begin{array}{c}
(U, c^* j^* M_{\overline{P}_u}) \\
\downarrow \beta \\
\mathrm{Spec}(N_\iota \to \mathbb{Z}[N_\iota]) \longrightarrow \mathrm{Spec}(N \to \mathbb{Z}[N]),
\end{array}
\tag{6.3.50.3}
$$

where N_ι is as in 6.3.47.

Corollary 6.3.51 *The map 6.3.50.1 is a closed immersion (in the logarithmic sense 2.3.12). In particular the underlying morphism of stacks is a closed immersion.*

6.3.52. Write $\overline{P}_u^{\rho,\log}$ for the underlying stack of $(\overline{P}_u, M_{\overline{P}_u})^\rho$ and $M_{\overline{P}_u^{\rho,\log}}$ for the log structure so by definition we have

$$(\overline{P}_u, M_{\overline{P}_u})^\rho = (\overline{P}_u^{\rho,\log}, M_{\overline{P}_u^{\rho,\log}}). \tag{6.3.52.1}$$

Let

$$i : (\overline{P}_u^{\rho,\log}, M_{\overline{P}_u^{\rho,\log}}) \to (\overline{P}_u, M_{\overline{P}_u}) \tag{6.3.52.2}$$

be the inclusion so we have a commutative diagram

$$(\overline{P}_u^{\rho,\log}, M_{\overline{P}_u^{\rho,\log}}) \xrightarrow{\ z\ } (\overline{P}_u^\rho, j^* M_{\overline{P}_u}) \xrightarrow{\ j\ } (\overline{P}_u, M_{\overline{P}_u}). \tag{6.3.52.3}$$

$$i$$

6.3.53. If S is a scheme, the groupoid $\overline{P}_u^{\rho,\log}(S)$ is the groupoid of data

$$(G, M_S, P, M_P, L, \sigma, t : M_S \to M_S', p), \tag{6.3.53.1}$$

where

(i) $(G, M_S, P, M_P, L, \sigma)$ is an object of $\overline{\Sigma}_{g,\delta}$;

(ii) $t : M_S \to M'_S$ is a morphism of fine log structures on S;

(iii) $p : (S, M'_S) \to (P, M_P)$ is a morphism of log schemes such that the diagram

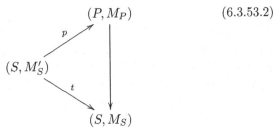

(6.3.53.2)

commutes. In addition we require that $\rho \circ p = p$ and that the map $p^* M_P \to M'_S$ identifies M'_S with the colimit (in the category of integral log structures) of the diagram

$$p^* M_P \overset{\rho}{\underset{\mathrm{id}}{\rightrightarrows}} p^* M_P.$$

(6.3.53.3)

The log structure $M_{\overline{P}^{\rho,\log}_u}$ on $\overline{P}^{\rho,\log}_u$ is the log structure defined by M'_S.

6.3.54. Note in particular that any object of $\overline{P}^{\rho,\log}_u$ admits an action of $\mathbb{G}_m$ (acting on the line bundle). Let $\overline{\Theta}_{g,\delta}$ denote the rigidification of $\overline{P}^{\rho,\log}_u$ with respect to this $\mathbb{G}_m$-action. As in 5.11.11 the log structure $M_{\overline{P}^{\rho,\log}_u}$ descends to a log structure $M_{\overline{\Theta}_{g,\delta}}$ on $\overline{\Theta}_{g,\delta}$.

We can now state the main result of this section:

Theorem 6.3.55 *(i) The stack $\overline{\Theta}_{g,\delta}$ is proper over $\mathbb{Z}[1/2d]$;*

(ii) The log stack $(\overline{P}^{\rho,\log}_u, M_{\overline{P}^{\rho,\log}_u})$, and hence also $(\overline{\Theta}_{g,\delta}, M_{\overline{\Theta}_{g,\delta}})$, is log smooth over $\mathbb{Z}[1/2d]$;

(iii) The natural map $\Theta_{g,\delta} \to \overline{\Theta}_{g,\delta}$ is a dense open immersion, and identifies $\Theta_{g,\delta}$ with the maximal open substack over which $M_{\overline{\Theta}_{g,\delta}}$ is trivial.

The proof is in several steps 6.3.56–6.3.64.

6.3.56. That $\overline{\Theta}_{g,\delta}$ is proper over $\mathbb{Z}[1/2d]$ can be seen as follows. Let

$$\overline{\Sigma}_{g,\delta}//\mathbb{G}_m \quad \text{and} \quad \overline{\mathscr{T}}_{g,\delta}[1/2d]//\mathbb{G}_m$$

(6.3.56.1)

denote the rigidifications with respect to $\mathbb{G}_m$ of $\overline{\Sigma}_{g,\delta}$ and $\overline{\mathscr{T}}_{g,\delta}[1/2d]$ respectively, where $\overline{\Sigma}_{g,\delta}$ is defined as in 6.3.45. Let $\mathscr{H}$ denote the finite flat group scheme over $\overline{\mathscr{A}}_{g,\delta}[1/2d]$ defined in 5.11.11. Then $\overline{\mathscr{T}}_{g,\delta}[1/2d]//\mathbb{G}_m$ is a $\mathscr{H}$-gerbe over

$\overline{\mathscr{A}}_{g,\delta}[1/2d]$, and in particular is proper over $\mathbb{Z}[1/2d]$. The properness of $\overline{\Theta}_{g,\delta}$ over $\mathbb{Z}[1/2d]$ then follows from the observation that there is a diagram

$$\overline{\Theta}_{g,\delta} \xrightarrow{\;v\;} \overline{\Sigma}_{g,\delta}/\!/\mathbb{G}_m \xrightarrow{\;w\;} \overline{\mathscr{T}}_{g,\delta}[1/2d]/\!/\mathbb{G}_m, \qquad (6.3.56.2)$$

where v is proper and w is finite. It follows that $\overline{\Theta}_{g,\delta}$ is proper over $\mathbb{Z}[1/2d]$.

It is also clear that the map $\Theta_{g,\delta} \to \overline{\Theta}_{g,\delta}$ identifies $\Theta_{g,\delta}$ with the maximal open substack where $M_{\overline{\Theta}_{g,\delta}}$ is trivial. If $(\overline{\Theta}_{g,\delta}, M_{\overline{\Theta}_{g,\delta}})$ is log smooth over $\mathbb{Z}[1/2d]$ then this open substack is also dense. To prove 6.3.55 it therefore suffices to prove that $(\overline{P}_u^{\rho,\log}, M_{\overline{P}_u^{\rho,\log}})$ is log smooth over $\mathbb{Z}[1/2d]$.

6.3.57. To verify the log smoothness of $(\overline{P}_u^{\rho,\log}, M_{\overline{P}_u^{\rho,\log}})$ we verify the infinitesimal lifting property.

Let $R' \to R$ be a surjection of artinian local rings and let R_0 be the residue field of R. Assume that the kernel $J \subset R'$ of $R' \to R$ is annihilated by the maximal ideal of R' and hence can be viewed as a R_0-module. Fix data as in 5.2.1 over R' and a map $\beta : H_S^{\mathrm{sat}} \to R'$ sending all nonzero elements to the maximal ideal, and let

$$(G', M_{R'}, P', M_{P'}, L') \qquad (6.3.57.1)$$

denote the resulting standard family over $\mathrm{Spec}(R')$. Let (G, M_R, P, M_P, L) (resp. $(G_0, M_{R_0}, P_0, M_{P_0}, L_0)$) denote the reduction to R (resp. R_0). Using 6.3.18 we may assume that the line bundle $\mathscr{M}$ on A is symmetric and fix an isomorphism $\iota_{\mathscr{M}} : \iota^* \mathscr{M} \to \mathscr{M}$.

Fix also the following data:

(i) A morphism of log structures $t' : M_{R'} \to M'_{R'}$. Let $t : M_R \to M'_R$ (resp. $t_0 : M_{R_0} \to M'_{R_0}$) be the reduction of t' to R (resp. R_0).

(ii) An isomorphism $\sigma : \mathscr{G}(\delta)_R \to \mathscr{G}_{(P,M_P,L)}$ inducing the identity on $\mathbb{G}_m$. By 6.3.44 this defines a morphism $\rho : (P, M_P, L) \to (P, M_P, L)$.

(iii) A morphism $p : (\mathrm{Spec}(R), M'_R) \to (P, M_P)$ over $(\mathrm{Spec}(R), M_R)$ such that $\rho \circ p = \rho$ and such that the map to M'_R identifies M'_R with the colimit of the diagram

$$p^* M_P \;\overset{\rho}{\underset{\mathrm{id}}{\rightrightarrows}}\; p^* M_P. \qquad (6.3.57.2)$$

This data defines a commutative diagram of solid arrows

$$\begin{array}{ccc}
(\mathrm{Spec}(R), M'_R) & \xrightarrow{\;a\;} & (\overline{P}_u^{\rho,\log}, M_{\overline{P}_u^{\rho,\log}}) \\
\downarrow{\scriptstyle b} & \overset{m}{\nearrow} & \downarrow{\scriptstyle d} \\
(\mathrm{Spec}(R'), M'_{R'}) & \xrightarrow{\;c\;} & (\overline{\mathscr{T}}_{g,\delta}, M_{\overline{\mathscr{T}}_{g,\delta}}).
\end{array} \qquad (6.3.57.3)$$

We show that there exists a morphism m filling in the diagram.

6.3.58. By 6.3.7, the automorphism group scheme of $\mathscr{G}(\delta)$ is étale over $\mathbb{Z}[1/2d]$, and therefore the isomorphism σ lifts uniquely to an isomorphism

$$\sigma' : \mathscr{G}(\delta)_{R'} \to \mathscr{G}_{(P', M_{P'}, L')}. \tag{6.3.58.1}$$

By 6.3.44 this in turn determines an automorphism

$$\rho' : (P', M_{P'}) \to (P', M_{P'}). \tag{6.3.58.2}$$

6.3.59. Let $\widetilde{P}' \to P'$ be the covering used in the standard construction, and let $\pi : \widetilde{P}' \to A'$ be the projection. As before let $\widetilde{P}$ (resp. $\widetilde{P}_0$) denote the reduction of $\widetilde{P}$ to R (resp. R_0).

By the proof of 6.3.33 the automorphism ρ lifts to an automorphisms

$$\tilde{\rho} : (\widetilde{P}, M_{\widetilde{P}}) \to (\widetilde{P}, M_{\widetilde{P}}) \tag{6.3.59.1}$$

over the isomorphism $\iota : A \to A$ (since we assumed $\mathscr{M}$ was symmetric).

Since $\widetilde{P}' \to P'$ is étale, we can after possibly replacing R' by an étale covering find a lifting

$$\tilde{p} : (\mathrm{Spec}(R), M_R') \to (\widetilde{P}, M_{\widetilde{P}}) \tag{6.3.59.2}$$

of p. Let $a \in A(R)$ denote $\pi(\tilde{p})$ (where $\pi : \widetilde{P} \to A$ is the projection). Since $\rho \circ p = p$ we have $\tilde{\rho} \circ \tilde{p} = t_y(\tilde{p})$ for some $y \in Y$. Applying π we get that $\iota(a) = a + c^t(y)$. Equivalently

$$-2a = c^t(y). \tag{6.3.59.3}$$

Choose a point $a' \in A'(R')$ reducing to a such that $-2a' = c^t(y)$ in $A'(R')$.

Let $(Z_{a'}, M_{Z_{a'}})$ denote the fiber product

$$Z_{a'} := \mathrm{Spec}(R') \times_{a', A'} \widetilde{P}' \tag{6.3.59.4}$$

with $M_{Z_{a'}}$ the pullback of the log structure on $\widetilde{P}'$, and let $(Z_{a'+y}, M_{Z_{a'+y}})$ denote the fiber product

$$Z_{a'+y} := \mathrm{Spec}(R') \times_{a'+c^t(y), A} \widetilde{P}' \tag{6.3.59.5}$$

with the pullback of $M_{\widetilde{P}'}$. The map $\tilde{\rho}$ induces an isomorphism (which we again denote by $\tilde{\rho}$)

$$\tilde{\rho} : (Z_{a'}, M_{Z_{a'}}) \to (Z_{a'+y}, M_{Z_{a'+y}}). \tag{6.3.59.6}$$

Translation by y also induces a map

$$t_y : (Z_{a'}, M_{Z_{a'}}) \to (Z_{a'+y}, M_{Z_{a'+y}}) \tag{6.3.59.7}$$

and we let

$$\alpha : (Z_{a'}, M_{Z_{a'}}) \to (Z_{a'}, M_{Z_{a'}}) \tag{6.3.59.8}$$

denote the automorphism $\tilde{\rho}^{-1} \circ t_y$. The point $\tilde{p}$ then defines a map

$$w : (\mathrm{Spec}(R), M_R') \to (Z_{a'}, M_{Z_{a'}}) \qquad (6.3.59.9)$$

such that $\alpha \circ w = w$. We need to show that we can lift this to a map

$$w' : (\mathrm{Spec}(R'), M_{R'}') \to (Z_{a'}, M_{Z_{a'}}) \qquad (6.3.59.10)$$

with $\alpha \circ w' = w'$.

6.3.60. Let $\mathcal{R}_{a'}$ denote the ring

$$\mathcal{R}_{a'} = \oplus_{(d,x) \in P} \mathcal{M}(a')^{\otimes d} \otimes L_x(a') \qquad (6.3.60.1)$$

with algebra structure defined by the maps 4.1.10.2. Then

$$Z_{a'} = \mathrm{Proj}(\mathcal{R}_{a'}). \qquad (6.3.60.2)$$

By the construction of t_y and the map $\tilde{\rho}$ (see for example the proof of 6.3.33), the automorphism α is induced by maps

$$\alpha_x : L_x(a') \to L_{-x}(a') \qquad (6.3.60.3)$$

such that for every $x, z \in X$ the diagram

$$
\begin{array}{ccc}
L_x(a') \otimes L_z(a') & \xrightarrow{\alpha_x \otimes \alpha_z} & L_{-x}(a') \otimes L_{-z}(a') \\
{\scriptstyle \mathrm{can}} \downarrow & & \downarrow {\scriptstyle \mathrm{can}} \\
L_{x+z}(a') & \xrightarrow{\alpha_{x+z}} & L_{-x-z}(a')
\end{array}
\qquad (6.3.60.4)
$$

commutes.

Lemma 6.3.61 *After replacing R' by an étale covering, there exists trivializations $\xi_x \in L_x(a')$ such that for every $x, z \in X$ the diagram*

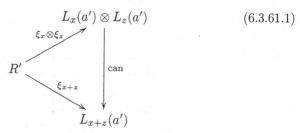

$$(6.3.61.1)$$

commutes, and such that for every $x \in X$ the diagram

$$(6.3.61.2)$$

commutes.

Proof. Choose first trivializations $\xi_x \in L_x(a')$ such that the diagrams 6.3.61.1 commute. This is possibly simply by fixing a basis $x_1, \ldots, x_r$ for X, choosing ξ_{x_i} for $i = 1, \ldots, r$, and then defining for $x = a_1 x_1 + \cdots + a_r x_r$ the trivialization ξ_x to be the element

$$\xi_1^{a_1} \otimes \cdots \otimes \xi_r^{a_r} \in L_{x_1}^{a_1}(a') \otimes \cdots L_{x_r}^{a_r}(a') = L_x(a'). \tag{6.3.61.3}$$

For every $x \in X$ we then get a unit $u_x \in R'^*$ characterized by the condition that

$$\alpha_x(\xi_x) = u_x \xi_{-x}. \tag{6.3.61.4}$$

The commutativity of 6.3.60.4 together with the commutativity of the diagrams 6.3.61.1 implies that for all $x, z \in X$ the diagrams

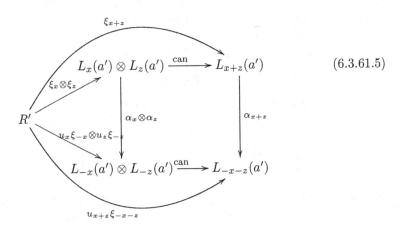

$$\tag{6.3.61.5}$$

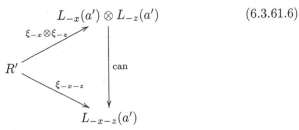

$$\tag{6.3.61.6}$$

commute. From this it follows that the u_x's define a homomorphism

$$u : X \to R'^*. \tag{6.3.61.7}$$

After replacing R' by an étale cover (since 2 is invertible in R'), there exists a homomorphism

$$\lambda : X \to R'^* \tag{6.3.61.8}$$

such that for all $x \in X$ we have

$$\lambda_x^2 = u_x^{-1}. \tag{6.3.61.9}$$

Then we have

$$\alpha_x(\lambda_x \xi_x) = \lambda_x u_x \xi_{-x} = \lambda_x^{-1} \xi_{-x} = \lambda_{-x} \xi_{-x}. \tag{6.3.61.10}$$

It follows that by replacing ξ_x by $\lambda_x \xi_x$ we obtain the desired trivializations of the $L_x(a')$. □

6.3.62. Choose trivializations $\xi_x \in L_x(a')$ as in 6.3.61. These choices define an isomorphism of log schemes

$$(Z_{a'}, M_{Z_{a'}}) \simeq \mathrm{Proj}(P \rtimes H_S \to R \otimes_{\mathbb{Z}[H_S]} \mathbb{Z}[P \rtimes H_S]) \tag{6.3.62.1}$$

such that the automorphism α is induced by the involution $\iota : P \rtimes H_S \to P \rtimes H_S$ described in the proof of 6.3.33 (recall that ι is the automorphism induced by multiplication by -1 on X).

For $p \in P$ let $\widetilde{U}_p \subset \mathrm{Spec}(R \otimes_{\mathbb{Z}[H_S]} \mathbb{Z}[P \rtimes H_S])$ be the open subset of points x where p maps to a unit in $k(x)$, and let $U_p \subset Z_{a'}$ denote the image of $\widetilde{U}_p$. Note that U_p is affine. If $Q_p \subset (P \rtimes H_S)_p$ denotes the submonoid of degree 0-elements in the localization then as a log scheme

$$U_p = \mathrm{Spec}(Q_p \to R \otimes_{\mathbb{Z}[H_S]} \mathbb{Z}[Q_p]). \tag{6.3.62.2}$$

For any $p \in P$ the open subset $U_{p+\iota(p)} \subset Z_{a'}$ is α-invariant, and the union of these open sets cover the set of points fixed by α.

We can therefore find a strict open immersion compatible with involutions

$$\mathrm{Spec}(Q \to R \otimes_{\mathbb{Z}[H_S]} \mathbb{Z}[Q]) \subset (Z_{a'}, M_{Z_{a'}}) \tag{6.3.62.3}$$

where $H_S \to Q$ is a morphism of fine monoids and $\iota : Q \to Q$ is an involution fixing H_S, such that the image of w is contained in this open set. Let Q_ι denote the colimit (in the category of integral monoids) of the diagram

$$Q \overset{\iota}{\underset{\mathrm{id}}{\rightrightarrows}} Q. \tag{6.3.62.4}$$

Then the map w (6.3.59.9) defines a strict morphism

$$\bar{w} : (\mathrm{Spec}(R), M_R') \to \mathrm{Spec}(Q_\iota \to R' \otimes_{\mathbb{Z}[H_S]} \mathbb{Z}[Q_\iota]), \tag{6.3.62.5}$$

and to define the lifting w' such that $\alpha \circ w' = w'$ it suffices to lift this morphism $\bar{w}$ to a morphism

$$\bar{w}' : (\mathrm{Spec}(R'), M_{R'}') \to \mathrm{Spec}(Q_\iota \to R' \otimes_{\mathbb{Z}[H_S]} \mathbb{Z}[Q_\iota]). \tag{6.3.62.6}$$

Note that such a lifting $\bar{w}'$ is automatically strict which ensures that the colimit condition in 6.3.53 (iii) holds.

Lemma 6.3.63 *(i) The kernel and the torsion part of the cokernel of the map $H_S^{gp} \to Q_\iota^{gp}$ are annihilated by a power of 2.*

(ii) The morphism

$$\text{Spec}(Q_\iota \to R' \otimes_{\mathbb{Z}[H_S]} \mathbb{Z}[Q_\iota]) \to (\text{Spec}(R'), M_{R'}) \tag{6.3.63.1}$$

is log smooth.

Proof. Part (ii) follows form (i) and 2.3.16.

To see (i), note that Q is obtained by taking the degree 0 part of a localization of $P \rtimes H_S$. It follows that if N^{gp} denotes the cokernel of $H_S^{gp} \to Q^{gp}$ then there is a commutative diagram

$$\begin{array}{ccccccccc}
0 & \longrightarrow & H_S^{gp} & \longrightarrow & Q^{gp} & \longrightarrow & N^{gp} & \longrightarrow & 0 \\
& & {=}\downarrow & & \downarrow & & \downarrow & & \\
0 & \longrightarrow & H_S^{gp} & \longrightarrow & (P \rtimes H_S)^{gp} & \longrightarrow & P^{gp} & \longrightarrow & 0,
\end{array} \tag{6.3.63.2}$$

where the rows are exact and the vertical arrows are inclusions. In particular $H_S^{gp} \to Q^{gp}$ is injective and N^{gp} is torsion free. Let G denote the group $\{\pm 1\}$. Then

$$Q_\iota^{gp} \simeq \mathbb{Z} \otimes_{\mathbb{Z}[G]} Q^{gp}, \tag{6.3.63.3}$$

where $\mathbb{Z}$ is viewed as a trivial G-module and Q as a G-module via the automorphism ι. To prove part (i) of the lemma it therefore suffices to show that

$$\mathbb{Z}[1/2] \otimes_{\mathbb{Z}[G]} H_S^{gp} \to \mathbb{Z}[1/2] \otimes_{\mathbb{Z}[G]} Q^{gp} \tag{6.3.63.4}$$

is injective and that $\mathbb{Z}[1/2] \otimes_{\mathbb{Z}[G]} N^{gp}$ is torsion free. This is clear because if M is a $\mathbb{Z}[1/2][G]$-module then $M \simeq M^+ \oplus M^-$, where G acts trivially on M^+ and by multiplication by -1 on M^- and

$$\mathbb{Z} \otimes_{\mathbb{Z}[G]} M \simeq M^+. \tag{6.3.63.5}$$

$\square$

6.3.64. This completes the proof of 6.3.55, for since

$$\text{Spec}(Q_\iota \to R' \otimes_{\mathbb{Z}[H_S]} \mathbb{Z}[Q_\iota]) \to (\text{Spec}(R'), M_{R'}) \tag{6.3.64.1}$$

is log smooth we can by the infinitesimal lifting property 2.3.12 (iii) find a lifting $\bar{w}'$ of $\bar{w}$. $\square$

6.3.65. Note that the proof in fact shows that the morphism

$$\pi : (\overline{\Theta}_{g,\delta}, M_{\overline{\Theta}_{g,\delta}}) \to (\overline{\mathscr{A}}_{g,\delta}[1/2d], M_{\overline{\mathscr{A}}_{g,\delta}[1/2d]}) \tag{6.3.65.1}$$

is log smooth and proper. Since the restriction

$$\Theta_{g,\delta} \to \mathscr{A}_{g,\delta}[1/2d] \tag{6.3.65.2}$$

is finite and étale we obtain:

Corollary 6.3.66 *The morphism 6.3.65.1 is proper and log étale.*

Remark 6.3.67. Let $\overline{\Theta}^{\text{tot}}_{g,\delta} \subset \overline{\Theta}_{g,\delta}$ be the closure of $\mathscr{M}^{\text{tot}}_{g,\delta}$ (defined as in 6.3.14). Since $\overline{\Theta}_{g,\delta}$ is locally integral, the stack $\overline{\Theta}^{\text{tot}}_{g,\delta}$ is open and closed in $\overline{\Theta}_{g,\delta}$. If all d_i are divisible by 8, then there is a natural map

$$P : \overline{\Theta}^{\text{tot}}_{g,\delta} \to \overline{\mathscr{M}}^{\text{tot}}_{g,\delta} \qquad (6.3.67.1)$$

restricting to the identity on $\mathscr{M}^{\text{tot}}_{g,\delta}$ (here $\overline{\mathscr{M}}^{\text{tot}}_{g,\delta}$ is as in 6.3.15). This map is defined as follows.

Since $\mathscr{M}_{g,\delta} \subset \overline{\Theta}^{\text{tot}}_{g,\delta}$ is a dense open substack and $\overline{\Theta}^{\text{tot}}_{g,\delta}$ is locally integral, it suffices to define a morphism

$$\overline{\Theta}^{\text{tot}}_{g,\delta} \to \mathbb{P}(V_\delta) \qquad (6.3.67.2)$$

extending the map 6.3.15.4. For this in turn it suffices to define a map

$$\tilde{P} : \overline{P}^{\rho,\log}_u \to \mathbb{P}(V_\delta). \qquad (6.3.67.3)$$

For this note that it follows from 5.2.6 that for any scheme S and object

$$(G, M_S, P, M_P, L, \sigma, t : M_S \to M'_S, p) \in P^{\rho,\log}_u(S) \qquad (6.3.67.4)$$

as in 6.3.53 the sheaf $f_* L$ on S is a locally free sheaf of rank d on S and its formation commutes with arbitrary base change. Furthermore the map $f^* f_* L \to L$ is surjective by 6.3.10 and 5.2.7 since each d_i is divisible by 4. We therefore get a surjection

$$f_* L = p^* f^* f_* L \to p^* L \qquad (6.3.67.5)$$

on S. On the other hand, by [36, Part II, Proposition 2] there exists a unique line bundle K on S and an isomorphism of $\mathscr{G}(\delta)$-representations $V_\delta \otimes_{\mathbb{Z}[1/2d]} K \simeq f_* L$ (which is unique up to multiplication by an element of $\mathscr{O}^*_S$). The surjection 6.3.67.5 therefore defines a map

$$S \to \mathbb{P}(V_\delta). \qquad (6.3.67.6)$$

In this way we obtain the desired morphism 6.3.67.3.

References

1. D. Abramovich, M. Olsson, and A. Vistoli, *Twisted stable maps to tame Artin stacks*, preprint (2008).
2. D. Abramovich and A. Vistoli, *Compactifying the space of stable maps*, J. Amer. Math. Soc. **15** (2002), 27–75.
3. V. Alexeev, *Complete moduli in the presence of semiabelian group action*, Annals of Math. **155** (2002), 611–708.
4. V. Alexeev and I. Nakamura, *On Mumford's construction of degenerating abelian varieties*, Tohoku Math. J. **51** (1999), 399–420.
5. M. Artin, *Algebraization of formal moduli. I.*, in Global Analysis (Papers in Honor of K. Kodaira), Tokyo Press, Tokyo (1969), 21–71.
6. _____, *Algebraic approximation of structures of complete local rings*, Inst. Hautes Études Sci. Publ. Math. **36** (1969), 23–58.
7. _____, *Versal deformations and algebraic stacks*, Inv. Math. **27** (1974), 165–189.
8. M. Artin, A. Grothendieck, and J.-L. Verdier, *Théorie des topos et cohomologie étale des schémas.*, Lecture Notes in Mathematics **269, 270, 305**, Springer-Verlag, Berlin (1972).
9. A. Ash, D. Mumford, M. Rapoport, and Y.S. Tai, *Smooth compactification of locally symmetric varieties*, Math. Sci. Press, Brookline (1975).
10. S. Bosch, W. Lutkebohmert, and M. Raynaud, *Néron models*, Ergebnisse der Mathematik und ihrer Grenzgebiete **21**, Springer-Verlag, Berlin, 1990.
11. L. Breen, *Fonctions thêta et théorème du cube*, Lecture Notes in Mathematics **980**, Springer-Verlag, Berlin (1983).
12. C. Cadman, *Using stacks to impose tangency conditions on curves*, Amer. J. Math. **129** (2007), 405–427.
13. C.-L. Chai and G. Faltings, *Degeneration of abelian varieties*, Ergebnisse der Mathematik **22**, Springer-Verlag (1990).
14. B. Conrad, *Keel-Mori theorem via stacks*, preprint.
15. P. Deligne and M. Rapoport, *Les schémas de modules de courbes elliptiques*, Lecture Notes in Math **349**, Springer-Verlag, Berlin (1973), 143–316.
16. M. Demazure and A. Grothendieck, *Schémas en Groupes*, Lecture Notes in Math. **151, 152, 153**, Springer–Verlag, Berlin (1970).
17. J. Dieudonné and A. Grothendieck, *Éléments de géométrie algébrique*, Inst. Hautes Études Sci. Publ. Math. **4, 8, 11, 17, 20, 24, 28, 32** (1961–1967).

18. G. van der Geer and B. Moonen, *Abelian varieties*, preliminary manuscript available at http://staff.science.uva.nl/~bmoonen/boek/BookAV.html.

19. A. Grothendieck, *Groupes de Monodromie en Géométrie Algébrique*, Lecture Notes in Math. **288**, Springer–Verlag, Berlin (1972).

20. R. Hartshorne, *Algebraic geometry*, Springer-Verlag, New York, 1977, Graduate Texts in Mathematics **52**.

21. L. Illusie, *An overview of the work of K. Fujiwara, K. Kato, and C. Nakayama on logarithmic étale cohomology*, Astérisque **279** (2002), 271–322.

22. T. Kajiwara, K. Kato, and C. Nakayama, *Logarithmic abelian varieties. Part I. Complex analytic theory*, preprint (2004).

23. F. Kato, *Log smooth deformation theory*, Tohoku Math. J. **48** (1996), 317–354.

24. K. Kato, *Logarithmic structures of Fontaine-Illusie*, Algebraic analysis, geometry, and number theory (Baltimore, MD, 1988), Johns Hopkins Univ. Press, Baltimore, MD, 1989, pp. 191–224.

25. _____, *Toric singularities*, American Jour. of Math. **116** (1994), 1073–1099.

26. S. Keel and J. Tevelev, *Chow quotients and grassmanians I and II*, preprints (2005).

27. S. Keel and S. Mori, *Quotients by groupoids*, Ann. Math. **145** (1997), 193–213.

28. S. Kleiman, *A note on the Nakai-Moisezon test for ampleness of a divisor*, Amer. J. Math. **87** (1965), 221–226.

29. D. Knutson, *Algebraic spaces*, Springer Lecture Notes in Math. **203** (1971).

30. L. Lafforgue, *Chirurgie des grassmanniennes*, CRM Monograph Series **19**, American Mathematical Society, Providence, RI, 2003, xx+170

31. G. Laumon and L. Moret-Bailly, *Champs algébriques*, Ergebnisse der Mathematik **39**, Springer-Verlag, Berlin (2000).

32. J.S. Milne, *Étale cohomology*, Princeton Mathematical Series **33**, Princeton University Press, Princeton, N.J., 1980.

33. L. Moret-Bailly, *Pinceaux de variétés abéliennes*, Astérisque **129** (1985).

34. D. Mumford, *Geometric invariant theory*, Ergebnisse der Mathematik und ihrer Grenzgebiete **34**, Springer-Verlag, Berlin-New York, 1965.

35. _____, *Abelian varieties*, Tata Institute of Fundamental Research Studies in Mathematics **5**, Published for the Tata Institute of Fundamental Research, Bombay; Oxford University Press, London, 1970.

36. _____, *On the equations defining abelian varieties I, II, III*, Inv. Math. **1** (1966), 287–354, and **3** (1967), 71–135 and 215–244.

37. C. Nakayama, *Logarithmic étale cohomology*, Math. Ann. **308** (1997), 365–404.

38. _____, *Nearby cycles for log smooth families*, Comp. Math. **112** (1998), 45–75.

39. Y. Namikawa, *A new compactification of the Siegel space and degeneration of Abelian varieties. I and II*, Math. Ann. **221** (1976), 97–141 and 201–241.

40. T. Oda, *Convex bodies and algebraic geometry*, Ergebnisse der Math., 3 Folge, Band 5, Springer-Verlag, 1989.

41. A. Ogus, *Lectures on logarithmic algebraic geometry*, Texed notes.

42. M. Olsson, *Logarithmic geometry and algebraic stacks*, Ann. Sci. École Norm. Sup. **36** (2003), 747–791.

43. _____, *On proper coverings of Artin stacks*, Advances in Mathematics **198** (2005), 93–106.

44. _____, *Sheaves on Artin stacks*, J. Reine Angew. Math. (Crelle's Journal) **603** (2007), 55–112.

45. _____, *Hom-stacks and restriction of scalars*, Duke Math. J. **134** (2006), 139–164.

46. _____, *Semi-stable degenerations and period spaces for polarized K3 surfaces*, Duke Math. J. **125** (2004), 121–203.

47. I. Satake, *On the compactification of the Siegel space*, J. Indian Math. Soc. **20** (1956), 259–281.

48. I. Vidal, *Morphismes log étale et descente part homéomorphismes universels*, C. R. Acad. Sci. Paris Ser. I Math. **332** (2001), 239–244.

49. A. Vistoli, *Grothendieck topologies, fibered categories and descent theory*, in "Fundamental algebraic geometry", Math. Surveys Monogr. **123** Amer. Math. Soc., Providence, RI (2005) 1–104.

Index of Terminology

S-groupoid, 8
$\Gamma_0(n)$-level structure, 240
d-th root construction, 225
(integral) paving (abelian case), 85
(integral) paving (toric case), 57
(integral) polytope, 57

abelian algebraic space, 31
algebraic space, 19
algebraic stacks, 22

biextension, 34
big Kummer étale topos, 234

chart for a log structure, 44
cleavage, 10
closed immersion of log schemes, 45
coarse moduli space, 26
coherent monoid, 43

Deligne-Mumford stack, 22
dimension of a polytope, 57

exact morphism (of monoids or log schemes), 232

face of a polytope, 57
fibered category, 8
fine log structure, 44
fine monoid, 43
fs log scheme, 50

generalized Tate family, 213
group associated to a monoid, 43
groupoid in a category $\mathcal{C}$, 14

integral monoid, 43
integral morphism of log schemes, 47

Kummer étale site, 232
Kummer morphism (of monoids or log schemes), 231

log étale, 45
log scheme, 43
log smooth, 45
log smooth deformation, 49
log structure, 43
log structure associated to pre-log structure, 44
log tangent bundle, 49
logarithmic differentials, 48

Poincare bundle, 33
pre-log structure, 43
presentation of a stack, 22
prestack, 17
pseudo-functor, 11
pullback, 7

regular paving (abelian case), 85
regular paving (toric case), 58
representable morphism of stacks, 22
rigidification of a stack, 28
rigidified line bundle, 32

saturated log scheme, 50
saturated log structure, 50
saturated monoid, 50
saturation of log scheme, 50
schematic morphism of stacks, 20
semi-abelian scheme, 31
seminormal scheme, 51
stable semiabelic pair, 53
stable semiabelic variety, 53
stack, 18
standard family (general case), 140
standard family (principally polarized case), 93
standard family (toric case), 60
strict closed immersion, 45
strict morphism, 45

symmetric biextension, 38
symmetric invertible sheaf, 241
symmetric theta level structure,
    243

theta group, 136, 144
totally symmetric invertible sheaf,
    245
type, 237

Index of Notation

$(X, M_X)_{\text{ket}}$, 232
$(fs/(X, M_X))_{\text{ket}}$, 234
BG, 8
$F(\{T_i \to T\})$, 17
H_S (abelian case), 86
H_S (toric case), 58
$I_{g,n}$, 236
$K(\delta), H(\delta)$, 242
$P \rtimes H_S$ (abelian case), 86
$P \rtimes H_S$ (toric case), 59
$T_{(X,M_X)/(Y,M_Y)}$, 49
$U(B_0)$, 172
V_δ, 243
$X^{(d)}$, 225
$[X/G]$, 8
$[X_0/X_1]$, 19
$\Omega^1_{(X,M_X)/(Y,M_Y)}$, 48
$\Sigma_{g,\delta}$, 249
$\text{Spec}(P \to R[P])$, 45
$\Theta_{g,\delta}$, 251
$\mathscr{A}_g$, 9
$\mathscr{A}_{g,d}$, 135
$\mathscr{A}^{\text{Alex}}_{g,d}$, 54

$\mathscr{G}(L)$, 241
$\mathscr{G}_{(A,P,L)}$, 136
$\mathscr{G}_{(\mathscr{P},M_{\mathscr{P}},L_{\mathscr{P}})}$, 144
$\mathscr{K}_Q$, 80
$\mathscr{K}_g$, 121
$\mathscr{M}_{g,\delta}$, 243
$\mathscr{M}^{\text{tot}}_{g,\delta}$, 245
$\mathscr{N}_{g,\delta}$, 250
$\mathscr{N}^{\text{sym}}_{g,\delta}$, 250
$\mathscr{T}_{g,d}$, 135
$\overline{\Sigma}_{g,\delta}$, 260
$\overline{\Theta}_{g,\delta}$, 264
$\overline{\Theta}^{\text{tot}}_{g,\delta}$, 271
$\overline{\mathscr{A}}_{g,\delta}[1/d]$, 237
$\overline{\mathscr{A}}_{g,d}$, 203
$\overline{\mathscr{T}}_{g,d}$, 141
$\overline{\mathscr{T}}_{g,\delta}$, $\mathscr{T}_{g,\delta}$, 237
$\text{KEt}(X, M_X)$, 232
$\text{TORS}_{\mathscr{P}}(X)$, 142
Tate_d, 229
$\underline{\text{Isom}}(-,-)$, 15
$\underline{\text{Pic}}(P), \underline{\text{Pic}}^0(P)$, 31
$\mathcal{TP}^{\text{fr}}[Q]$, 53

Lecture Notes in Mathematics

For information about earlier volumes
please contact your bookseller or Springer
LNM Online archive: springerlink.com

Vol. 1774: V. Runde, Lectures on Amenability (2002)

Vol. 1775: W. H. Meeks, A. Ros, H. Rosenberg, The Global Theory of Minimal Surfaces in Flat Spaces. Martina Franca 1999. Editor: G. P. Pirola (2002)

Vol. 1776: K. Behrend, C. Gomez, V. Tarasov, G. Tian, Quantum Comohology. Cetraro 1997. Editors: P. de Bartolomeis, B. Dubrovin, C. Reina (2002)

Vol. 1777: E. García-Río, D. N. Kupeli, R. Vázquez-Lorenzo, Osserman Manifolds in Semi-Riemannian Geometry (2002)

Vol. 1778: H. Kiechle, Theory of K-Loops (2002)

Vol. 1779: I. Chueshov, Monotone Random Systems (2002)

Vol. 1780: J. H. Bruinier, Borcherds Products on O(2,1) and Chern Classes of Heegner Divisors (2002)

Vol. 1781: E. Bolthausen, E. Perkins, A. van der Vaart, Lectures on Probability Theory and Statistics. Ecole d' Eté de Probabilités de Saint-Flour XXIX-1999. Editor: P. Bernard (2002)

Vol. 1782: C.-H. Chu, A. T.-M. Lau, Harmonic Functions on Groups and Fourier Algebras (2002)

Vol. 1783: L. Grüne, Asymptotic Behavior of Dynamical and Control Systems under Perturbation and Discretization (2002)

Vol. 1784: L. H. Eliasson, S. B. Kuksin, S. Marmi, J.-C. Yoccoz, Dynamical Systems and Small Divisors. Cetraro, Italy 1998. Editors: S. Marmi, J.-C. Yoccoz (2002)

Vol. 1785: J. Arias de Reyna, Pointwise Convergence of Fourier Series (2002)

Vol. 1786: S. D. Cutkosky, Monomialization of Morphisms from 3-Folds to Surfaces (2002)

Vol. 1787: S. Caenepeel, G. Militaru, S. Zhu, Frobenius and Separable Functors for Generalized Module Categories and Nonlinear Equations (2002)

Vol. 1788: A. Vasil'ev, Moduli of Families of Curves for Conformal and Quasiconformal Mappings (2002)

Vol. 1789: Y. Sommerhäuser, Yetter-Drinfel'd Hopf algebras over groups of prime order (2002)

Vol. 1790: X. Zhan, Matrix Inequalities (2002)

Vol. 1791: M. Knebusch, D. Zhang, Manis Valuations and Prüfer Extensions I: A new Chapter in Commutative Algebra (2002)

Vol. 1792: D. D. Ang, R. Gorenflo, V. K. Le, D. D. Trong, Moment Theory and Some Inverse Problems in Potential Theory and Heat Conduction (2002)

Vol. 1793: J. Cortés Monforte, Geometric, Control and Numerical Aspects of Nonholonomic Systems (2002)

Vol. 1794: N. Pytheas Fogg, Substitution in Dynamics, Arithmetics and Combinatorics. Editors: V. Berthé, S. Ferenczi, C. Mauduit, A. Siegel (2002)

Vol. 1795: H. Li, Filtered-Graded Transfer in Using Noncommutative Gröbner Bases (2002)

Vol. 1796: J.M. Melenk, hp-Finite Element Methods for Singular Perturbations (2002)

Vol. 1797: B. Schmidt, Characters and Cyclotomic Fields in Finite Geometry (2002)

Vol. 1798: W.M. Oliva, Geometric Mechanics (2002)

Vol. 1799: H. Pajot, Analytic Capacity, Rectifiability, Menger Curvature and the Cauchy Integral (2002)

Vol. 1800: O. Gabber, L. Ramero, Almost Ring Theory (2003)

Vol. 1801: J. Azéma, M. Émery, M. Ledoux, M. Yor (Eds.), Séminaire de Probabilités XXXVI (2003)

Vol. 1802: V. Capasso, E. Merzbach, B. G. Ivanoff, M. Dozzi, R. Dalang, T. Mountford, Topics in Spatial Stochastic Processes. Martina Franca, Italy 2001. Editor: E. Merzbach (2003)

Vol. 1803: G. Dolzmann, Variational Methods for Crystalline Microstructure – Analysis and Computation (2003)

Vol. 1804: I. Cherednik, Ya. Markov, R. Howe, G. Lusztig, Iwahori-Hecke Algebras and their Representation Theory. Martina Franca, Italy 1999. Editors: V. Baldoni, D. Barbasch (2003)

Vol. 1805: F. Cao, Geometric Curve Evolution and Image Processing (2003)

Vol. 1806: H. Broer, I. Hoveijn. G. Lunther, G. Vegter, Bifurcations in Hamiltonian Systems. Computing Singularities by Gröbner Bases (2003)

Vol. 1807: V. D. Milman, G. Schechtman (Eds.), Geometric Aspects of Functional Analysis. Israel Seminar 2000-2002 (2003)

Vol. 1808: W. Schindler, Measures with Symmetry Properties (2003)

Vol. 1809: O. Steinbach, Stability Estimates for Hybrid Coupled Domain Decomposition Methods (2003)

Vol. 1810: J. Wengenroth, Derived Functors in Functional Analysis (2003)

Vol. 1811: J. Stevens, Deformations of Singularities (2003)

Vol. 1812: L. Ambrosio, K. Deckelnick, G. Dziuk, M. Mimura, V. A. Solonnikov, H. M. Soner, Mathematical Aspects of Evolving Interfaces. Madeira, Funchal, Portugal 2000. Editors: P. Colli, J. F. Rodrigues (2003)

Vol. 1813: L. Ambrosio, L. A. Caffarelli, Y. Brenier, G. Buttazzo, C. Villani, Optimal Transportation and its Applications. Martina Franca, Italy 2001. Editors: L. A. Caffarelli, S. Salsa (2003)

Vol. 1814: P. Bank, F. Baudoin, H. Föllmer, L.C.G. Rogers, M. Soner, N. Touzi, Paris-Princeton Lectures on Mathematical Finance 2002 (2003)

Vol. 1815: A. M. Vershik (Ed.), Asymptotic Combinatorics with Applications to Mathematical Physics. St. Petersburg, Russia 2001 (2003)

Vol. 1816: S. Albeverio, W. Schachermayer, M. Talagrand, Lectures on Probability Theory and Statistics. Ecole d'Eté de Probabilités de Saint-Flour XXX-2000. Editor: P. Bernard (2003)

Vol. 1817: E. Koelink, W. Van Assche (Eds.), Orthogonal Polynomials and Special Functions. Leuven 2002 (2003)

Vol. 1818: M. Bildhauer, Convex Variational Problems with Linear, nearly Linear and/or Anisotropic Growth Conditions (2003)

Vol. 1819: D. Masser, Yu. V. Nesterenko, H. P. Schlickewei, W. M. Schmidt, M. Waldschmidt, Diophantine Approximation. Cetraro, Italy 2000. Editors: F. Amoroso, U. Zannier (2003)

Vol. 1820: F. Hiai, H. Kosaki, Means of Hilbert Space Operators (2003)

Vol. 1821: S. Teufel, Adiabatic Perturbation Theory in Quantum Dynamics (2003)

Vol. 1822: S.-N. Chow, R. Conti, R. Johnson, J. Mallet-Paret, R. Nussbaum, Dynamical Systems. Cetraro, Italy 2000. Editors: J. W. Macki, P. Zecca (2003)

Vol. 1823: A. M. Anile, W. Allegretto, C. Ringhofer, Mathematical Problems in Semiconductor Physics. Cetraro, Italy 1998. Editor: A. M. Anile (2003)

Vol. 1824: J. A. Navarro González, J. B. Sancho de Salas, $\mathscr{C}^\infty$ – Differentiable Spaces (2003)

Vol. 1825: J. H. Bramble, A. Cohen, W. Dahmen, Multiscale Problems and Methods in Numerical Simulations, Martina Franca, Italy 2001. Editor: C. Canuto (2003)

Vol. 1826: K. Dohmen, Improved Bonferroni Inequalities via Abstract Tubes. Inequalities and Identities of Inclusion-Exclusion Type. VIII, 113 p, 2003.

Vol. 1827: K. M. Pilgrim, Combinations of Complex Dynamical Systems. IX, 118 p, 2003.

Vol. 1828: D. J. Green, Gröbner Bases and the Computation of Group Cohomology. XII, 138 p, 2003.

Vol. 1829: E. Altman, B. Gaujal, A. Hordijk, Discrete-Event Control of Stochastic Networks: Multimodularity and Regularity. XIV, 313 p, 2003.

Vol. 1830: M. I. Gil', Operator Functions and Localization of Spectra. XIV, 256 p, 2003.

Vol. 1831: A. Connes, J. Cuntz, E. Guentner, N. Higson, J. E. Kaminker, Noncommutative Geometry, Martina Franca, Italy 2002. Editors: S. Doplicher, L. Longo (2004)

Vol. 1832: J. Azéma, M. Émery, M. Ledoux, M. Yor (Eds.), Séminaire de Probabilités XXXVII (2003)

Vol. 1833: D.-Q. Jiang, M. Qian, M.-P. Qian, Mathematical Theory of Nonequilibrium Steady States. On the Frontier of Probability and Dynamical Systems. IX, 280 p, 2004.

Vol. 1834: Yo. Yomdin, G. Comte, Tame Geometry with Application in Smooth Analysis. VIII, 186 p, 2004.

Vol. 1835: O.T. Izhboldin, B. Kahn, N.A. Karpenko, A. Vishik, Geometric Methods in the Algebraic Theory of Quadratic Forms. Summer School, Lens, 2000. Editor: J.-P. Tignol (2004)

Vol. 1836: C. Năstăsescu, F. Van Oystaeyen, Methods of Graded Rings. XIII, 304 p, 2004.

Vol. 1837: S. Tavaré, O. Zeitouni, Lectures on Probability Theory and Statistics. Ecole d'Eté de Probabilités de Saint-Flour XXXI-2001. Editor: J. Picard (2004)

Vol. 1838: A.J. Ganesh, N.W. O'Connell, D.J. Wischik, Big Queues. XII, 254 p, 2004.

Vol. 1839: R. Gohm, Noncommutative Stationary Processes. VIII, 170 p, 2004.

Vol. 1840: B. Tsirelson, W. Werner, Lectures on Probability Theory and Statistics. Ecole d'Eté de Probabilités de Saint-Flour XXXII-2002. Editor: J. Picard (2004)

Vol. 1841: W. Reichel, Uniqueness Theorems for Variational Problems by the Method of Transformation Groups (2004)

Vol. 1842: T. Johnsen, A. L. Knutsen, K_3 Projective Models in Scrolls (2004)

Vol. 1843: B. Jefferies, Spectral Properties of Noncommuting Operators (2004)

Vol. 1844: K.F. Siburg, The Principle of Least Action in Geometry and Dynamics (2004)

Vol. 1845: Min Ho Lee, Mixed Automorphic Forms, Torus Bundles, and Jacobi Forms (2004)

Vol. 1846: H. Ammari, H. Kang, Reconstruction of Small Inhomogeneities from Boundary Measurements (2004)

Vol. 1847: T.R. Bielecki, T. Björk, M. Jeanblanc, M. Rutkowski, J.A. Scheinkman, W. Xiong, Paris-Princeton Lectures on Mathematical Finance 2003 (2004)

Vol. 1848: M. Abate, J. E. Fornaess, X. Huang, J. P. Rosay, A. Tumanov, Real Methods in Complex and CR Geometry, Martina Franca, Italy 2002. Editors: D. Zaitsev, G. Zampieri (2004)

Vol. 1849: Martin L. Brown, Heegner Modules and Elliptic Curves (2004)

Vol. 1850: V. D. Milman, G. Schechtman (Eds.), Geometric Aspects of Functional Analysis. Israel Seminar 2002-2003 (2004)

Vol. 1851: O. Catoni, Statistical Learning Theory and Stochastic Optimization (2004)

Vol. 1852: A.S. Kechris, B.D. Miller, Topics in Orbit Equivalence (2004)

Vol. 1853: Ch. Favre, M. Jonsson, The Valuative Tree (2004)

Vol. 1854: O. Saeki, Topology of Singular Fibers of Differential Maps (2004)

Vol. 1855: G. Da Prato, P.C. Kunstmann, I. Lasiecka, A. Lunardi, R. Schnaubelt, L. Weis, Functional Analytic Methods for Evolution Equations. Editors: M. Iannelli, R. Nagel, S. Piazzera (2004)

Vol. 1856: K. Back, T.R. Bielecki, C. Hipp, S. Peng, W. Schachermayer, Stochastic Methods in Finance, Bressanone/Brixen, Italy, 2003. Editors: M. Fritelli, W. Runggaldier (2004)

Vol. 1857: M. Émery, M. Ledoux, M. Yor (Eds.), Séminaire de Probabilités XXXVIII (2005)

Vol. 1858: A.S. Cherny, H.-J. Engelbert, Singular Stochastic Differential Equations (2005)

Vol. 1859: E. Letellier, Fourier Transforms of Invariant Functions on Finite Reductive Lie Algebras (2005)

Vol. 1860: A. Borisyuk, G.B. Ermentrout, A. Friedman, D. Terman, Tutorials in Mathematical Biosciences I. Mathematical Neurosciences (2005)

Vol. 1861: G. Benettin, J. Henrard, S. Kuksin, Hamiltonian Dynamics – Theory and Applications, Cetraro, Italy, 1999. Editor: A. Giorgilli (2005)

Vol. 1862: B. Helffer, F. Nier, Hypoelliptic Estimates and Spectral Theory for Fokker-Planck Operators and Witten Laplacians (2005)

Vol. 1863: H. Führ, Abstract Harmonic Analysis of Continuous Wavelet Transforms (2005)

Vol. 1864: K. Efstathiou, Metamorphoses of Hamiltonian Systems with Symmetries (2005)

Vol. 1865: D. Applebaum, B.V. R. Bhat, J. Kustermans, J. M. Lindsay, Quantum Independent Increment Processes I. From Classical Probability to Quantum Stochastic Calculus. Editors: M. Schürmann, U. Franz (2005)

Vol. 1866: O.E. Barndorff-Nielsen, U. Franz, R. Gohm, B. Kümmerer, S. Thorbjønsen, Quantum Independent Increment Processes II. Structure of Quantum Lévy Processes, Classical Probability, and Physics. Editors: M. Schürmann, U. Franz, (2005)

Vol. 1867: J. Sneyd (Ed.), Tutorials in Mathematical Biosciences II. Mathematical Modeling of Calcium Dynamics and Signal Transduction. (2005)

Vol. 1868: J. Jorgenson, S. Lang, $Pos_n(R)$ and Eisenstein Series. (2005)

Vol. 1869: A. Dembo, T. Funaki, Lectures on Probability Theory and Statistics. Ecole d'Eté de Probabilités de Saint-Flour XXXIII-2003. Editor: J. Picard (2005)

Vol. 1870: V.I. Gurariy, W. Lusky, Geometry of Müntz Spaces and Related Questions. (2005)

Vol. 1871: P. Constantin, G. Gallavotti, A.V. Kazhikhov, Y. Meyer, S. Ukai, Mathematical Foundation of Turbulent Viscous Flows, Martina Franca, Italy, 2003. Editors: M. Cannone, T. Miyakawa (2006)

Vol. 1872: A. Friedman (Ed.), Tutorials in Mathematical Biosciences III. Cell Cycle, Proliferation, and Cancer (2006)

Vol. 1873: R. Mansuy, M. Yor, Random Times and Enlargements of Filtrations in a Brownian Setting (2006)

Vol. 1874: M. Yor, M. Émery (Eds.), In Memoriam Paul-André Meyer - Séminaire de Probabilités XXXIX (2006)

Vol. 1875: J. Pitman, Combinatorial Stochastic Processes. Ecole d'Eté de Probabilités de Saint-Flour XXXII-2002. Editor: J. Picard (2006)

Vol. 1876: H. Herrlich, Axiom of Choice (2006)

Vol. 1877: J. Steuding, Value Distributions of L-Functions (2007)

Vol. 1878: R. Cerf, The Wulff Crystal in Ising and Percolation Models, Ecole d'Eté de Probabilités de Saint-Flour XXXIV-2004. Editor: Jean Picard (2006)

Vol. 1879: G. Slade, The Lace Expansion and its Applications, Ecole d'Eté de Probabilités de Saint-Flour XXXIV-2004. Editor: Jean Picard (2006)

Vol. 1880: S. Attal, A. Joye, C.-A. Pillet, Open Quantum Systems I, The Hamiltonian Approach (2006)

Vol. 1881: S. Attal, A. Joye, C.-A. Pillet, Open Quantum Systems II, The Markovian Approach (2006)

Vol. 1882: S. Attal, A. Joye, C.-A. Pillet, Open Quantum Systems III, Recent Developments (2006)

Vol. 1883: W. Van Assche, F. Marcellán (Eds.), Orthogonal Polynomials and Special Functions, Computation and Application (2006)

Vol. 1884: N. Hayashi, E.I. Kaikina, P.I. Naumkin, I.A. Shishmarev, Asymptotics for Dissipative Nonlinear Equations (2006)

Vol. 1885: A. Telcs, The Art of Random Walks (2006)

Vol. 1886: S. Takamura, Splitting Deformations of Degenerations of Complex Curves (2006)

Vol. 1887: K. Habermann, L. Habermann, Introduction to Symplectic Dirac Operators (2006)

Vol. 1888: J. van der Hoeven, Transseries and Real Differential Algebra (2006)

Vol. 1889: G. Osipenko, Dynamical Systems, Graphs, and Algorithms (2006)

Vol. 1890: M. Bunge, J. Funk, Singular Coverings of Toposes (2006)

Vol. 1891: J.B. Friedlander, D.R. Heath-Brown, H. Iwaniec, J. Kaczorowski, Analytic Number Theory, Cetraro, Italy, 2002. Editors: A. Perelli, C. Viola (2006)

Vol. 1892: A. Baddeley, I. Bárány, R. Schneider, W. Weil, Stochastic Geometry, Martina Franca, Italy, 2004. Editor: W. Weil (2007)

Vol. 1893: H. Hanßmann, Local and Semi-Local Bifurcations in Hamiltonian Dynamical Systems, Results and Examples (2007)

Vol. 1894: C.W. Groetsch, Stable Approximate Evaluation of Unbounded Operators (2007)

Vol. 1895: L. Molnár, Selected Preserver Problems on Algebraic Structures of Linear Operators and on Function Spaces (2007)

Vol. 1896: P. Massart, Concentration Inequalities and Model Selection, Ecole d'Été de Probabilités de Saint-Flour XXXIII-2003. Editor: J. Picard (2007)

Vol. 1897: R. Doney, Fluctuation Theory for Lévy Processes, Ecole d'Été de Probabilités de Saint-Flour XXXV-2005. Editor: J. Picard (2007)

Vol. 1898: H.R. Beyer, Beyond Partial Differential Equations, On linear and Quasi-Linear Abstract Hyperbolic Evolution Equations (2007)

Vol. 1899: Séminaire de Probabilités XL. Editors: C. Donati-Martin, M. Émery, A. Rouault, C. Stricker (2007)

Vol. 1900: E. Bolthausen, A. Bovier (Eds.), Spin Glasses (2007)

Vol. 1901: O. Wittenberg, Intersections de deux quadriques et pinceaux de courbes de genre 1, Intersections of Two Quadrics and Pencils of Curves of Genus 1 (2007)

Vol. 1902: A. Isaev, Lectures on the Automorphism Groups of Kobayashi-Hyperbolic Manifolds (2007)

Vol. 1903: G. Kresin, V. Maz'ya, Sharp Real-Part Theorems (2007)

Vol. 1904: P. Giesl, Construction of Global Lyapunov Functions Using Radial Basis Functions (2007)

Vol. 1905: C. Prévôt, M. Röckner, A Concise Course on Stochastic Partial Differential Equations (2007)

Vol. 1906: T. Schuster, The Method of Approximate Inverse: Theory and Applications (2007)

Vol. 1907: M. Rasmussen, Attractivity and Bifurcation for Nonautonomous Dynamical Systems (2007)

Vol. 1908: T.J. Lyons, M. Caruana, T. Lévy, Differential Equations Driven by Rough Paths, Ecole d'Été de Probabilités de Saint-Flour XXXIV-2004 (2007)

Vol. 1909: H. Akiyoshi, M. Sakuma, M. Wada, Y. Yamashita, Punctured Torus Groups and 2-Bridge Knot Groups (I) (2007)

Vol. 1910: V.D. Milman, G. Schechtman (Eds.), Geometric Aspects of Functional Analysis. Israel Seminar 2004-2005 (2007)

Vol. 1911: A. Bressan, D. Serre, M. Williams, K. Zumbrun, Hyperbolic Systems of Balance Laws. Cetraro, Italy 2003. Editor: P. Marcati (2007)

Vol. 1912: V. Berinde, Iterative Approximation of Fixed Points (2007)

Vol. 1913: J.E. Marsden, G. Misiołek, J.-P. Ortega, M. Perlmutter, T.S. Ratiu, Hamiltonian Reduction by Stages (2007)

Vol. 1914: G. Kutyniok, Affine Density in Wavelet Analysis (2007)

Vol. 1915: T. Bıyıkoğlu, J. Leydold, P.F. Stadler, Laplacian Eigenvectors of Graphs. Perron-Frobenius and Faber-Krahn Type Theorems (2007)

Vol. 1916: C. Villani, F. Rezakhanlou, Entropy Methods for the Boltzmann Equation. Editors: F. Golse, S. Olla (2008)

Vol. 1917: I. Veselić, Existence and Regularity Properties of the Integrated Density of States of Random Schrödinger (2008)

Vol. 1918: B. Roberts, R. Schmidt, Local Newforms for GSp(4) (2007)

Vol. 1919: R.A. Carmona, I. Ekeland, A. Kohatsu-Higa, J.-M. Lasry, P.-L. Lions, H. Pham, E. Taflin, Paris-Princeton Lectures on Mathematical Finance 2004.

Editors: R.A. Carmona, E. Çinlar, I. Ekeland, E. Jouini, J.A. Scheinkman, N. Touzi (2007)

Vol. 1920: S.N. Evans, Probability and Real Trees. Ecole d'Été de Probabilités de Saint-Flour XXXV-2005 (2008)

Vol. 1921: J.P. Tian, Evolution Algebras and their Applications (2008)

Vol. 1922: A. Friedman (Ed.), Tutorials in Mathematical BioSciences IV. Evolution and Ecology (2008)

Vol. 1923: J.P.N. Bishwal, Parameter Estimation in Stochastic Differential Equations (2008)

Vol. 1924: M. Wilson, Littlewood-Paley Theory and Exponential-Square Integrability (2008)

Vol. 1925: M. du Sautoy, L. Woodward, Zeta Functions of Groups and Rings (2008)

Vol. 1926: L. Barreira, V. Claudia, Stability of Nonautonomous Differential Equations (2008)

Vol. 1927: L. Ambrosio, L. Caffarelli, M.G. Crandall, L.C. Evans, N. Fusco, Calculus of Variations and Non-Linear Partial Differential Equations. Cetraro, Italy 2005. Editors: B. Dacorogna, P. Marcellini (2008)

Vol. 1928: J. Jonsson, Simplicial Complexes of Graphs (2008)

Vol. 1929: Y. Mishura, Stochastic Calculus for Fractional Brownian Motion and Related Processes (2008)

Vol. 1930: J.M. Urbano, The Method of Intrinsic Scaling. A Systematic Approach to Regularity for Degenerate and Singular PDEs (2008)

Vol. 1931: M. Cowling, E. Frenkel, M. Kashiwara, A. Valette, D.A. Vogan, Jr., N.R. Wallach, Representation Theory and Complex Analysis. Venice, Italy 2004. Editors: E.C. Tarabusi, A. D'Agnolo, M. Picardello (2008)

Vol. 1932: A.A. Agrachev, A.S. Morse, E.D. Sontag, H.J. Sussmann, V.I. Utkin, Nonlinear and Optimal Control Theory. Cetraro, Italy 2004. Editors: P. Nistri, G. Stefani (2008)

Vol. 1933: M. Petkovic, Point Estimation of Root Finding Methods (2008)

Vol. 1934: C. Donati-Martin, M. Émery, A. Rouault, C. Stricker (Eds.), Séminaire de Probabilités XLI (2008)

Vol. 1935: A. Unterberger, Alternative Pseudodifferential Analysis (2008)

Vol. 1936: P. Magal, S. Ruan (Eds.), Structured Population Models in Biology and Epidemiology (2008)

Vol. 1937: G. Capriz, P. Giovine, P.M. Mariano (Eds.), Mathematical Models of Granular Matter (2008)

Vol. 1938: D. Auroux, F. Catanese, M. Manetti, P. Seidel, B. Siebert, I. Smith, G. Tian, Symplectic 4-Manifolds and Algebraic Surfaces. Cetraro, Italy 2003. Editors: F. Catanese, G. Tian (2008)

Vol. 1939: D. Boffi, F. Brezzi, L. Demkowicz, R.G. Durán, R.S. Falk, M. Fortin, Mixed Finite Elements, Compatibility Conditions, and Applications. Cetraro, Italy 2006. Editors: D. Boffi, L. Gastaldi (2008)

Vol. 1940: J. Banasiak, V. Capasso, M.A.J. Chaplain, M. Lachowicz, J. Miękisz, Multiscale Problems in the Life Sciences. From Microscopic to Macroscopic. Będlewo, Poland 2006. Editors: V. Capasso, M. Lachowicz (2008)

Vol. 1941: S.M.J. Haran, Arithmetical Investigations. Representation Theory, Orthogonal Polynomials, and Quantum Interpolations (2008)

Vol. 1942: S. Albeverio, F. Flandoli, Y.G. Sinai, SPDE in Hydrodynamic. Recent Progress and Prospects. Cetraro, Italy 2005. Editors: G. Da Prato, M. Röckner (2008)

Vol. 1943: L.L. Bonilla (Ed.), Inverse Problems and Imaging. Martina Franca, Italy 2002 (2008)

Vol. 1944: A. Di Bartolo, G. Falcone, P. Plaumann, K. Strambach, Algebraic Groups and Lie Groups with Few Factors (2008)

Vol. 1945: F. Brauer, P. van den Driessche, J. Wu (Eds.), Mathematical Epidemiology (2008)

Vol. 1946: G. Allaire, A. Arnold, P. Degond, T.Y. Hou, Quantum Transport. Modelling, Analysis and Asymptotics. Cetraro, Italy 2006. Editors: N.B. Abdallah, G. Frosali (2008)

Vol. 1947: D. Abramovich, M. Mariño, M. Thaddeus, R. Vakil, Enumerative Invariants in Algebraic Geometry and String Theory. Cetraro, Italy 2005. Editors: K. Behrend, M. Manetti (2008)

Vol. 1948: F. Cao, J-L. Lisani, J-M. Morel, P. Musé, F. Sur, A Theory of Shape Identification (2008)

Vol. 1949: H.G. Feichtinger, B. Helffer, M.P. Lamoureux, N. Lerner, J. Toft, Pseudo-Differential Operators. Quantization and Signals. Cetraro, Italy 2006. Editors: L. Rodino, M.W. Wong (2008)

Vol. 1950: M. Bramson, Stability of Queueing Networks, Ecole d'Eté de Probabilités de Saint-Flour XXXVI-2006 (2008)

Vol. 1951: A. Moltó, J. Orihuela, S. Troyanski, M. Valdivia, A Non Linear Transfer Technique for Renorming (2008)

Vol. 1952: R. Mikhailov, I.B.S. Passi, Lower Central and Dimension Series of Groups (2008)

Vol. 1953: K. Arwini, C.T.J. Dodson, Information Geometry (2008)

Vol. 1954: P. Biane, L. Bouten, F. Cipriani, N. Konno, N. Privault, Q. Xu, Quantum Potential Theory. Editors: U. Franz, M. Schuermann (2008)

Vol. 1955: M. Bernot, V. Caselles, J.-M. Morel, Optimal transportation networks (2008)

Vol. 1956: C.H. Chu, Matrix Convolution Operators on Groups (2008)

Vol. 1957: A. Guionnet, On Random Matrices: Macroscopic Asymptotics, Ecole d'Eté de Probabilités de Saint-Flour XXXVI-2006 (2008)

Vol. 1958: M.C. Olsson, Compactifying Moduli Spaces for Abelian Varieties (2008)

Recent Reprints and New Editions

Vol. 1702: J. Ma, J. Yong, Forward-Backward Stochastic Differential Equations and their Applications. 1999 – Corr. 3rd printing (2007)

Vol. 830: J.A. Green, Polynomial Representations of GL_n, with an Appendix on Schensted Correspondence and Littelmann Paths by K. Erdmann, J.A. Green and M. Schoker 1980 – 2nd corr. and augmented edition (2007)

Vol. 1693: S. Simons, From Hahn-Banach to Monotonicity (Minimax and Monotonicity 1998) – 2nd exp. edition (2008)

Vol. 470: R.E. Bowen, Equilibrium States and the Ergodic Theory of Anosov Diffeomorphisms. With a preface by D. Ruelle. Edited by J.-R. Chazottes. 1975 – 2nd rev. edition (2008)

Vol. 523: S.A. Albeverio, R.J. Høegh-Krohn, S. Mazzucchi, Mathematical Theory of Feynman Path Integral. 1976 – 2nd corr. and enlarged edition (2008)

Vol. 1764: A. Cannas da Silva, Lectures on Symplectic Geometry 2001 – Corr. 2nd printing (2008)

LECTURE NOTES IN MATHEMATICS   🐎 Springer

Edited by J.-M. Morel, F. Takens, B. Teissier, P.K. Maini

Editorial Policy (for the publication of monographs)

1. Lecture Notes aim to report new developments in all areas of mathematics and their applications - quickly, informally and at a high level. Mathematical texts analysing new developments in modelling and numerical simulation are welcome.

   Monograph manuscripts should be reasonably self-contained and rounded off. Thus they may, and often will, present not only results of the author but also related work by other people. They may be based on specialised lecture courses. Furthermore, the manuscripts should provide sufficient motivation, examples and applications. This clearly distinguishes Lecture Notes from journal articles or technical reports which normally are very concise. Articles intended for a journal but too long to be accepted by most journals, usually do not have this "lecture notes" character. For similar reasons it is unusual for doctoral theses to be accepted for the Lecture Notes series, though habilitation theses may be appropriate.

2. Manuscripts should be submitted either to Springer's mathematics editorial in Heidelberg, or to one of the series editors. In general, manuscripts will be sent out to 2 external referees for evaluation. If a decision cannot yet be reached on the basis of the first 2 reports, further referees may be contacted: The author will be informed of this. A final decision to publish can be made only on the basis of the complete manuscript, however a refereeing process leading to a preliminary decision can be based on a pre-final or incomplete manuscript. The strict minimum amount of material that will be considered should include a detailed outline describing the planned contents of each chapter, a bibliography and several sample chapters.

   Authors should be aware that incomplete or insufficiently close to final manuscripts almost always result in longer refereeing times and nevertheless unclear referees' recommendations, making further refereeing of a final draft necessary.

   Authors should also be aware that parallel submission of their manuscript to another publisher while under consideration for LNM will in general lead to immediate rejection.

3. Manuscripts should in general be submitted in English. Final manuscripts should contain at least 100 pages of mathematical text and should always include

   - a table of contents;
   - an informative introduction, with adequate motivation and perhaps some historical remarks: it should be accessible to a reader not intimately familiar with the topic treated;
   - a subject index: as a rule this is genuinely helpful for the reader.

For evaluation purposes, manuscripts may be submitted in print or electronic form, in the latter case preferably as pdf- or zipped ps-files. Lecture Notes volumes are, as a rule, printed digitally from the authors' files. To ensure best results, authors are asked to use the LaTeX2e style files available from Springer's web-server at:

ftp://ftp.springer.de/pub/tex/latex/svmonot1/ (for monographs).

Additional technical instructions, if necessary, are available on request from: lnm@springer.com.

4. Careful preparation of the manuscripts will help keep production time short besides ensuring satisfactory appearance of the finished book in print and online. After acceptance of the manuscript authors will be asked to prepare the final LaTeX source files (and also the corresponding dvi-, pdf- or zipped ps-file) together with the final printout made from these files. The LaTeX source files are essential for producing the full-text online version of the book (see www.springerlink.com/content/110312 for the existing online volumes of LNM).
   The actual production of a Lecture Notes volume takes approximately 12 weeks.

5. Authors receive a total of 50 free copies of their volume, but no royalties. They are entitled to a discount of 33.3% on the price of Springer books purchased for their personal use, if ordering directly from Springer.

6. Commitment to publish is made by letter of intent rather than by signing a formal contract. Springer-Verlag secures the copyright for each volume. Authors are free to reuse material contained in their LNM volumes in later publications: a brief written (or e-mail) request for formal permission is sufficient.

Addresses:

Professor J.-M. Morel, CMLA,
École Normale Supérieure de Cachan,
61 Avenue du Président Wilson, 94235 Cachan Cedex, France
E-mail: Jean-Michel.Morel@cmla.ens-cachan.fr

Professor F. Takens, Mathematisch Instituut,
Rijksuniversiteit Groningen, Postbus 800,
9700 AV Groningen, The Netherlands
E-mail: F.Takens@math.rug.nl

Professor B. Teissier, Institut Mathématique de Jussieu,
UMR 7586 du CNRS, Équipe "Géométrie et Dynamique",
175 rue du Chevaleret
75013 Paris, France
E-mail: teissier@math.jussieu.fr

For the "Mathematical Biosciences Subseries" of LNM:

Professor P.K. Maini, Center for Mathematical Biology,
Mathematical Institute, 24-29 St Giles,
Oxford OX1 3LP, UK
E-mail: maini@maths.ox.ac.uk

Springer, Mathematics Editorial I, Tiergartenstr. 17
69121 Heidelberg, Germany,
Tel.: +49 (6221) 487-8410
Fax: +49 (6221) 4876-8259
E-mail: lnm@springer.com